"[*Solving the Climate Crisis*] is an important and engaging book that will have broad appeal not just to the millions concerned about climate change but to readers interested in new scientific and technological breakthroughs and in ways of sparking a national economic renewal."
—JOHN H. ADAMS, ESQ.
Founding Director, Past President, and Trustee,
The Natural Resources Defense Council

"*Solving the Climate Crisis* is a vital tool and voice that is urgently needed right now. We do not need more books about doom-and-gloom and dysfunction. We do need to inspire decision-makers with a path forward and let wavering politicians and bureaucrats know that we can solve the climate crisis with a combination of personal, local, state, regional, national and global actions."
—BRUCE HAMILTON,
former National Policy Director, Sierra Club

"John is a good storyteller. His entertaining book is filled with inspiring and memorable stories and characters while also being authoritative, comprehensive, and well-researched."
—PETER G. JOSEPH, MD
Group Leader, Marin Chapter
Citizens' Climate Lobby

"John Berger has produced a critically important book at just the right time. In a moment of immense urgency on climate, John's book walks us through the solutions and strategies the world needs to address the most important challenge of our time. His book is a ray of light in a challenging time."
—DAVID HOCHSCHILD
Chairman, California Energy Commission

"Comprehensive. Everyone will find at least one thing useful in here that they can do to help."
—BILL MCKIBBEN, author of *The End of Nature*

"John Berger combines authoritative science, accessible writing, and persuasive logic to provide a vital book for creating a future worthy of the next generations. His work is a gift to our best possibilities on this planet."

—NORMAN SOLOMON
National Director, RootsAction Education Fund
Author of *War Made Invisible*

"A scientifically reliable overview of currently available climate solutions, *Solving the Climate Crisis* provides thought-provoking coverage of proven, cost-effective technologies and strategies that could be scaled up to hugely reduce greenhouse-gas emissions and protect the climate."

—JOHN HARTE
Distinguished Professor of the Graduate School, Division of Ecosystem Sciences, University of California, Berkeley

PRAISE FOR *CLIMATE PERIL*

"A brilliant book, and one that might just change the world. By far the best overview of climate science and its implications for our planet that I've ever read."

—TIM FLANNERY
Chief Councilor, Australian Climate Council
Author of *The Weather Makers*

"John Berger has produced a critically needed and eloquent compendium of not only the crisis of global warming, but of the many interdependent shortages and crises—of financial risks; of land, water, and cultural preservation; and of economic opportunity for the poor—that all exist today and that conspire to challenge us to be better stewards of the planet."

—DANIEL M. KAMMEN
Distinguished Professor of Energy,
University of California, Berkeley

ALSO BY JOHN J. BERGER

*Beating the Heat: Why and How We Must
Combat Global Warming*

*Charging Ahead: The Business of Renewable Energy and
What It Means for America*

Climate Myths: The Campaign Against Climate Science

*Climate Peril: The Intelligent Reader's Guide to
Understanding the Climate Crisis*

*Ecological Restoration Directory in the San Francisco
Bay Area* (editor)

*Environmental Restoration: Science and Strategies
for Restoring the Earth* (editor)

Forests Forever: Their Ecology, Restoration, and Protection

*Nuclear Power, the Unviable Option: A Critical Look
at Our Energy Alternatives*

*Restoring the Earth: How Americans Are Working to
Renew Our Damaged Environment*

Understanding Forests

Solving the Climate Crisis

FRONTLINE REPORTS FROM THE RACE TO SAVE THE EARTH

John J. Berger, PhD

introduction by
Senator Russ Feingold

Seven Stories Press
new york • oakland • london

Seven Stories Press
140 Watts Street
New York, NY 10013
www.sevenstories.com

Library of Congress Cataloging-in-Publication Data

Names: Berger, John J., 1945- author.
Title: Solving the climate crisis : frontline reports from the race to save
    the Earth / John J. Berger.
Description: New York : Seven Stories Press, [2023] | Includes
    bibliographical references and index.
Identifiers: LCCN 2023013473 | ISBN 9781644213223 (trade paperback) | ISBN
    9781644213230 (ebook)
Subjects: LCSH: Clean energy--Government policy--United States. | Clean
    energy investment--United States. | Climatic changes--Government
    policy--United States.
Classification: LCC HD9502.5.C543 B47 2023 | DDC
    333.790973--dc23/eng/20230801
LC record available at https://lccn.loc.gov/2023013473

College professors and high school and middle school teachers may order free examination copies
of Seven Stories Press titles. Visit https://www.sevenstories.com/pg/resources-academics or email
academic@sevenstories.com.

Printed in the United States of America

9 8 7 6 5 4 3 2 1

In memory of George Helmholtz
(1942-2021)

We are now faced with the fact that tomorrow is today. We are confronted with the fierce urgency of now. In this unfolding conundrum of life and history there is such a thing as being too late . . . We may cry out desperately for time to pause in her passage, but time is deaf to every plea and rushes on. Over the bleached bones and jumbled residue of numerous civilizations are written the pathetic words: "Too late."

—MARTIN LUTHER KING JR.

It always seems impossible until it's done.

—NELSON MANDELA

Contents

Part I: Mega-threats and Breakthrough Solutions

Part II: Clean Technology, Inexhaustible Resources

Part III: Beyond the City: Natural Climate Solutions

Part IV: Political and Legal Reform and Other Challenges

Foreword

If you're like me, you've read a plethora of materials about how dire the climate crisis is and are often frustrated by the lagging conversation about practical, real-world solutions. This is where this book comes in. John Berger spends little time convincing us that the climate crisis is grave and, instead, focuses on the more pressing question: How do we solve it? This is the book for readers who are interested not in pointing the finger, but in enacting practical and achievable solutions.

Berger combines real-world examples, explanations of cutting-edge technologies, and sound energy policy to provide a practical road map for how we can effectively combat climate change and move to 100 percent clean energy. As someone from a political and policy background, I particularly appreciate Berger's commitment to grounding his book in the realities of U.S. and global politics. Solutions to climate change that rely on sudden and widespread unity among political parties or countries are just not realistic given the times in which we live. If we are to solve climate change, we need a plan that acknowledges political realities rather than denies them.

The U.S. Supreme Court, in its decision in *West Virginia v. EPA*, has kneecapped the executive branch's ability to regulate greenhouse gas emissions. This decision is a galling example of the political challenges to combating climate change and underscores the growing need for solutions that do not rely exclusively on one level of government. Now more than ever, solutions to climate change need to involve every level of government and every sector of the economy, from agriculture to tech.

The climate crisis is an existential crisis because, if left unabated, it will impact every facet of our lives and our institutions. Already we are witnessing unprecedented apathy and loss of trust in government to solve our problems. As the climate crisis worsens, political grievances will only intensify as government repeatedly fails to meet the moment. The climate crisis is a test for government, and Berger explains how government, at the national, state, and local levels, can and should partner with the private sector to solve this shared existential crisis.

This book could not be timelier for many reasons, including economic ones. Right now, we are confronting economic upheaval with a labor market that seems as unfriendly to workers as it does to employers. Gasoline prices soared during the pandemic, and fuel costs are still consuming more of people's paychecks as energy companies take advantage of current conditions to profit handsomely at the public's expense.

While we've been told for decades that shifting to clean energy could provide thousands of new jobs, there has been a glaring lack of detail regarding how this actually gets done. Berger provides those missing details, demonstrating how his proposed action plan would be a boost to the labor market and provide a sustainable source of economic growth. Not relying on hypotheticals, he instead details how the clean-energy transition has already achieved job and economic growth for towns and localities that are leading the way.

Whether you are a concerned citizen, an environmental scientist, an elected official, a CEO, or an engineer, this book offers scalable, profitable business solutions and practical lessons about how we can combat climate change. Just as important, if you do not consider yourself an environmentalist and, instead, are focused on economics or technological innovation, put this book at the top of your reading list. As Berger points out, the challenge of combating the climate crisis is as much an economics and technology challenge as it is an environmental one. Just as the climate crisis impacts all of us, the solution to the crisis requires all of us.

—RUSS FEINGOLD,
president, American Constitution Society,
former U.S. senator

Author's Introduction

No one would have suspected, early in his career, that a staunchly conservative Republican mayor of a small Texas city would catapult his community from relative obscurity to clean-energy superstardom. Yet, when I met Dale Ross in 2017, the round-faced, blond mayor of Georgetown, Texas, was on a victory lap at a Las Vegas clean-energy conference and was being videotaped by a German film crew as we spoke. Ross was fifty-eight, and his claim to fame was that, in 2015, he had convinced a pro-Trump oil patch Texas town to switch from fossil fuel power to 100 percent wind and solar.

At first, Ross's bold advocacy was very popular. (Later, as we'll see, it backfired, before eventually regaining support.) A certified public accountant and self-described "right-wing fiscal conservative," Ross makes financial decisions based on numbers. So, instead of being drawn to the environmental benefits of clean power, he liked the price stability and long-term contracts that renewable-energy companies had to offer. "We solved that math problem," he declared. "The natural gas providers would only guarantee us a seven-year contract. The new contracts guarantee us the same rates in year one as they do in year twenty-five," while reducing regulatory risks. Initially, the switch to clean power was hugely popular in Georgetown because it saved money. "If you win the economic argument," said Ross, "then you're going to win the environmental argument by default."

The decision to give up fossil fuel power catapulted this Trump-voting Texas tightwad into the limelight, turning him into an environmental celebrity who hobnobbed with former vice president Al Gore and appeared both in the Bloomberg Philanthropies–funded documentary, *From the Ashes*, about the transition from coal to renewables, and in a 2017 HBO documentary, *Happening: A Clean Energy Revolution*, produced by James Redford (Robert Redford's son). Ross was, even then, highly skeptical of Trump's energy policies and had staked out an iconoclastic position within his party: "[President Trump] keeps on saying the coal industry is coming back. The coal industry is not coming back, okay? It's dead," Ross told me.

Georgetown's switch to renewables altered Ross's own lifestyle as well as his image. He took to riding around town on an electric BMW scooter and boosted solar panels onto his roof. Renewable energy is the future, he concluded, because it makes economic sense. Renewables, he told me, were "a huge economic development tool" that helped make Georgetown "the greatest city on planet Earth," even more attractive to large corporations looking to do business there.

Ross appears to have had the proverbial wind at his back. Texas—headquarters of the oil and gas industry and the state that dominates U.S. oil and gas production—is now also by far the nation's largest producer of wind power. In 2022, the state got more power from wind than from coal and two and a half times as much as from nuclear power. In fact, Texas will get more power from wind and solar in 2023 even than from natural gas. And it will add far more clean-energy projects than any other state in the union.

✝   ✝   ✝

Decades ago, and also in America's heartland, another bold experiment was beginning that would eventually benefit the climate. In the mid-1990s, after three years of devastating hailstorms and a year of drought, North Dakota rancher-farmer Gabe Brown was going bankrupt and facing foreclosure. Recently married, he was desperate to protect his land and provide for his young family.

To save the farm, he and his wife took second jobs and did the farm chores at night, working until 2 a.m. They mowed hay in roadside ditches to get free feed for their cattle. But no matter how hard they worked, Brown and his wife could not get out of debt. Hail and drought hammered their land; it felt to the Browns like being struck by a series of plagues. Yet, while those years were "hell to go through," Brown now says they were actually the best thing that ever happened to him, as they forced him to completely change the way he approached farming.

Farmers in North Dakota tend to concentrate on the most profitable single crop, allowing the land to lie bare between seedings and dousing the soil liberally with pesticides and herbicides to kill weeds and insects. This kind of farming—the energy-hungry, ecologically costly, factory farm way—leads to spiraling costs for farmers, charged against ever-diminishing returns. Convinced there had to be a better, more economical way to farm, Brown went to the local library.

There, in the diaries of explorers Meriwether Lewis and William Clark, who had passed through North Dakota between 1804 and 1806, he discovered long-forgotten agricultural practices used by the native Mandan people. About that time, he also heard a talk by Alberta rancher Don Campbell, in which Campbell said, "If you want to make small changes, change the way you do things, but if you want to make major changes, change the way you *see* things." After thinking about this for a day or two, Gabe Brown realized that he had to look at his whole operation through the eyes of nature, to better understand how nature functions, so that he could work in harmony with it in a way that would best serve him and, more important, the land. Following this practice, which included observing how hailstorms armored his soil by hammering plants onto the surface, Brown discovered the value of cover crops—vegetation planted after a harvest to cover and enrich the soil with nutrients, protecting it from erosion and increasing its ability to hold water. Gradually, he also found other ways to replenish his exhausted soil, all without pesticides, herbicides, fertilizers, or plowing.

His neighbors laughed at him at first. But, with startling speed, his new farming practices reduced his expenses and any need for the fossil fuel–based chemicals on which other farmers and ranchers relied. Even more important to Brown, his new approach dramatically increased the fertility and health of his soil. In turn, his crops and forage soon grew strong and healthy, as did the livestock that grazed the newly restored, drought-resistant prairie. At last, Gabe Brown and his family prospered.

✦ ✦ ✦

Like Dale Ross, Gabe Brown hadn't intended to tackle climate change. Yet the alterations he made to his farming practices not only created new income streams and a highly profitable operation, but also contributed to a healthier climate by sucking tons of carbon out of the atmosphere and reducing greenhouse gas (GHG) emissions. Both Mayor Ross and Rancher Brown were "accidental climate savers," people who became climate champions through insights born of necessity—or opportunism.

Other climate savers in this book are far more intentional, having little in common with the accidental kind, save for their passion, determination, and resourcefulness. But they are all highly effective and transformational agents of change for our planet. Attorney Dan Reicher, for example, is a Stanford University energy scholar, former U.S. Department of Energy chief of staff, and former chief climate officer for Google; he also cofounded New Energy Capital, the nation's first renewable-energy project finance company. Much of Reicher's professional life in government, academia, and business has been spent trying to put fossil fuels in the rearview mirror. By exerting his influence in Congress as a private citizen, for example, he managed to save a multibillion-dollar federal energy research program from extinction.

With his understanding of energy finance, Reicher is a strong advocate of using public-private partnerships to pay for the clean-energy transition. Such partnerships use federal loan subsidies and guarantees to attract private capital and multiply federal clean-energy investments many times over. For example, the \$465 million low-interest, federally subsidized loan to Tesla in 2009 that Reicher supported while at the Department of Energy helped Tesla build its award-winning Model S electric sedan and, in turn, supercharged the electric vehicle (EV) industry. Tesla repaid the loan with interest in 2013. As of 2022, it was the world's most valuable car manufacturer and the eighth most valuable company overall, with a market cap of \$904 billion.

The roots of solicitude for the planet and the climate sometimes sprout early. Reicher's interest in climate issues arose from his boyhood concerns about water quality on badly polluted Onondaga Lake (later a Superfund site) around his hometown of Syracuse, New York. Then there's Professor Mark Z. Jacobson, whose youthful concern over the acrid air he was breathing while playing competitive tennis spurred him to see if his talents for science and math could lead to a career doing something about air pollution. For his doctorate at UCLA, he created the first computer model of the complex chemistry of Los Angeles's air pollution. Then he integrated an existing meteorology model into his work, and the resulting program became one of the world's first computer models of global air pollution and climate. In the process, Jacobson made an important scientific discovery: black carbon, a by-product of fossil fuel combustion, plays a powerful role in heating the atmosphere; it's the second largest contributor to global warming after carbon dioxide (CO_2).

Jacobson is now an engineering professor at Stanford and the world's twelfth most-cited scientist in meteorology and atmospheric sciences. As we'll see in chapter 5, his work has demonstrated how the United States and the world could run smoothly and cost-effectively entirely on renewable energy from wind, water, and solar power, while creating 24 million net new jobs.

✢   ✢   ✢

This book is an account of visionary scientists starting new clean-steel, "green" cement, and CO_2-reuse companies and of great engineers building new electric airplanes, semi-trucks, and buses. It's also the story of transportation experts envisioning cities that accommodate the fully automated electric vehicles of the future and of forward-looking city leaders, from San Francisco to Stockholm, who are dedicated to making their cities carbon-free. Finally, the book describes how, in this moment of climate crisis, the same transformational policies needed to eliminate GHG emissions flooding the atmosphere can also reduce long-standing environmental and social injustice. That's important, because climate disruption disproportionately hurts the world's poorest and most vulnerable people, who not only are least able to defend themselves against the ravages of extreme weather, but who also are least responsible for the emissions that cause it in the first place.

The diverse and wide-ranging narratives in this book all point toward a fundamental truth: in any serious, high-level strategic thinking about a solution to climate change, technological, ecological, and social elements must all be deeply considered, and integrated in any comprehensive strategic plan to deal with it. No competent climate strategy can rely solely on only one or two of these elements. Technologically well-versed thinkers, for example, often focus narrowly on kilowatt-hours saved and clean-energy production curves. They advocate massive new research-and-development investments to produce better machines and buildings and new ways of generating energy. But if sound energy policy, law, and finance don't support the scaling of new clean technologies, they cannot become mainstream and reach their full potential. And if ecosystems are neglected or abused due to an underappreciation of their critical role in removing and storing carbon, their GHG emissions can rise and negate gains made in technology.

Part of a true climate solution is thus in the ecological realm, especially in the agricultural and forestry sectors and even the food industry. Healthy natural and restored ecosystems—not only forests, but also wetlands, peatlands, rangelands, and agricultural soils—will help us normalize the concentration of GHGs in the air. Restoring ecosystems will also help to sustain their benefits and yields—such as wild fish, forest products, biodiversity, and natural services like those rendered by pollinators. We also need to alter how we grow, process, consume, and dispose of food, so as to reduce emissions from all these activities. Even the most assiduous attention to the health of ecosystems, however, will not be sufficient without a tightly woven lattice of laws and policies to undergird and financially support it.

The human and social dimension, including values and systems of governance, is therefore also essential to any successful and comprehensive program to combat climate change. Our values guide us in determining how we allocate resources to address climate threats and impacts and the extent to which technological and ecological solutions get funded. Thorny issues like environmental justice, income inequality, and intergenerational equity also enter the mix. The social dimension is indeed virtually as complex as society itself: It embraces politics, diplomacy, information and communication systems (including media), finance and economics, and urban planning. Most saliently, it also involves political and economic power—which, when wielded to spread misinformation about climate science, creates public resistance to needed climate policies. Proponents of a full-court press to protect the climate will thus need to acquire influential media outlets of their own and generate countervailing power through savvy messaging and persistent political organizing to build powerful public and stakeholder support and strategic coalitions. Along with legal and electoral efforts and relentless public education, more direct action in the form of marches, rallies, and demonstrations will also be needed. Those who hold power do not relinquish it without a struggle. Understanding how to gain substantial power is thus of paramount importance if a successful campaign is to be mounted in time against the world's relentless drift toward a climate cataclysm far worse than what we've seen so far.

It is thus in this broad context, and with these understandings, that the chapters ahead—with their accounts of breakthrough new technologies, hard-nosed entrepreneurship, and wise public policies—make their claim on your attention. This book is the story of energetic, creative people applying tourniquets to stem the flood of CO_2 pouring into the atmosphere. You'll get glimpses in the pages ahead of visionary frontline efforts in homes and factories, in laboratories and legislatures, on farms and rangelands, and in forests. Collectively, the climate savers' pioneering work forms a tapestry of comprehensive efforts—a holistic technological, ecological, and social approach to reducing climate change. This book's fundamental premise is that nothing other than such a holistic approach stands a chance of success, but also that it must be quickly scaled up to powerfully counteract the global climate emergency—because while we're parsing nicely turned phrases about climate solutions, a real-world climate time bomb is ticking.

✦   ✦   ✦

None of this social, technological, and ecological work will be simple or easy. Fortunately, each person who is committed to helping to advance a swift and fair energy transition can make a contribution. Homeowners can insulate their homes, put up solar panels, and use heat pumps instead of gas furnaces. Some of us may bike to work, make our next vehicle an EV, or jump on public transit. Beyond this, each of us has different inclinations, abilities, and financial resources; each of us must look within, find our inner activist, and then commit to making climate activism as broad a social movement today as the antiwar and civil rights struggles of the 1960s.

The complexity of the climate crisis, however, transcends the individual's sphere of influence. So, we also need to join and support climate-protection organizations. Whereas this is not a book about political organizing, it nonetheless needs to be said that such collective political action is needed now more than at any previous time in human history to pivot society as quickly as possible from fossil fuels to clean energy and ecosystem repair. Armed with knowledge of the many solutions available, and with the inspiration of climate role models, we, the people, must mobilize ourselves in collective efforts to compel policymakers to make the right decisions, even though many now receive political contributions from the very industries they need to put in check. By contrast, stressing individual action and responsibility to the exclusion of collective action will ensure that the climate-protection movement remain always on the fringes of power, lacking enough of it to institute a climate agenda that will transform the economy as fast as possible, rather than as slowly as the fossil fuel industry wants.

As I'll also explain in the book, the climate crisis will demand a larger, more active role for government than ideological opponents of "big government" might like. It will therefore be challenging to develop a social consensus to take the necessary steps. The suggestion that government needs to be more engaged in combating climate change might discomfit some readers by raising the specter of socialism or of crushing financial burdens. If so, all the more reason to read on: I believe you'll see in the pages ahead that the ensemble of public and private measures needed to save the climate will be in the nation's long-term economic interest and will yield more net employment and long-term economic security than any other path. By contrast, foot-dragging by politicians based on an ideological predisposition to minimalist government will lead to terrible humanitarian consequences and crushing financial costs— between 6 and 29 percent of global economic output by 2100. It will also leave us vulnerable to the full force of climate disruption. To preserve the opportunity to enjoy a good and prosperous life in the future, doing all we can now to head off a ruinous climate catastrophe makes perfect sense. Not protecting the climate now, however, will *definitely* lead to more government and more sacrifices later—under much harsher living conditions, on a greatly degraded planet, and in a reactive, disaster-response mode.

The job calls for "us," in the United States and abroad—not just you and me as individuals. Only you, your friends, your community, your country, and other nations all together can do this. It must be an all-of-society effort to face an "all hands on deck" emergency; an intense, coordinated national and global effort to greatly accelerate the clean-energy transition; a political movement that consciously strives for holistic technological, ecological, and socioeconomic solutions—and one that remains concerned about environmental justice, so that no nation and no person is left behind or disadvantaged in the transition.

Speed and decisive, assertive action are urgent in this enterprise. We don't need to rely just on climate scientists anymore to warn us that we are in a Code Red emergency. Millions of people have already died, been sickened, or been displaced by climate change. We can see the change outside our windows and feel it in searing droughts, heat waves, and enormous wildfires; in record-breaking rains and unprecedented floods; and in hurricanes that hit with the force of a bomb, flattening everything in their path.

Yet all these impacts represent only the first waves of climate devastation. So, now we *must* make the space, in our busy and distracted lives, not just to recycle a few cans and bottles or buy a veggie burger, but also to become advocates and activists for the earth, linking arms in wise, farsighted, strategic collective action. We need no further reminder of what will happen if even we, who recognize the problem and understand the solutions, somehow just don't manage to find the weekends, evenings, or even at least a few hours to join in some collective action.

Clearly, avoiding further climate tragedies is paramount. But, as you're about to see in this book, the good news is that replacing the fossil fuel system in a timely manner with a sustainable clean-energy system will be akin to a heart transplant for the global economy. The newly invigorated, modernized clean-energy economy will produce tens of millions of new jobs and save trillions of dollars, even as it avoids tens or hundreds of trillions of dollars more in climate damage. Protecting the climate is thus potentially the greatest economic opportunity of our time. The climate savers' progress in their pursuit is cause for hope and inspiration.

PART I

Mega-threats and Breakthrough Solutions

I.

Daunting Challenges,
Rays of Hope

On July 29, 2018, at 7:29 p.m., fire inspector Jeremy Stoke was driving his pickup in northwest Redding, California. Ominous reddish-gray clouds of smoke darkened the sky around him. The thirty-seven-year-old family man and community volunteer had returned to duty from vacation and was on his way to protect and evacuate residents in the path of the Carr Fire.

One minute later, the unpredictable fire exploded into a towering one-thousand-foot-wide fire tornado. The swirling funnel of flame, with scorching 2,700°F winds of up to 165 mph, ripped roofs off houses and sent cars and power lines flying through the air. In an instant, the flames burned over Stoke, killing him. He left behind a grieving wife and two children—just one of many tragedies that vast wildfires, sparked by drought-related climate change, have left in their wake in the United States and globally.

Monster wildfires are now just another all-too-common climate nightmare—like the "five-hundred-year" flood arriving for the second year in a row, or the three feet of water Hurricane Harvey dropped on Houston in 2017, the most extreme precipitation event in U.S. history. So many hurricanes developed in 2020 that the National Hurricane Center ran out of letters in the Roman alphabet to name them. Meanwhile, around the world, ten-thousand-year-old glaciers melt, polar ice caps dwindle, and regions of the Amazon rain forest die from drought or fires, releasing millions of tons of greenhouse gases (GHG). Wildlife extinctions soar. Coral reefs perish from heat and rising ocean acidity.

It's no secret that because of climate disruption, heat waves and droughts become more common. Crops parch. Grain yields slip. Livestock perishes. Food prices rise. People go hungry. Globally, tens of millions of people are already fleeing the impacts of extreme weather. Moreover, while just 1 percent of the earth today is too hot for human habitation, by 2070, if current trends continue, 19 percent of the earth's land area—home to billions of people—will be uninhabitably hot. This will be catastrophic for people who live in these and neighboring regions.

Disasters are happening so frequently now that it's hard to track them all. In 2022, for example, an unprecedented flood submerged a third of Pakistan, killing more than 1,700 people and disrupting tens of millions of lives. A high-pressure summer "heat dome" lingered over parts of Europe, the Middle East, Africa, and Asia, bringing triple-digit temperatures, droughts, and fires. Parts of Asia's largest river, the Yangtze, dried up. Gradually, a kind of "disaster fatigue" sets in as a barrage of such news desensitizes us to our new reality. So dims our collective memory of a healthy planet and a normal climate.

After experiencing just the first 1.2°C (about 2°F) of heating, and with far more likely to come, the earth's climate is already gravely tormented. That average 1.2°C increase may not seem like much. Yet, obviously, it is more than enough to throw the climate out of whack, especially because the heating is uneven. Land heats up one and a half times as much as the average for the entire surface of the earth (more than 70 percent of which is ocean). But the Arctic is heating up *four times* as fast as the global average. And it's an integral part of the global climate system, sometimes referred to as the planet's air conditioner. As higher temperatures melt snow and ice over vast areas of the Arctic, they reduce the reflection of solar energy back into space. Instead, that energy is absorbed, further heating the earth. The heat also gradually thaws Arctic permafrost, which contains huge stores of frozen carbon. Meanwhile, higher Arctic temperatures also allow immense, carbon-rich northern forests to burn more frequently than at any time in the past ten thousand years. So, local temperature increases in the Arctic can destabilize the entire climate system.

The United States and many other governments have long referred to 2°C (3.6°F) of warming above pre-industrial temperatures as "safe," although this dogma is without sound scientific foundation. Consider, for example, what would happen to the animals and plants of the world. The Intergovernmental Panel on Climate Change has warned that with an increase of 2°C, nearly one in five land animals would likely go extinct, and almost a third of all insect species would also be "at high risk of extinction," while some of our most precious biodiverse regions, including tropical rain forests and tropical coral reefs, would be irreversibly damaged. The Arctic would be particularly hard hit. Another problem is that if we continue on our current business-as-usual trajectory, the world will not stop warming in 2100. Neither will sea level rise halt, even if we halted all our emissions; nor will various other climate impacts cease. So, do we just keep on doing what has brought us to the current climate precipice and plunge over it, or do we do something else?

Climate System Lags and the Long View

Some further climate change and additional global warming are now inevitable. By way of explanation, consider the following three alternative scenarios that would, to varying degrees, begin reducing the rate of future climate change.

In the first scenario, imagine that the nations of the world are able only to stop *increasing* their greenhouse gas emissions but nonetheless continue emitting at their current elevated level of about

37 billion metric tons of CO_2 annually. Much of these continued annual emissions would accumulate in the atmosphere. Although this would be less disruptive to the climate than a continuing *increase* in emissions every year, the concentrations of long-lived GHGs like CO_2 would continue to climb rapidly, which would result in further accelerated heating and further climate change, with risks of still further acceleration in GHG releases from highly dangerous "positive feedbacks" arising from heat-perturbed natural systems, such as drier, more flammable forests; methane clathrates, a form of frozen methane found in seafloor sediments; and thawing Arctic permafrost.

Now consider a second, less disruptive scenario. Imagine that in a magical event, all human CO_2 emissions (but *only* CO_2 emissions) suddenly dropped from 37 billion metric tons a year to zero, where they remained year after year. That would be a big improvement over the first case, and CO_2 concentrations in the atmosphere would begin falling. Some scientific uncertainties exist as to what precisely would then happen. One might suppose that this would begin cooling the earth. However, for complex reasons, it appears most plausible that the earth's temperature would remain roughly steady at about whatever elevated level it was when emissions halted. This appears to be the case because, first, the vast amounts of heat stored in the oceans would serve to reduce or prevent atmospheric cooling. Second, emissions of other GHGs (such as methane and nitrous oxide) would still be occurring.*

To not only stop heating the planet but also cause the earth's temperature to begin falling, *all* human-induced GHG emissions (not just CO_2) would need to cease. In this third scenario, climate science models tell us that temperatures would begin falling immediately or a few years after cessation, depending on the complexities of what happened concurrently to all other non-CO_2 greenhouse gases and

---

* This assumption is made only for argument's sake. In reality, if CO_2 emissions ceased, the "upstream" emissions of methane and nitrogen oxides associated with fossil fuel extraction and combustion would be eliminated, although those associated with agriculture would not.

aerosols.* Notably, the resulting fall in temperature could also be accelerated if humanity not only curtailed its emissions but went a step farther, physically and biologically removing excess GHGs that had accumulated in the atmosphere by applying some of the techniques described in this book.

None of these three scenarios describes what's happening nowadays, however, as annual global emissions are *still rising*, although there are signs they finally may be starting to flatten.** Without major emissions reductions, the world's temperature is expected to rise by 3°C to 5°C (5.4°F to 9°F) by 2100, with catastrophic consequences.*** Moreover, in 2022, a new draft paper released by renowned climate scientist James Hansen and colleagues projected that "climate sensitivity"—the climate's response to a doubling of GHG—is not on the order of 3°C, as most estimates had found, but is possibly much higher. If true, this would suggest that without drastic action to greatly accelerate the transition to clean energy, those catastrophic impacts will arrive sooner. In other words, our goose will be cooked faster than we've feared.

In addition to global heating caused by GHG emissions, extra heat now stored in the ocean will also gradually emerge and will not only elevate global temperature but also melt ice and drive a more vigorous water cycle, causing heavier rains and snows, more drying, and more wildfires. The Greenland Ice Sheet has already lost more than four trillion tons of ice, and additional major losses are all but certain.

That's dangerous, but not as risky as the destabilization of the Ant-

---

* Aerosols are tiny reflective particles of sulfur or nitrogen that are generated during combustion and that tend to cool the earth. As industrial emissions decline due to a reduction in combustion, a corresponding reduction in aerosols occurs. When their atmospheric concentration falls, their cooling influence is likewise reduced, and atmospheric temperature rises.

** Global industrial GHG emissions swelled by more than two billion metric tons (6 percent) in 2021, as I mention later in the book, but they rose by only 1 percent in 2022. Although this is a promising development, as the saying goes, "one swallow doesn't make a spring," and it would be premature to declare that emissions have peaked unless we see a plateau in emissions of some years' duration.

*** It's impossible to pinpoint exactly how much the earth's temperature will rise, because no one knows how long it will take the world to reduce its GHG emissions to net zero. If the emission-reduction commitments made by most nations of the world at the 2015 Paris climate conference were fully attained—a far-from-forgone conclusion—the world's average surface temperature would still be expected to rise by at least 2.5°C.

arctic polar ice cap. Antarctica has been aptly called the "wild card" of sea level rise. With ten times the ice mass of Greenland, its already accelerating ice loss will become irreversible by 2060 if current emissions rates continue. The moments at which a climate impact becomes irreversible are known as tipping points, and that would be a big one.

If global temperature rises more than 2°C above pre-industrial temperatures, the pace of sea level rise will then accelerate "abruptly." The melting of polar land ice and expansion of the oceans will lead to sea level rise that will affect hundreds of millions of people in low-lying cities and nations. Dr. Hansen has said that a rise of even 6.5 feet—not impossible by 2100, if high emissions persist—would produce "a different planet." The world's coastlines will not be transformed overnight, of course. But longer-range impacts are too devastating to ignore: just ninety years after a 2060 tipping point, in 2150, the world's oceans would be rising almost a foot every five years.

If you have a home or apartment on a bay or ocean, you may have already noticed the water creeping up toward your house and sometimes perhaps even flooding in when a storm surge occurs. Miami now sometimes experiences "sunny day" flooding, when king tides plus sea level rise send water ashore. And some coastal forests are already dying from saltwater intrusion, like the "ghost forests" of Alligator River National Wildlife Refuge, in North Carolina.

Inland, if you visit national parks and wilderness areas, you may have noticed that vast swaths of the western forests from the Yukon along the Rockies to Mexico and all along the West Coast have already been destroyed by pine bark beetles, which, before global heating, were killed off by the cold winter freezes that are now increasingly rare. The loss of the pine trees deprives bears, birds, squirrels, and other species of their habitat and their dietary staple, pine nuts. Infestation also makes the trees highly susceptible to wildfires, an example of how climate impacts cascade and produce domino effects. Climate change is like a sledgehammer to the delicate ecosystems, natural resources, and agricultural systems on which we all depend. It's a multiplier of virtually all ecosystem threats and many national security threats, including risks to critical infrastructure, supply chains, international stability, and the global economy.

Even if you are not directly in the path of a climate-related disaster and are not planning to be around in 2100, let alone 2150, you will still experience some impacts of the climate disruption we have already locked in. (The longer you plan on living, the more of these impacts you'll experience.) Probably you'll have to endure longer, more frequent heat waves and higher summer temperatures as the local climate of places like New York begins to resemble that of Georgia or South Carolina. Maybe you'll experience more violent storms and floods or "bad-air days," when pollution is chemically fomented by overheated air. Perhaps your asthma or allergies will be significantly worse. I hope not, but maybe you'll catch a tropical disease that was previously unknown in your area but moved north as the Northern Hemisphere warmed.

Maybe the price of foods you love will spike, or perhaps their taste will change when they are grown in a hotter climate with more carbon in the air. Some foods may not be available at all, due to heat or new diseases, overexploitation, or habitat alteration. Not only have we already fished out 95 percent of all large fish in the sea, but climate change and ocean acidification are harming other species (oysters, lobsters, and crabs, for example) right down to the microscopic plankton at the bottom of complex food webs, whose populations are plummeting.

Although this isn't a book about nature and climate impacts, such impacts, large and small, cannot be ignored. Collectively, they imperil our homes, jobs, security, health, and future, and, increasingly, they diminish the quality of our daily lives, our working environments, and our recreational opportunities.

Awakening to the Climate Challenge

A survey by the Pew Research Center shows that three fifths of all Americans now regard climate change as a major threat. Nearly two thirds of us believe the federal government is still doing too little about climate disruption. The American public is understandably growing impatient for more effective climate action. No doubt you, too, realize that the situation is dire and want to know what can be done. Perhaps you've been told that things are hopeless, or that the solutions we have are too expensive, or won't work, or would change your life in unpleasant ways. However, the committed people described in this book, who are on the front lines of the effort to solve the climate crisis, do not accept this negative, fatalistic outlook. They have found many ways to put out fossil fuel fires and get our clean-energy act together.

Two thirds of the U.S. population now live in areas participating in significant climate-protection efforts. Thousands of cities, businesses, universities, health care facilities, and cultural organizations worldwide are already working to reduce their emissions. A thousand cities, a thousand educational institutions, and more than four hundred financial institutions have committed to slashing global emissions in half by 2030. More than eight hundred cities in more than a hundred regions plan to cut their emissions to net zero by 2050. These municipal actions are extremely important: cities are responsible for 70 percent of the world's greenhouse gases.

Corporations are getting involved as well. More than 2,287 American firms and investors—including Amazon, Gap, Twitter, Target, and Nike—have joined the "We Are Still In" coalition to help fulfill the United States' 2015 Paris Agreement pledge. The coalition of nearly 4,000 signatories includes numerous mayors and governors who represent 173 million Americans and 58 percent of U.S. GDP. Many Fortune 500 companies have already achieved net-zero carbon or have committed to it, including Google, American Express, Apple, Bank of America, Citigroup, Facebook, and Ikea. Walmart, the world's largest retailer, along with its suppliers, aims to reduce its carbon emissions by a globally significant billion metric tons by 2030. That would be the equivalent of taking 211 million passenger cars off America's roads for a year.

But we are off to a very late start, and thus, timely success in stabilizing the climate is far from guaranteed. The scale and urgency of the problems are daunting. We need to pick up the pace of climate action, advance distant timetables, and raise our level of ambition while finding ways to restrain large energy companies, which are still compounding our problems by producing all the fossil fuels they can. Unless their energy is made uncompetitive by cheaper alternatives, or they are forced to reduce production through regulatory policy, they will continue on their present course, because it brings them billions in profits. And as long as their fuel is priced lower than cleaner alternatives, consumers will buy as much of the dirtier fuel as they need. The United States today still counts on fossil fuels for nearly four fifths of its energy needs, and many influential politicians remain opposed to decisive action to wean the economy from them.

This book will explain how fossil fuels can be made obsolete and uncompetitive without sparking energy shortages and higher prices. It will also show how all our energy services can be provided without fossil fuels through clean, efficient, profitable, and increasingly affordable new energy technologies.

One major, very positive development is that the clean-energy economy has already blasted off its launching pad and is rapidly gaining momentum. By 2025, solar and wind power will be the world's largest source of electricity. By 2027, renewable energy will account for more than half the world's electricity-generating capacity, while the share of coal and natural gas will have declined to 40 percent.* For wind and solar technologies that only a few years ago produced so little power that they were scarcely on the radar, this development is truly revolutionary and offers myriad economic benefits.

Three million Americans and counting now work in clean-energy jobs all over the country. At least 350 occupations will expand in the years ahead as demand grows for more workers in energy efficiency, environmental planning and management, energy finance, and

---

* Coal production hit an all-time high in 2022, in part because of natural-gas shortages due to Russia's war in Ukraine. However, the International Energy Agency experts expect its use to drop by almost a third by 2025.

environmental regulation. Today, nearly one in every five new construction jobs is a clean-energy job. Millions more new jobs will be created for American workers, technicians, scientists, engineers, and managers to build new solar panels and power plants, wind turbines, electric and hydrogen fuel cell vehicles, advanced batteries, charging stations, and electrical components.

When a large, new renewable-energy power plant, such as an offshore wind farm, is built, thousands of new jobs are created in construction, operation, and maintenance—and on the payrolls of all the companies in that wind farm's supply chain. The same is true when a major new EV manufacturing plant is constructed. Ford, for example, along with SK Innovation, is building a $5.6 billion mega-plant on a nearly six-square-mile site called BlueOval City in rural Stanton, Tennessee. The company will manufacture its next-generation electric F-Series pickups and advanced batteries there for electric Fords and Lincolns, creating approximately six thousand new jobs. Ford and its partner will also open two advanced lithium-ion battery manufacturing plants in Glendale, Kentucky. All told, the new factories will represent an investment of $11.4 billion and employ nearly eleven thousand workers.

Advanced electric and fuel cell vehicles like those described in chapter 6, with advanced sensors and controls and artificial intelligence, are already much more efficient than conventional vehicles. Moreover, electric vehicles are today only a few years away from being as cheap as gasoline vehicles, and by 2025, they may even be cheaper, as batteries continue their steep declines in price—86 percent from 2010 to 2022. Virtually all major automakers are making large commitments to EVs, with Ford planning to spend fifty billion dollars by 2026.

The clean-energy job boom is not confined to automakers or even to multinational corporations. More than a thousand companies, large and small, are involved just in the solar power and solar storage industry. In an interview with the environmental research and advocacy group E2, Richard Harkrader, founder of Carolina Solar Energy in Durham, North Carolina, said, "When I started my company in

2004, we were one of the first to build solar in the Southeast. Now the solar and wind industry employs over nine thousand people and is a robust part of our region's economy, especially in rural areas. North Carolina leads the nation in rural clean-energy jobs. There is so much potential for clean energy to bring continued economic growth and job creation to our state in communities that need it most."

Civil engineer Carla Walker-Miller, the founder and CEO of Walker-Miller Energy Services, has built her firm into one of the country's largest Black- and female-owned energy efficiency companies. "I feel blessed to employ more than a hundred Michiganders in a growing field and one of the foundational fields as we fight climate change," she told Michigan governor Gretchen Whitmer in a recent interview reported by E2. "It's a doing-well, doing-good industry, and we need more women, more Black people, more brown people . . . [so we can] learn and help not just our environment but our economy. . . . We were able to weather the pandemic without laying off one person, without decreasing one salary."

A Broad Energy Transformation

Rapid progress is also being made in the development of all kinds of new clean-energy technologies, from hydrogen production systems to new wind turbines and solar panels, fuel cells, and energy storage. Biofuels, hydrogen, and synthetic fuels (called synfuels) offer the prospect of carbon-neutral or zero-carbon propulsion for ships and aircraft and heavy equipment that currently burns fossil fuels and would be hard to propel electrically with today's technology. Shipping industry leader Maersk in 2023, for example, launched the world's first oceangoing carbon-neutral container ship, fueled with cleanly produced methanol. (More details on how such technologies work and their implications are in chapter 7.) Affordable energy storage will help expand the use of clean power from wind and solar power plants on the power grid and, in the case of rooftop solar, on homes and office buildings. Various well-known companies—Generac, Sonnen, and SunPower, to name a few—provide battery-electric storage that already makes it possible for consumers to become more, or even totally, energy independent.

Advances in hydrogen technology also hold great promise. Hydrogen is a carbon-free fuel that releases only water vapor when burned. It can be produced by using an electrolyzer.* Hydrogen-powered passenger vehicles are already available and, as hydrogen fueling stations become more common and hydrogen technology advances, hydrogen will increasingly be a suitable and affordable fuel for long-haul vehicles, powering a fuel cell to propel a container ship, locomotive, or semi-tractor truck or being burned directly in a vehicle's engine. It can also cleanly provide high-temperature heat for a steel mill furnace or a cement factory, something few other clean fuels can do. The oxygen co-produced during electrolysis is valuable as an industrial gas.

Hydrogen production facilities can be sited at solar and wind power plants. When a clean power plant produces surplus electricity, this surplus can then be used to electrolytically produce hydrogen, which can be stored until needed to reverse the reaction and generate electricity again in a fuel cell. Thus, more and more of the power grid can be based on renewable energy, without backup from fossil fuel plants. As a bonus, more renewably produced hydrogen will be available to power transportation. And, as its production scales up, costs will come down.**

Hydrogen is just one of many rapidly advancing and lucrative clean-energy technologies. The optimistic projections the investment community is already making about it give an inkling of the future size of the larger global renewable-energy market: Goldman Sachs projects that hydrogen production, shipping, storage, and use

---

* In an electrolyzer, electricity splits water into hydrogen and oxygen. By recombining hydrogen with oxygen in a fuel cell, the cell produces electricity and water.
** Today, most hydrogen is still produced from natural gas, a fossil fuel, rather than from water. That's because, in the absence of widespread or significant carbon taxes, "green" hydrogen costs a lot more. But green hydrogen costs are dropping.

will be a $12 *trillion* business by 2050, and Bank of America analysts think it could lead to the construction of $11 trillion worth of hydrogen-related infrastructure, along with $2.5 trillion in other direct investment.

Technology advances are also bringing major improvements in energy efficiency. Smart, energy-efficient homes and even multistory office buildings, for example, are making it possible for people to live and work in more comfortable, healthier spaces—as we'll see in chapter 4. In addition, in a powerful technological synergy, energy system sensors and controls in new buildings will make those buildings more efficient to operate, saving money and fuel and making on-site renewable-power systems more affordable. And, of course, saving fuel means reducing pollution and greenhouse gases.

You can now see the outlines of a new clean-energy economy emerging. Increasingly, millions of workers will be able to find jobs in clean-energy and energy efficiency industries. The rest of us will be able to live and work in more comfortable, healthier buildings. We'll drive (or be driven in) more energy-efficient cars and trucks, and our lives will create smaller carbon footprints.

A powerful clean-energy revolution is also occurring in industry and manufacturing. You've likely heard about the increased efficiency of home air-conditioning and heating systems. The same revolution is occurring in industry, where more efficient engines, pumps, motors, and ducting are becoming commonplace, along with "green chemistry" technology, which makes it possible to use safer chemicals to synthesize many everyday products, like plastics, with less toxic waste and emissions. The concept of green chemistry is used more broadly in this book to include ways of decarbonizing industries—even heavy ones like steel and cement manufacturing—that until recently were thought of as almost impossible to decarbonize. In the steel industry, for example, a new process makes it possible to produce steel and other metals with clean electricity instead of in a coal-fired furnace.

Great strides have also been made in carbon recycling, capture, and storage and in energy-saving resource recycling and recovery. Your next set of tires, for example, may include ultra-tough fiber made from used plastic bags, reducing the volume of waste plastic making its way into our oceans and, ultimately, through the marine food chain, back into our bodies. The jet you take across the country may soon operate on fuel that microbes helped create from captured CO_2.

Collectively, new technologies like these will enable us to leave the smoke, slag, and noxious emissions of the fossil fuel age behind—permanently. A new generation of ultra-long-lived sodium-sulfur batteries using relatively inexpensive materials is now showing considerable promise in the lab. Designed to provide energy storage for large, grid-connected renewable energy systems, the technology packs four times the energy density of lithium batteries but is potentially much cheaper because, unlike lithium, which is in relatively short supply and has risen thirteenfold in price in the past two years, sodium is abundant and can be refined cheaply from seawater.

Dramatic emissions reductions in agriculture and the food industry are also possible and are beginning to be achieved. Indoor agricultural systems using vertical agricultural and hydroponic systems, for example, use far less land and water, and they produce less waste and greenhouse gas. New aquaculture and aquaponics systems are also under development to efficiently produce both plants and fish. On more traditional farms, "regenerative agriculture" (described in chapter 12) prioritizes the health of the soil, and of the whole agricultural ecosystem, over short-term farm profits, producing more healthful food in more humane, sustainable ways, with less climate impact.

Livestock currently release seven billion tons of CO_2-equivalent greenhouse gases a year. Fortunately, plenty of plant-based meat substitutes are available. You can order a McPlant burger at McDonald's, an Impossible Whopper at Burger King, an Impossible Slider at White Castle, a Beyond Sausage at Dunkin' Donuts, and a Beyond Meatball Marinara sandwich at Subway. Start-up Nature's Fynd is working on a meat substitute produced by microbial fermentation, and Wildtype is using cultured salmon cells to grow "sushi."

In addition to sushi, other tissue culture meats are also in the works. (Theoretically, muscle cells originally taken from an animal can be grown indefinitely under controlled conditions to produce immortal cell lines and create meat without slaughtering animals.)

Although this "lab meat" is not going to appeal to everyone (yours truly included), researchers have demonstrated that Americans who just eat *less* meat would be a lot healthier, and most of us could easily do so. Pound for pound, cultured meats and plant-based substitutes will result in far less methane released from animal waste products, less nitrous oxide (another powerful GHG from chemical fertilizers), less bacterial contamination from animal waste, and less CO_2 from farm equipment. If 340 million of the earth's more affluent people, here in the United States, reduced their meat consumption, it would lead to a significant drop in GHG emissions.

All told, the scientific and technological advances now occurring in energy production, industry, green chemistry, and regenerative agriculture mean that we can do everything we now do with fossil fuels, conventional industry, and extractive industrial-style agriculture, but in cleaner, healthier, more modern ways.

Clean Power, Clean Profits

When a new electric-power technology can be had at lower cost than existing technology, that new upstart will spread like wildfire. That's exactly what we're seeing with clean energy today. The economic advantages explain the massive growth in clean renewables mentioned earlier; to the marketplace, the environmental benefits are icing on the cake—or, in policyspeak, "co-benefits."

In 2018, revenue from advanced energy systems and technologies was more than a quarter of a trillion dollars, growing four times faster than the rest of the U.S. economy. That's a trend from which U.S. investors could profit, and doubtless some already do. Global investment in clean energy, even excluding large hydro, is exceeding that in oil and gas. In another telltale sign of the unstoppable energy transition now in progress, more than a thousand institutions have divested or committed to divest $11 trillion from fossil fuel industries, which are also facing the likelihood of increased financial regulation.

Solar-electric panel costs have dropped by an almost unbelievable 99 percent since 1976.* Not surprisingly, therefore, solar power is the fastest-growing new power source, and solar jobs in the United States are increasing seventeen times faster than employment in general. They also pay more than the median annual wage. The industry is a big engine of economic growth: in 2020, it employed more than 230,000 workers at more than 10,000 companies throughout the United States (and in every state) and generated $25 billion in new investment.

The solar industry is growing by almost 42 percent annually and is poised for even more massive growth. It could provide 40 percent of the nation's power by 2035, according to the Biden administration, and will play a major role in fulfilling the administration's plan to decarbonize the power sector by 2035. By then, it could provide enough power for every home in the United States, says U.S. energy secretary Jennifer Granholm.

Wind power costs, though a little higher than solar, are nose-diving—and stimulating global demand, too. Onshore wind power fell 60 percent in cost in just over ten years. That's thanks to big improvements in technology that make enormous, powerful, and efficient turbines possible. In 2019, land-based wind machines averaged about 3 million watts (MW) in size. Just a year later, they were over 4 MW—and almost the height of a football field. Some newly financed offshore wind projects are now set to use turbines of up to 12 MW in size.

Not only are turbines getting bigger, but so are the wind projects, creating major economies of scale. Not surprisingly, growth in renewable-energy capacity spiked over 50 percent in 2020 compared

---

* In prime locations, utility-scale solar power can now be produced for an astonishingly low two cents per kilowatt-hour.

to 2019. An advanced U.S. wind industry could create new American jobs for workers to meet U.S. turbine demand and to create turbines for export abroad.

New renewable power is now cheaper than fossil power for two thirds of the people on earth, who use 85 percent of the world's energy. So, it's not hard to understand why wind and solar now provide roughly twice as much new U.S. capacity as rival natural-gas generation. Indeed, renewables are now so competitive that they are also attracting enormous amounts of investment capital, beating out fossil fuel competition for new power plant orders as old fossil plants are retired.

With the exception of shale oil and tar sands oil, coal is the dirtiest fossil fuel in terms of emissions of CO_2 and other dangerous pollutants, such as mercury. Coal produces 30 percent of the world's CO_2 emissions and, in 2018, was responsible for the largest increase in emissions of any source. Eighty percent of U.S. coal plants now cost more *just to operate* than the cost of replacing them with brand-new, clean solar and wind energy plants.

Given their high costs and the wariness of investors, not to mention utility regulators, many fossil fuel plants are now closing at an accelerating pace. More than 100 gigawatts (i.e., 100 billion watts) of older, and hence relatively inefficient, U.S. coal and natural-gas plants have been shut down since 2002. More than 300 U.S. coal plants, out of 530, were closed just in the decade ending in 2020, and all remaining U.S. coal plants will likely be gone by 2030. Globally, trillions of dollars in future fossil fuel investments have been canceled.

Renewable-power sources are also challenging nuclear power, whose costs are high, especially for older plants. Despite perennial efforts to revive it, the nuclear industry has stagnated for decades due to these cost concerns and to safety issues.* From 2005 to 2022, ninety-six nuclear power plants around the world were shut down

---

* Safety concerns arise throughout the entire nuclear fuel cycle: in uranium mining and milling, fuel enrichment and fabrication, power generation, waste storage, waste transportation, and nuclear proliferation. Also of concern are routine emissions during operations and the potential for catastrophic accidents due to human error, earthquakes, tsunamis, simultaneous multisystem failures, and terrorist attacks.

permanently, and a quarter of the nuclear plants in advanced industrial economies are likely to close by 2025 due to their advanced age. The decline of fossil fuel and nuclear technologies has created a unique demand opportunity for clean-energy pioneers and financiers to fill.

The World Turns to Renewables

While still very heavily committed to burning coal,* China—the world's largest greenhouse gas emitter—is currently increasing its renewable capacity at a blistering pace and is already at 27 percent renewable power. It plans to get a quarter of its total energy from renewables by 2030, while peaking its carbon emissions in 2026 and sharply increasing its energy efficiency. It is currently covering the equivalent of one soccer field every hour with solar panels and, in 2017, it averaged more than one wind turbine installation every hour of every day. The Chinese invest almost twice as much in renewables as the United States—from 2010 to 2020, $918 billion, compared to $528 billion. China's heavy investments have made it the world's largest manufacturer of solar, wind, battery, and EV technology and the world's biggest financier of state-of-the art renewable-power plants in developing countries.

The United States is second only to China in renewable-energy development, but, as noted earlier in this chapter, our nation is still heavily dependent on fossil fuels. About 78 percent of all U.S. energy and, in 2022, 58 percent of U.S. electricity still come from fossil fuels. Despite all our national polemics about clean energy, we're still the world's third largest coal producer and user. However, as previously described, the United States has also made tremendous progress in renewables, doubling its renewable capacity over the past decade. Domestic wind power capacity has tripled, and solar capacity has

---

* China approved more coal plants in 2022 than at any time since 2015 and, in the first quarter of 2023, approved more than in all of 2022. Poornima Weerasekara, "China approves coal power surge despite emissions pledge," AFP/Energy Daily, April 24, 2023.

grown eightyfold, over the past ten years. In the first six months of 2022, 24 percent of U.S. electrical power came from nonnuclear renewables. Some U.S. solar power plants now even include energy storage, making them better able to provide around-the-clock power.

Despite the already remarkable drop in solar energy costs, the Biden administration in 2021 announced a goal of driving down the cost of large solar power plants by another 60 percent by 2030. That's in service to the administration's plans to see the whole U.S. electricity-generating system decarbonized by 2035. Although power sector emissions have already fallen by more than half just in the past fifteen years, the administration in 2021 proposed a ten-year extension of clean-energy production and investment tax credits to keep up the momentum and accelerate it if possible. The administration also plans to use clean energy in all federal buildings and has proposed an investment tax credit specifically to provide incentives for the creation of long-distance, high-capacity power lines, so that solar power from remote, hot deserts and wind power from the windy plains of the Midwest can reach power-hungry cities elsewhere.

Vanguard Communities

For decades, opponents of government support for clean energy have argued that shifting to it would be too costly and would hurt workers and the economy. But energy efficiency, clean-energy policy, and economic prosperity are fully compatible. California provides a glimpse of how clean energy can help grow an economy. The state has the world's fourth largest economy; is the country's largest producer of solar, wind, biomass, and geothermal electricity; gets two thirds of its power from non-carbon sources; and had an estimated $38 billion state budget surplus in 2021, despite the COVID-19 pandemic—yet it had the fourth lowest energy consumption per person in the United States. Meanwhile, according to data available in 2021, the state's climate policies had cumulatively pumped more than $102 billion in public and private investments into its economy, helping almost half a million Californians obtain jobs in clean-energy industries and cutting emissions deeply—equivalent to taking 14 million cars off the road for a year.

California is not alone in the nation in expanding its reliance on renewable energy. Iowa, Kansas, and North Dakota, for example, each gets more than half its electricity from wind power. Maine, Oregon, South Dakota, Vermont, and Washington get over 70 percent of their power from renewable sources.

Abroad, some 220 subnational governments have voluntarily committed to bringing their emissions to net zero by 2050, in the hope of keeping the rise in global temperature well below 2°C. The signatories of the Under2 Coalition collectively represent 1.3 billion people and 43 percent of the global economy. Like the United States, dozens of other nations are planning on a future with 100 percent clean power. (Worldwide, more than 140 cities already use 70 to 100 percent clean power.) Countries with unique resources—such as Iceland, with its large hydro and geothermal power resources—already obtain virtually 100 percent of their power from renewable energy. Norway gets 98 percent of its power from renewables, mainly from hydropower. Sweden gets essentially 100 percent of its power without fossil fuels. More than 80 percent of Denmark's electric power comes from renewables (most of it from bioenergy), and the country plans to achieve 100 percent by 2035. Costa Rica is already at 100 percent; Uruguay is at 95 percent. On April 5, 2020, even the United Kingdom, cradle of the Industrial Revolution, got 80 percent of its power from non–fossil fuel sources.

One community has been able to achieve not just 100 percent clean power, but the more challenging goal of providing 100 percent clean *energy*. Rural Samsø, an island off the Danish mainland, was once a nondescript farming and fishing region of twenty-two villages, home to 3,800 people. Now, however, it has become well known as a community where people not only rely on but enthusiastically support, invest in, and profit from renewable energy.

Since 1998, the island's transformation has been led by Søren Hermansen, who previously headed the Samsø Energy and Environment Organization and now leads the Samsø Energy Academy.* Hermansen is an idealistic yet pragmatic vegetable farmer turned high school environmental studies teacher who knocked around the world for years on Norwegian fishing boats before returning to Samsø, where he was born and raised. When he first got involved there, the island was dependent on increasingly expensive imported oil for heat and on electricity generated in coal-fired plants. Fishing and farming were in decline. Jobs were scarce. Something had to be done to revitalize the economy.

Thanks in large part to Hermansen's efforts, and to supportive government policies, Samsø today gets 100 percent of its power from the sun, the wind, and the earth, save for the hybrid electric/natural-gas ferry to the island (soon to be converted to biofuel) and Samsø's remaining internal combustion engine vehicles. Each citizen of Samsø today accounts for *minus* 12 metric tons of CO_2 emissions a year, whereas the average Dane produces more than 6 tons, and Americans produce 15.

Much of the clean power on Samsø comes from onshore and offshore wind turbines. Samsø exports a surplus 80,000 megawatt-hours of solar and wind power a year, and owners of the turbines—local farmers, co-ops, and the municipality—enjoy millions in profits as co-investors in the island's energy projects. Instead of heating with imported oil, Samsø now gets heat from geothermal heat pumps and from boilers in four local district heating biomass plants fired with local straw and forestry wastes. Farmers get paid for the nearly carbon-neutral straw, which burns far more cleanly than oil, and the ash goes back to farms as fertilizer. Local plumbers won contracts to install the island's geothermal heat pumps and the piping that sends hot water from the district heating plants to radiators around

---

* Whereas the former organization still exists as an NGO, Samsø Energy Academy is doing the majority of the island's climate and renewable energy work and international outreach.

the community. Danish kroner are invested in the community, not in imported oil.

The Samsø economy is carbon neutral, and its carbon emissions today are close to zero. By 2030, it plans to be totally independent of all fossil fuels, but the community is benefiting economically now. As of 2015, the little island's renewable-energy investments had already exceeded eighty million dollars. Samsø consumers enjoy cheaper, cleaner electricity, while heating costs on the island have fallen 40 percent.

The Samsø case suggests what we, in the United States, could accomplish—thanks in part to the efforts of the climate savers we'll be meeting. Of course, the transformation of Samsø could not be precisely replicated in Manhattan. But it's not hard to imagine that these and similar sources of income might be appreciated in parts of depressed rural America now hurting from the nation's loss of manufacturing jobs.

Make No Little Plans

A range of ever more economical, clean technologies being advanced by the climate savers in this book is gaining widespread support and will eventually be able to meet all our energy needs sustainably and profitably. At the same time, the climate savers' work highlights that the pace of the clean-energy transition must move more quickly. It also shows that climate saving means not only turning off the CO_2 faucet but also, ultimately, putting the genie back in the bottle by removing from the air excess atmospheric greenhouse gases already emitted.

Trillions of public and private dollars a year will need to be invested to create a clean-energy world. Millions of new solar and wind energy devices will need to be built and paid for. More than one hundred million buildings in the United States will need to be made more energy efficient, with new energy-efficient windows, better insulation, electric heating and cooling, heat pumps, solar panels, and high-efficiency appliances. Hundreds of millions of new electric and fuel cell vehicles will be needed, along with hundreds of thousands of charging stations and hydrogen fueling stations. In addition, if carbon is to be widely captured at cement plants and other factories—an arguable proposition, but one often heard—then thousands of new carbon capture devices will need to be installed at all these facilities, and tens of thousands of miles of CO_2 pipelines will be needed to transport the captured CO_2 to injection wells. Finally, if airplanes, ships, and engines for heavy construction equipment are to operate on net-zero-carbon fuels, enormous new sustainable-fuel industries will need to arise from what are now small companies.

In short, lots of problems will have to be solved and lots of opposition surmounted. The size and scale of the challenges ahead are daunting. Where, too, is the money for all this to come from? In the next chapter, you will see how the United States can readily pay for its clean-energy transformation while earning trillions of dollars in new wealth and putting millions of Americans profitably to work.

2

Show Me the Money: Paying for the Clean-Energy Transformation

In an NPR interview in the summer of 2021, U.S. Special Presidential Envoy for Climate John Kerry noted, unsurprisingly, that trillions of dollars were going to be needed for a rapid clean-energy transition. Kerry made news, however, when he told NPR that U.S. representatives had been in discussions about clean-energy finance with the six largest banks in the nation and that these banks would now set aside more than four trillion dollars over the next ten years to accelerate the clean-energy transition. "And that," said Kerry, "is a floor, not a ceiling." Those commitments are an important step forward, but they still cover only a fraction of the total amount needed globally to stabilize the climate.* That total is, nonetheless, definitely well within reach. The United States and other nations simply have not been trying hard enough—yet.

For example, at the 2009 UN Climate Change Conference, held in Copenhagen, the developed nations of the world committed to providing the world's developing nations with one hundred billion dollars a year to help fend off climate change. Their commitment was reaffirmed at the 2015 UN Climate Change Conference, in Paris.

---

* The International Renewable Energy Agency (IRENA) has estimated that $110 trillion will be needed between 2016 and 2050 for the world to have a chance of keeping temperature rise to 1.5°C.

Yet, on a global scale, it's a tiny sum, especially given the urgency and magnitude of the need. Even so, at the time of the Glasgow-held UN Climate Change Conference, in November 2021, more than a decade later, the sources of even that manifestly inadequate funding still had not yet been fully identified. Fortunately, Special Envoy Kerry seized the occasion to pressure some donor nations and was able to secure a major funding commitment from Japan that at last promised to fulfill the long-awaited financing commitment.

Apart from the four trillion dollars in commitments by the U.S. banking sector and federal loan guarantees (as will be discussed in the next chapter), many other financial mechanisms exist for advancing the clean-energy transition. During the financial downturn initially created by the COVID-19 pandemic, the Federal Reserve Bank went on one of its periodic bond-buying sprees (known as "quantitative easing"), to hold down interest rates. It purchased $120 billion a month in Treasury bonds and mortgage-backed securities on the open market, at a cost of almost $1.5 trillion a year, all without a net increase to the U.S. debt—just one indication of the U.S. government's enormous financial strength.*

It's not difficult to think of many ways that the federal government could use this kind of financial might to supercharge a rapid clean-energy transition. It could, for example, establish a generously funded "green bond" program. Green bonds are similar to ordinary bonds, in which the issuer borrows from investors for a fixed period at a set interest rate to finance a project or ongoing activities. The difference is that, with a green bond, the capital is earmarked for projects aligned with the clean-energy transition.

Currently, there is an explosion of growth in green bonds in the private sector. Green bonds offer both greater safety and greater convenience to investors who might be too risk averse to directly finance a renewable-energy project. That's because green bonds spread project

---

* Because bonds are essentially a loan from the buyer to the seller, these funds are eventually repaid with interest to the Fed, so there is no long-term increase in the national debt.

risk across more investors, can be bought in convenient denominations, and offer liquidity, as they can be traded on the bond market. To increase investors' confidence, the creditworthiness of green bonds, like that of other bonds, is rated by credit-rating agencies. In addition, if the bond is issued by a government, the interest rate can be lower, because the bond is backed by the government's full faith and credit. The government can then use the funds raised to inexpensively finance investments in the transition to clean energy.

Investment in green bonds has grown very rapidly and, since their introduction in 2006, is now cumulatively in excess of two trillion dollars. The prestigious Climate Bonds Initiative, an expert international financial organization, has called for the annual issuance of five trillion dollars in green bonds by 2025. "We have adequate global capital available," the organization asserts, "and the rapid market growth to date has demonstrated the appetite for capital to move." As one analyst put it, "The green bond phenomenon has taken the investing world by storm." Governments, multinational organizations, financial institutions, and corporations are all getting in on the act: Canada has announced the launch of a green bond program with a five-billion-dollar first issue. Germany and twenty-two other nations have now also issued green bonds—but, regrettably, the United States has not. Ultimately, the concept could generate a much-needed global flood of capital for energy transition projects. My guess is that it's just a matter of time before the United States falls in line with other nations and issues green bonds, too.

One possibility is that qualified banks around the country could issue their own green bonds at low interest rates. The banks would then be able to lend the capital locally or regionally at slightly higher, but still low, rates to qualifying clean-energy projects.* There's ample precedent for "single-purpose bonds" such as green bonds: during World War II, 84 million patriotic Americans bought war bonds to

---

* Various international standards already exist, including the Green Bond Principles and the Climate Bonds Standard, supplemented by third-party certifiers such as Moody's and Sustainalytics.

help finance the war effort, even though war bonds paid less interest than ordinary commercial bonds. Green bonds, or "climate bonds," would help us fight a different kind of war.

A "green Treasury bond" might be just what's needed to help win the international campaign to save the climate, says financial analyst Andrew Poreda. "Treasuries account for a staggering 38 percent of the investment-grade U.S. bond market, which currently means that a large percentage of investment dollars are doing nothing unique from a sustainability perspective." But all stakeholders would stand to gain from a green Treasury bond, Poreda observed. Investors could fill the Treasury portion of their fixed-income portfolio with interest-bearing green bonds, and the U.S. government could use the proceeds to combat climate change.

Clean-Energy Transition Costs and Savings

Various studies have found that a clean-energy transition would cost no more than 2 percent of gross domestic product (GDP) in the United States. I would suggest that we maintain a healthy agnosticism about that figure at this point, because some macroeconomic effects on the economy as a whole—stemming from the large increase in both private and government investment, tax credits, subsidies, and new energy regulations—might ultimately entail larger costs that have not yet been fully analyzed or foreseen. I will take that 2 percent of GDP as a reasonable starting point, however, as many analysts have converged around that early estimate.

In 2021, the U.S. GDP was about $23 trillion, so 2 percent of GDP is about $460 billion. Yet, historically, the United States has spent 6 to 13 percent of GDP on energy, mostly on fossil fuels that would largely no longer be needed in the clean-energy economy. So, if the nation were to pay 2 percent of GDP for an accelerated clean-energy transition—and, therefore, eventually save at least a net 4 percent of GDP by not buying fossil fuels and even more through deep efficiency and clean, renewable electric energy—it would actually be *making* a lot of money, long term, in freeing its economy from fossil fuels.

That's a pretty good deal, given all the other economic, environmental, and health benefits the United States would also receive, essentially, as free bonuses—or co-benefits. Moreover, the infrastructure costs for the clean-energy economy might turn out to be *less* than 2 percent. The studies that came up with that 2 percent estimate are based on older costs for clean energy, but because the costs of renewable energy and energy storage have been plummeting for decades now, the upfront costs of the clean-energy transition could be significantly less than 2 percent of GDP, especially in real (that is, inflation-adjusted) dollars.

One reason the true net costs to the economy would ultimately be lower than the initial costs is because many current societal costs would be reduced or eliminated with the elimination of fossil fuel use and combustion. For example, by weight, we in the United States transport more coal, crude petroleum, gasoline, kerosene, ethanol, diesel, and other fossil fuel products than any other category of bulk commodity. Just moving 6.3 billion tons of fossil fuels by rail, ship, pipeline, and truck each year consumes a huge amount of energy that would no longer be needed—not to mention all the energy we would no longer need to expend in mining or drilling for these fuels and then processing and refining them; nor the money spent by consumers to buy them.

Public health and environmental costs would drop significantly, too: Nearly 20 percent of all deaths worldwide—those of some eight million people a year—are due to fossil fuel air pollution. It causes myriad health problems, including heart and lung diseases, stroke, cancer, and acute respiratory illnesses. If, on an actuarial basis, a human life is worth one million dollars—and some estimates have placed the value five to ten times higher—then avoiding eight million deaths is actually worth at least eight trillion dollars a year. Of those deaths, about sixty thousand are in the United States, so the domestic mortality costs of air pollution alone are sixty billion dollars, by conservative estimates, and that doesn't include costs for illnesses and property and crop damage.

In addition, the environmental costs of fossil fuel production, transportation, and consumption are staggering in terms of land degradation, wildlife and habitat destruction, and water pollution—all from the impacts of exploring, drilling, mining, refining, and transporting coal, oil, and natural gas. Those avoidable costs are significant. The Natural Resources Defense Council puts it succinctly:

> Unearthing, processing, and moving underground oil, gas, and coal deposits take an enormous toll on our landscapes and ecosystems. The fossil fuel industry leases vast stretches of land for infrastructure, such as wells, pipelines, and access roads, as well as facilities for processing, waste storage, and waste disposal. In the case of strip mining, entire swaths of terrain—including forests and whole mountaintops—are scraped and blasted away to expose underground coal or oil.

It's hard to put a price tag on such damage, which is done each year, much of it on public lands that are leased to oil and gas companies for royalties as low as two dollars an acre.* On top of the unpriced damage costs, the industry has left the public at least 57,000 abandoned oil and gas wells (some say ten times as many) and tens of thousands of abandoned coal mines, for which the public must pay the untallied tab. So, even without accounting for all the economic and job benefits of the clean-energy transition and the totally unacceptable costs of uncontrolled climate change, halting fossil fuel use as quickly as clean-energy alternatives can be made available looks like a sensible course of action.

Whereas the U.S. government arguably might need to rearrange some priorities to come up with between $400 and $500 billion a

---

* Royalty rates haven't been raised in a hundred years.

year for the clean-energy transition, if the government raised that sum in Treasury bonds at 3.6 percent (the rate on the thirty-year U.S. Treasury in 2023), the interest cost on $500 billion a year would be a very manageable $18 billion a year. At the early 2023 inflation rate of 6.5 percent, however, $500 billion would lose $32.5 billion a year in purchasing power, so the U.S. Treasury would actually *make* $14.5 billion a year in real dollars by borrowing the money. Thus, the investment would actually be profitable for the government while benefiting the climate.

Misplaced Priorities

Of course, $460 (2 percent of our GDP) or even $500 billion would still be a lot of money to come up with if funded out of tax revenue—if, for some reason, it were not financed at least to some significant degree through federal borrowing. However, the Fed and the U.S. Treasury have in recent times made *trillions* of dollars in revenue available to address domestic financial problems that ultimately have far less impact on public health, the environment, and national security than does global climate change. In addition, Congress has spent enormous sums on questionable foreign military campaigns with few clear benefits.

The 2001–21 War in Afghanistan, for example, cost $2.26 trillion. Ironically, the vast majority of that spending was unnecessary to the disruption of Al Qaeda and the elimination of Osama bin Laden, the original stated reasons for launching the war. All told, the United States since 2001 has spent $6.4 trillion on its "Global War on Terror" in Afghanistan, Iraq, Syria, Pakistan, and Yemen. Yet it failed to achieve its goals for *any* of these foreign missions, leaving tragedy in its wake while also virtually ignoring the towering global and domestic risks posed by climate change. The Iraq War, launched on the false assertion that Saddam Hussein had weapons of mass destruction, cost taxpayers over $2 trillion outright. Just the interest incurred on the debt for that fighting has already exceeded $925 billion. In addition, the war produced a staggering toll in deaths, injuries, and displacements. Meanwhile, our annual military budget remains larger than those of the next nine countries combined.

Now consider a hypothetical: Suppose climate change historically had been a hot-button issue for voters, one that determined the outcomes of elections. Suppose that, instead of squandering its wealth on unnecessary or ineffectual military efforts, the United States had instead spent $6.4 trillion over the past twenty years combating the climate emergency. If this had been the case, the United States and the community of nations at large would today most likely be facing a far less severe climate threat.*

The disproportionate nature of the U.S. response to the climate emergency compared with the U.S. response to terrorism is but one demonstrative example of our not being very good at calibrating our response to risks, especially longer-term risks. Terrorism, which has attracted an avalanche of funding to counter it, has killed an average of 26,000 people a year *globally* for the decade 2009–19, accounting for less than 0.01 percent of all deaths in almost every nation outside Africa and the Middle East. If we have the wealth to squander trillions of dollars on numerically insignificant threats and on wars of choice, we should be able to find the sums needed to address a genuine and existential threat like the climate emergency.**

For a final example of how much financial firepower the nation can call on when it decides to do so, one need look no farther than the financial crisis of 2008, the "Great Recession." To support the economy, the U.S. government provided net bailouts (outlays less

---

* Some would also argue—correctly, I believe—that we would be facing a lesser terrorism threat as well, as the instability generated by the wars mentioned has been a seedbed for terrorism and because climate change impacts themselves are a frequent cause of political instability.

** Some people regard the military budget as sacrosanct, no matter how large it may be, not believing that any emergency justifies trimming it. Although combating the climate crisis does not *require* that we do so, it's worth considering that reducing the rising U.S. military budget ($741 billion in fiscal year 2022) by just $100 billion, as has been recommended by the Center for American Progress, would in itself finance more than a fifth of the nominal upfront gross capital needs of the clean-energy transition.

eventual repayments) totaling $4 trillion in 2009; $3.5 trillion on June 30, 2010; and another $3.4 trillion on September 30, 2010. On top of that, bailout guarantees by the U.S. Treasury, the Federal Reserve, and other federal agencies, including to some of the wealthiest financial services companies in the country, totaled more than $16 trillion in those periods. The nation's ability to spend on this enormous scale demonstrates unequivocally that the United States has more than enough money to expedite a clean-energy transformation. By comparison, $460 billion a year is chump change. But, again, public understanding and support are required to loosen those purse strings. If public support is to be expected, it's critical that this larger financial context be well explained to the public. If the public does not recognize adequate climate funding as an urgent necessity, or believes that we can't afford to fully address it, then it will not be of deep concern to politicians and decision makers, either.

Raising Revenue for the Climate

If Congress did decide to raise additional tax revenues to combat climate change, many analysts have found that upper-income earners and corporations could easily and painlessly afford to pay a little more. The wealthiest 1 percent of all Americans holds $25 trillion in wealth, more than the combined wealth held by 80 percent of the rest of the public. During the COVID-19 pandemic, these super-rich individuals actually *added* another $6.5 trillion to the value of their corporate equities and mutual funds. In addition, according to a wealth tax plan released by Sen. Ron Wyden of the Senate Finance Committee, "the average family worth more than $100 million has never paid tax on more than half of its wealth."

A wealth tax of only 2 percent on these super-wealthy people would thus scarcely be felt, but it would provide $500 billion a year, likely entirely covering the annual upfront costs of a clean-energy transition. And $25 trillion is but the icing on the financial cake that the wealthy are enjoying: while the bottom half of households possesses only 1.6 percent of the nation's wealth, the top 10 percent has *$74.5 trillion*. A 2 percent slice of that number would generate $1.5 trillion annually—which could make for a pretty snappy clean-energy transition. The question is, how do we convince our fellow citizens to care enough to mobilize political support for this plan? And how do we convince the wealthy to care enough to pay a little more in taxes if that's what it takes to protect the planet, the environment, and our way of life?

Some may object to or even be outraged at the notion that wealth should be taxed. Yet taxing wealth is not a new or radical idea: Norway has had a wealth tax for more than a century. Moreover, we regularly tax homes and buildings via property taxes, and many homeowners liable for those taxes are far from wealthy. We also tax vehicles, another form of property and one that is by no means reserved for the elite. A sensible wealth tax would not apply to the assets held by ordinary working people in their savings accounts or 401(k) plans. It would simply assess a fee on assets like very large stock and bond holdings, yachts, mansions, and others luxuries held by the superrich.

One co-benefit of a wealth tax would be a reduction in the extreme inequality in wealth in the United States I've just described. While the top 5 percent of U.S. households have increased their wealth 500 percent since 1985, ordinary people hold little, if any, wealth and have seen their real wages largely stagnate since then. Tax reform measures could collectively provide up to *$10 trillion* for the clean-energy transition without burdening ordinary working people. Moreover, the wealthy among us nowadays can already leave their offspring $11 million in tax-free inheritance and can further circumvent taxes by setting up trusts to pass even more tax-free assets to their heirs, creating hereditary financial dynasties of super-wealthy offspring. Without touching the modest inheritances that middle- and low-income people receive, additional tens of billions or more could be painlessly harvested for a clean-energy transition just by firmly closing these and other large tax loopholes for the wealthy.

Many other tax mechanisms of varying complexity, scalability, and political feasibility could be used to fund the clean-energy transition and climate restoration. These range from charging wealthy investors higher capital gains taxes to adding fees to the trillions of securities transactions that occur every year in the stock market. A $0.02-per-transaction tax on these trades, for example, would raise $50 billion a year.

Raising the capital gains tax rate for those with incomes over $500,000 a year is appealing to some. The idea is controversial, however, as it might perversely lead to a *loss* in tax revenue if investors responded by holding their capital gains longer before realizing them, in order to avoid the tax. Instead of taxing capital gains on stocks and bonds only when they're sold, taxing them on an accrual basis each year (meaning, according to their annual gain in value) would be more effective and would raise an extra $1.7 trillion in tax revenue over a decade—$170 billion a year—according to the Brookings Institution. As the saying goes, "A billion here, a billion there, and pretty soon you're talking about real money."

Not only could the nation capture big bucks for the clean-energy transition, and make the tax system more equitable simply by closing gaping tax loopholes, as explained earlier, it could also do so by collecting delinquent taxes from wealthy taxpayers. A 2021 expose published by ProPublica documented that many of the nation's ultra-rich, including well-known billionaires like Jeff Bezos, Elon Musk, Michael Bloomberg, and George Soros, scandalously (but legally) pay little or nothing in taxes. In addition, in the decade between 2010 and 2019, some of the nation's largest corporations, like Apple, Facebook (now Meta), Google (now Alphabet), Microsoft, and Netflix, managed to legally avoid more than $100 billion in taxation. In remarks at the White House on August 24, 2021, President Biden noted that "fifty-five of our largest companies in America paid zero dollars in taxes on more than $40 billion in profits last year." Finally, wealthy executives of the $4.5 trillion private equity industry, and other rich people who use private partnerships as investment vehicles, are able to use various tax dodges to avoid $75 billion in taxes per year. Over a decade, that's three quarters of a trillion dollars lost to the Treasury, a colossal windfall bestowed on people who have little need for public largesse—while the climate emergency remains neglected.

Public-Private Partnerships

The clean-energy transition could readily be paid for even without new taxes on wealth or the raising of other taxes. Currently, trillions of dollars in private capital sits on the sidelines of the economy while its owners scout for promising investments. The nation has something like $3.6 trillion in checking accounts, $12 trillion in time deposits (i.e., interest-bearing accounts with dates of maturity) and savings deposits, and $3 trillion in money market funds. Not all this capital, of course, is readily available for clean-energy investment, but prevailing interest rates in the economy in recent years were at historic lows. With interest rates of under 0.6 percent for the ten-year Treasury note in 2020, under 1 percent for the twenty-year Treasury bond, and not much more for the thirty-year Treasury bond, holders of private capital would have been delighted instead to invest torrents of money in federally guaranteed and sound renewable-energy and energy efficiency projects, even at very modest rates of return that would nonetheless have been above Treasury yields.* If a mere 2 percent of the trillions in private capital just mentioned flowed into clean-energy investments annually, that itself would practically pay for the estimated gross costs of the transition.

U.S. corporate profits—let alone corporate borrowing power—are another important source of investment capital: more than $2.5 tril-

---

* The exceptionally low Treasury interest rates mentioned here were below the typical rate of inflation from 2016 to 2019. When the inflation rate is greater than the interest rate, the government actually *makes* money on its borrowing, because the purchasing power of the money it has to return plus interest due is less than the purchasing power of the money it borrowed. This, therefore, would have been a particularly opportune time for the government to finance long-term public investments of all kinds, especially in clean-energy infrastructure.

lion in the third quarter of 2022 alone. U.S. corporate deals (another indicator of U.S. wealth) in 2022 totaled $1.73 trillion just for the first seven months of the year. Again, with proper tax inducements, loan guarantees, and a national commitment to meet bold emission-reduction targets, large amounts of capital from the private sector could be attracted to renewable-energy and energy efficiency investment.

Throwing Bad Money After Bad

In addition to enacting some of the many measures discussed for generating public and private revenue to funnel toward clean-energy initiatives, counterproductive direct and indirect subsidization of the fossil fuel industry has to be eliminated. For example, in the years 2016 to 2020, following the Paris climate summit, twelve major central banks "failed to prevent" or directly financed the flow of $3.8 trillion into fossil fuel industry coffers. Some of this money went directly to fund the exploration and development of new fossil fuel sources. Despite some high-minded rhetoric about the climate emergency as an existential crisis, "central banks are not using their influence as regulators of commercial banks to stem the flow of capital to polluting activities," according to environmental news outlet DeSmog. The 2021 report *Unused Tools: How Central Banks Are Fueling the Climate Crisis*, authored by Oil Change International, provides clear guidelines that banks ought to follow to deter further fossil fuel industry financing. Efficiently combating climate change entails not only strategically financing the "goods," but also refraining from funding the "bads."

This is even more the case when you consider the vast and reckless subsidies that the fossil fuel industry presently enjoys. The G20 nations, for example, have provided $3.3 trillion in subsidies for fossil fuels just since the Paris Agreement in 2015. Although the G20 nations officially agreed in 2009 to phase out "inefficient" fossil fuel subsidies, those subsidies remain robust today, with 60 percent going to fossil fuel companies and 40 percent to consumers in the form of lower fuel prices. Doing away with these subsidies could eliminate at least 5.5 billion metric tons of CO_2 emissions per year. It would also make trillions available for funding solar and wind power plants, electric vehicles, biofuels, and energy efficiency in buildings.

Care must be taken in crafting these policies.* The climate case against subsidies, however, is nonetheless stark. A recent working paper by the International Monetary Fund projected that in 2017 alone, global fossil fuel subsidies amounted to $5.2 trillion. It concluded that ending these subsidies in 2015 would have had tremendous benefits by "lower[ing] global carbon emissions by 28 percent and fossil fuel air pollution deaths by 46 percent, and increase[ing] government revenue by 3.8 percent of GDP." Subsidies for fossil fuels are not only causing mortal damage to the planet; they are also helping to "create and maintain fossil fuel infrastructure with a 30- to 50-year operating lifetime, thereby locking in future carbon emissions." Simply by redirecting existing capital flows away from the fossil fuel industry and toward clean energy, the transition could be funded on a "champagne budget."

We have now shown that the United States and its enormous economy have more than sufficient financial resources to fund a robust and rapid energy transition once we make a serious decision to go full speed ahead to create a clean, just energy future for ourselves and for posterity. The political will—both of the public and of government at home and abroad—must be mobilized to do so.

---

* Simply cutting off all subsidies to consumers would unintentionally inflict hardships on ordinary people, but there is little harm in reducing subsidies going to fossil fuel companies, many of which seized on the supply disruption caused by Russia's 2022 invasion of Ukraine as an opportunity to further raise prices. One solution is to provide generous but means-tested public energy bill assistance to lower-income consumers, while refraining from actions like lowering fossil fuel prices across the board, which would perversely stimulate consumption.

3.

More Smart Ways to Finance the Transformation

The 2021 UN Climate Change Conference, held in Glasgow, Scotland, opened with a dire warning from UN secretary-general António Guterres: "Our addiction to fossil fuels is pushing humanity to the brink. We face a stark choice: Either we stop it—or it stops us. It's time to say: enough. . . . Enough of burning and drilling and mining our way deeper. We are digging our own graves."

The world, said Guterres, is now on track to experience at least 2.7°C of heating over pre-industrial temperatures—making our planet hotter than at any time in the past three million years. This would mean a fearsome intensification of the devastating climate impacts we're already experiencing at the current 1.2°C. But Guterres held out hope: if the world invests heavily in a net-zero-carbon, climate-resilient economy, he said, that would "create feedback loops . . . virtuous circles of sustainable growth, jobs, and opportunity."

Often lost, as public figures like Guterres call for a rapid clean-energy transition, is not only the question of how much this will cost, but also sober discussion of how to pay for it and whether we can afford to do so. Defenders of the status quo predictably point to the $3.3 to $5 trillion annually that may be needed across the globe as if that were an insuperable roadblock. Yet we saw in the previous chapter how readily the United States could afford the transition, and that argument can easily be extended to other developed nations. Moreover, the trillion-dollar scale of the nascent international green bond market is an indication that the world as a whole also can afford it.

The Glasgow Financial Alliance for Net Zero (GFANZ) announced at the 2021 summit that GFANZ's 450 members from forty-five nations were committing to reducing their own emissions to net zero by 2050 or sooner. Those members control $130 trillion in assets, so they are financially capable of funding the clean-energy transition. It therefore seemed highly encouraging that GFANZ was officially committed to transitioning the entire world's financial system, including all major economic sectors, to net-zero emissions by 2050, at a pace ostensibly consistent with keeping global heating to no more than 1.5°C. It sounded too good to be true—and it was.

Staying below 1.5°C of warming when the world is already at 1.2°C is now extremely challenging given that at the current warming rate of 0.2°C per decade, we are likely to hit that mark by 2038.* Remaining below 1.5°C would require that the nations of the world collectively slash their emissions 50 percent by 2030, less than a decade from now. GFANZ's pledge, and the resources and instruments the group could mobilize, would mean that the alliance could be the *most important development ever* in global climate transition finance—if only its actions were aligned with its rhetoric.

Since joining GFANZ, however, major financial institutions, such as Mitsubishi UF3, have continued to invest hundreds of billions of dollars in fossil fuel expansion. A new report, "Throwing Fuel on the

---

* New studies indicates we may hit 1.5°C in the 2020s: https://insideclimatenews.org/news/26052023/james-hansen-climate-change-2-degrees-2050/. Separately, the UN World Meteorological Organization has found that there's an even chance that the world's average temperature will temporarily exceed 1.5°C in the 2020s. "Climate: World getting 'measurably closer' to 1.5-degree threshold," UN News, May 9, 2023; another new study by Noah S. Diffenbaugh and Elizabeth A. Barnes ("Data-driven predictions of the time remaining until critical global warming thresholds are reached," Proceedings of the National Academy of Sciences, January 30, 2023, https://doi.org/10.1073/pnas.2207183120) projects that durable warming will reach 1.5°C by the early 2030s.

Fire: GFANZ Financing of Fossil Fuel Expansion," found that the 161 GFANZ members studied have been funding 229 of the world's largest fossil fuel developers to build new coal-fired power plants, develop new coal mines, and open new oil and gas fields, pipelines, and terminals. All this activity is likely to lock in decades of additional massive carbon emissions. Commenting on the behavior of these financial institutions, Lucie Pinson, the director of Reclaim Finance, which led the study, declared, "It is business as usual for most banks and investors who continue to support fossil fuel developers without any restrictions, despite their high-profile commitments to carbon neutrality."

This kind of virtue-signaling hypocrisy by big banks might be even more dangerous than outright villainy. The dissembling disarms opponents and creates a false sense of security instead of bringing to the fore the critical question: "What can be done to stop these powerful financial organizations from undermining the world's climate?" Perhaps only governments are powerful enough to proscribe the banks' irresponsible activities or, alternatively, to provide sufficient financial incentives to make green and sustainable investing more profitable for them, so they forgo their fossil fuel investments out of self-interest.

Apart from bankers who can't resist delicious fossil fuel profits, another concern is whether funds intended to support a clean-energy transition will actually be strategically or soundly invested; money designated for protecting the climate could instead be wasted on unwise investments in direct air capture, bioenergy, costly nuclear plants, or "blue" hydrogen. Former U.S. vice president Al Gore, now the cofounder of Generation Investment Management, a sustainable asset management firm, has expressed concerns that climate investment is not generally being targeted at those projects with the greatest potential climate impact. Instead, investors typically and predictably seek to maximize their "risk-adjusted returns." Thus, Gore and his colleagues found that some of the most important sectors of the economy, producing half of all global emissions, are receiving only 10 percent of climate finance capital. Similarly, they found that relatively little climate capital is making its way into the developing world or to hard-to-decarbonize sectors of the economy. In 2021, the firm founded a new subsidiary, Just Climate, to focus on investing for climate impact and on hard-to-decarbonize industries.

Windfalls to Come from a Speedy Transition

As this chapter will reaffirm, affordability is not a major hurdle on the path to a clean-energy economy globally, and certainly not in the United States. In fact, the levelized cost of operating the United States entirely on clean energy in 2050, for example, will actually be *less* per unit of energy than it is today with the current energy system.* Moreover, transforming even the most difficult-to-decarbonize sectors of the economy (discussed in part II of this book) will ultimately be profitable. Yet, while the nation and Congress have the political will to enact large tax cuts—the Bush, Obama, Trump, and Biden tax cuts, which predominantly benefited corporations and the wealthy, collectively cost the Treasury $10 trillion without causing so much as a national financial hiccup—the will to spend comparable amounts on preventing the impending global climate meltdown has historically been lacking. For example, the component of the 2021 Biden infrastructure bill that is actually devoted to reducing greenhouse emissions, rather than adapting to climate change, amounts to only $89 billion a year over five years ($445 billion), a wholly insufficient amount given the gravity of the climate emergency. Biden's far larger signature Build Back Better Bill, which originally included $3.5 trillion in funding for climate and social programs, was reduced

---

* The levelized cost of energy (LCOE) is the present average monetary cost of energy from an energy system, a useful standard for comparing the costs of different systems. The LCOE is computed by taking the sum of all capital, operating, maintenance, and interest costs and dividing it by the system's total lifetime energy output. The result is expressed as a monetary cost of energy per unit of energy output—for example, as cents per kilowatt-hour or dollars per megawatt-hour.

in size to $2.2 trillion before ultimately failing to win congressional approval, as Senate critics charged it was still too expensive.

Yet, rather than being a drain on our economy, a massive clean-energy transformation will generate millions of net new energy system jobs. New construction and manufacturing jobs in the solar, wind, geothermal, hydro, transmission, and efficiency industries will greatly outnumber job losses incurred when coal mines and coal power plants close. The transition will also avoid tens of trillions of dollars in cumulative climate, environmental, and public health costs. The International Renewable Energy Agency has found, for example, that by 2050, the clean-energy transition would create savings in those areas worth three to seven times its costs—not a bad return on investment! Yet another major report, by an arm of the prestigious Global Commission on the Economy and Climate, estimated that "bold climate action could deliver at least US $26 trillion in economic benefits" through 2030.

These rosy forecasts make sense for at least three reasons. The first is the rapid decline in the costs of renewable generating and storage technology, as discussed in chapter 1. The second is that tomorrow's clean-energy system will be almost entirely electrical, involving near-zero fossil fuels. We'll therefore be able to meet almost all our energy needs with the least costly energy sources—clean, renewable electrical power.* Moreover, because electrical technology is inherently more efficient compared with fossil fuel combustion processes (a result of fundamental laws of physics), the clean-energy economy overall will be much more efficient than today's economy, requiring less than half the energy now needed per unit of output. This will help reduce our overall need for energy, making all supplies go farther and cost less and cutting emissions proportionately.

---

* The remaining few percentages of our energy will come from renewably produced biofuels, ammonia, and hydrogen. Energy *services* delivered in an almost all-electric economy will be economical for two reasons: First, the power will be generated mostly by the least costly modes of generation today—wind, solar, and hydro. Second, electric motors are thermodynamically much more efficient than internal combustion engines, so less energy will be needed per unit of delivered useful service.

The third reason the clean-energy economy looks so promising relates to energy efficiency, the capacity of getting more useful work and services for each energy unit used. Society in general has a vast abundance of unutilized and cost-effective energy efficiency opportunities. In fact, as the world's leading energy expert, physicist Amory B. Lovins, notes, "Efficiency is already the world's biggest energy 'source,' bigger than oil, and its vast further potential outcompetes any supply."

Lovins likes to say that not only is energy efficiency low-hanging fruit ripe for plucking, but it also has an endearing way of growing back after you've picked it: making one upgrade for efficiency often opens the door to further changes that also reduce energy use. Moreover, technology improves continually, creating new efficiency opportunities, and knowledge of the technology keeps spreading, enabling more people to put it to work. It is also invariably cheaper than new sources of energy supply, as Lovins has been pointing out since the 1970s. Moreover, Lovins notes, efficiency is a resource that is not only expanding in quantity but also declining in cost. For all these reasons, he has concluded that climate protection depends on "seeing and deploying the entire efficiency resource."

All these important factors—low cost, abundant supplies, and high efficiency—working in tandem—determine that an energy system maximally reliant on electricity will actually be more efficient and less costly than one based on fossil fuels. However, even though fossil fuels will ultimately be priced out of the market—as is already happening with coal, nuclear, and natural-gas power—it is legitimate to ask where the money will come from to pay for the first costs of the energy transition.

Cash Up Front

Those first costs are for building and installing all needed generating and efficiency equipment, retrofitting existing buildings, creating new energy infrastructure, and decommissioning obsolete power plants. Production of new electric vehicles also requires investments to modify manufacturing plants and create new battery factories. To figure out specifically from where the money for all these first costs will come, we'll consider where the nation's energy dollars are now being spent and how to strategically redirect them and other capital flows as cheaply and efficiently as possible. Doing so will require the thoughtful use of several innovative and underutilized financial tools and institutions so as to make clean-transition dollars appear quickly, in large quantities, and at the least cost to taxpayers.

To start, let's consider the financial hurdles standing in the way of cost-effectively liberating the vast reservoir of energy efficiency silently lurking in the building sector of the economy. At present, more than one hundred million buildings, and the billions of energy-using devices they contain, consume nearly 40 percent of all end-use energy in the United States and 70 percent of all electricity.* Yet *half* that energy is being wasted. No massive decarbonization of the country will be possible without cleaning up our act here. Entrepreneur and resource policy expert Rob Harmon thinks he knows how to do that, with a system he's developed to get capital flowing on a grand scale to finance building sector energy efficiency. Its widespread deployment would be an important piece of the puzzle of how to pay for the broader clean-energy transition, as similar opportunities exist economy-wide.**

Before he became a champion of energy efficiency in buildings,

---

* End-use energy is energy expended at the point of use—for example, it's the power used to cook your food in the microwave, not the primary energy used in the power plant to generate that electricity. Because of thermodynamics, for each unit of electricity produced for any use, something on the order of three units of primary energy is combusted in a conventional power plant. But that's not true in a renewable-energy facility.

** These opportunities are estimated to have been worth $1.2 trillion in energy savings from 2008 to 2020. Had they been financed, they would have eliminated roughly 23 percent of projected energy demand and avoided over a billion tons of GHG emissions annually. See Hannah Choi Granade et al., "Unlocking Energy Efficiency in the U.S. Economy, Executive Summary," McKinsey Global Energy and Materials, McKinsey & Company, July 2009, https://www.mckinsey.com/~/media/mckinsey/dotcom/client_service/epng/pdfs/unlocking%20energy%20efficiency/us_energy_efficiency_exc_summary.ashx.

Harmon focused on reducing the environmental impacts of large water consumers. As chief innovation officer for the Bonneville Environmental Foundation, he co-created the foundation's Water Restoration Certificate, which led to the return of fifty billion gallons of water to degraded streams. He also created the first retail Renewable Energy Certificates, which polluters can purchase to defray their greenhouse gas emissions.

Starting in 2011, Harmon began working on finding a way to make what he calls "deep energy efficiency" attractive to building owners and utilities. By "deep" efficiency, he meant reductions of 30 to 50 percent in energy use; by contrast, he had noticed that typical utility-sponsored energy efficiency investments provided comparatively paltry savings of only 5 to 15 percent. In order to achieve much larger energy savings he knew that costly changes would often be needed to upgrade a building's "envelope" (its walls and windows), its furnace, its air conditioner, or all the above. Harmon believed that the nation's massive stock of commercial buildings would therefore never be upgraded using traditional financial mechanisms, because the latter simply were never designed for achieving deep energy efficiency. Their intent, he said, was merely to reduce gross energy waste, so that utilities would "build four coal power plants instead of eight"—clearly not the right formula for an accelerated energy transition.

At the time these thoughts were germinating, Harmon was also parenting a small child, and he was concerned about the future. "We're cooking the atmosphere. If we want that to end, we've got to do energy efficiency at depth," he told me. As he saw it, the utility sector and its regulators "actually have to do new things and take some risk, [but] it's a very risk-averse industry.... We can't afford [that utility behavior] anymore." This realization spurred him to tackle the thorny and complicated problem of financing deep efficiency.

Meet MEETS

Harmon created and directs an international consortium he founded in Seattle to tackle the problem and pursue a new approach to energy efficiency. The group includes real estate developers, architects, engineers, large energy companies, and consultants; he named it MEETS, for "Metered Energy Efficiency Transaction Structure." MEETS's big idea is that it converts the dollars saved through energy efficiency improvements in buildings into cash flow streams for building owners, investors, and utilities—all while improving the building tenants' comfort, at no cost to them. It does so by aggregating the value of energy savings over time into a here-and-now financing instrument for making buildings more efficient; in the world of energy finance, it's "a way of leaping tall buildings at a single bound." (Hat tip to Superman.)

Until MEETS, large-scale energy efficiency investments in buildings were hampered by financing bottlenecks that paralyzed building owners, tenants, utilities, and investors and froze the inefficiency status quo in place. Tenants had next to no incentive to invest in energy efficiency to improve a building they did not own and might not occupy long enough to recoup their outlay through savings on their utility bills. Landlords also had little incentive to invest in energy efficiency, because tenants usually are responsible for paying utility bills. Utility companies, another seemingly logical funding prospect, usually are not eager to help deliver energy efficiency for customers, because normally when customers reduce their energy consumption, the utility then loses revenue.*

What about private investors? Those interested in providing capital for building retrofits usually don't have a simple way of investing directly in the energy savings produced by their investments while,

---

* To overcome that reluctance, some public utility commissions have set up programs to encourage utilities to implement customer energy efficiency programs. The programs, however, are neither universally available nor effective, and the goal of realizing *all* cost-effective energy efficiency gains is rarely, if ever, approached.

at the same time—and this is critically important—*accepting responsibility for how the investments perform.* Retrofitting for efficiency is pointless if the upgrade is merely window dressing, so that someone makes money, but little energy is actually saved. Another challenge is that private investors tend to expect to realize their returns quickly, and when lending to parties who may be seen as high credit risks, the investors tend to demand high rates of return. That could be a deal-breaker for energy efficiency retrofits, which may require long payback periods to "pencil out" and may not produce high enough returns to satisfy the average private investor or building owner looking for a large and early return.

To some people, this mishmash of misaligned financial incentives seemed a hopeless mess. But Harmon waded into the tangle, and when he was done, MEETS beautifully cured all these difficulties simultaneously for landlords, tenants, private investors, and utilities. It did so by creating a new financial transaction model that could be applied all over the country to finance capital improvements for energy efficiency in buildings.

The essence of how MEETS works is that it is performance based and built around a long-term contract. Its structure is similar to that of a traditional utility "power purchase agreement," or PPA, in which a provider agrees to furnish power to a utility or other customer over a set period at a predetermined price. With MEETS, however, there's one important difference: thanks to major advances in computer technology, the efficiency is metered. This allows energy efficiency to be treated like the electricity or heat sold by the utility. "We have wind energy and solar energy," Harmon noted. "Now we have 'efficiency energy.'"

Another major strength of MEETS is that it avoids what's known as the utility "death spiral." In this fatal dance of supply and demand, energy efficiency upgrades reduce the number of kilowatt-hours the utility sells. Therefore, to continue earning its maximal allowed revenue so as to cover its fixed costs on shrinking sales, a utility must raise its per-kilowatt-hour rates. Inevitably, the higher prices drive even more customers away; with fewer customers, the utility is once again obliged to raise its rates. The upshot? Expecting utilities operating under conventional financial arrangements to implement effective energy efficiency programs is an unnatural act, like requiring airlines to help their customers prioritize Zoom conferences over air travel. Under MEETS, however, while a utility does sell fewer traditional energy units, it replaces the revenue lost from them with "efficiency energy" units. The utility then charges customers for *all* units of power—including those saved through energy efficiency, not just units of traditional power used—so that its total sales are unchanged. The utility's customer, meanwhile, benefits by getting to occupy the upgraded premises, which are often more comfortable and pleasant.

While not a new approach to energy, MEETS is a radically new approach to energy *efficiency*. It therefore obliges all its contracting parties to depart from business-as-usual practices and policies, requiring them to change how they think about efficiency—a cultural shift on the part of utilities and other stakeholders. For some, that may be a big ask. But if this cultural shift can be widely accomplished through educational outreach and marketing, MEETS could have profound financial implications for the nation and even the world by supercharging the financing of building energy efficiency.

Many of the ideas that underpin MEETS were first articulated by a financier and private equity manager named Bill Campbell, who cofounded Equilibrium Capital (a sustainability-driven asset management firm), and two co-conspirators: Curtis Robinhold, then of BP Renewables, and Terry Egnor, who ran an energy efficiency program at PacifiCorp, a large electric utility in the Pacific Northwest.* Campbell's breakthrough insight was that MEETS could actually make net-zero-energy buildings commonplace by making them profitable. The key would be the deep-efficiency upgrades

---

* The impetus for MEETS arose when Terry Egnor was fired by OnSite Energy for doing too good a job implementing Oregon's mandatory energy efficiency regulations, according to Campbell. "They did not want PacifiCorp to be in the business of cannibalizing its own sales opportunities."

made financially attractive by MEETS contracts; with a building's energy demand minimized, its needs could be fully and economically met by affordable, on-site renewable-energy systems. Campbell saw that the logic of power purchase agreement pricing also applies cost-effectively "to *any* method of putting energy resources on customer premises," such as rooftop solar, energy storage, building controls, or combined heat-and-power systems.

The crux of what MEETS does is to turn energy efficiency "yield" (in the form of saved energy) into sellable energy, like any other energy commodity, instead of keeping it as savings "locked" to the building. Having successfully brought electricity to millions of consumers worldwide in a span of only forty years, utilities, Campbell observed, are "the most successful social finance, public-private construct in the history of mankind. MEETS uses the same transaction system [based on power purchase agreements] that lit up the planet a hundred years ago to green it today."*

With the intellectual tool kit thus devised by Campbell and associates, Harmon, by late 2014, succeeded in getting a contract signed with Seattle City Light to use MEETS to finance the unique energy efficiency features of a remarkable Seattle building, the Bullitt Center. Things went so well with the Bullitt Center pilot project that SCL subsequently asked for and received regulatory approval to expand MEETS to thirty other commercial buildings in Seattle. As of late 2020, five of those projects were already moving ahead. MEETS pilot projects are also planned in New York and Hawaii.

An analysis conducted for the Seattle 2030 District, a consortium of real estate interests in the city committed to cutting energy and water use in half by 2030, concludes that adopting MEETS in the district's commercial buildings would create nearly five thousand full-time jobs, save 30 to 50 percent of building energy use (an esti-

---

* As Campbell pointed out to me, if we saved half the five hundred billion dollars we spend on building sector energy every year, over twenty years we would have saved five *trillion*. Even discounting that sum to present value at utilities' current cost of capital, "you'd have two trillion to work with," he says. That would be enough for quite an impressive down payment on financing the clean-energy transition.

mated 500 million kWh per year), and attract investments of more than five hundred million dollars in private capital to improve Seattle's built environment. A massive *national* building retrofit program using MEETS would make millions of buildings across the nation highly energy efficient, generating huge energy savings across multiple decades and reducing greenhouse gas emissions. It would also create tens of billions of dollars in profitable, productive investment opportunities that could be a bonanza for U.S. investors in hundreds of cities. In so doing, it would also create hundreds of thousands of new, well-paying, full-time jobs that could not be outsourced. Finally, as these gainfully employed workers spent their wages on consumption, billions of dollars would surge through the larger U.S. economy. This model would work in Des Moines just as easily as in New York City or Seattle.

Harmon anticipates that MEETS contracts in Seattle will produce rates of return in the range of 9 percent, with building owners receiving 5 to 10 percent of the revenue stream as rent for use of their buildings as energy efficiency investment platforms. He's convinced that if MEETS were widely adopted, the promise of a 9 percent return would bring investment capital flooding into the market: "The biggest problem we would have would be finding enough qualified workforce to fix all the buildings." It's the kind of problem most cities would love to have.

Stretching Climate Investment Dollars

Recall that earlier in the chapter, I mentioned that an estimated $3.3 to $5 trillion per year would be needed worldwide to mount an aggressive international climate protection effort. Suppose that instead of world governments trying to come up with all that money themselves—which would mean using scarce tax revenues—they strategically used much smaller amounts of public investment capital to provide loan guarantees and revolving loans to mobilize public-private partnerships on a grand scale. Each public dollar invested could then easily leverage large multiples of that amount in public-private investment capital for a massive clean-energy transition effort. (Moreover, only a small percentage of each dollar provided in loan guarantees would need to be ring-fenced off every year to make good on bad loans.) And every time a government entered into such partnerships, it could insist on social equity and environmental justice contract provisions, which private investors would willingly accept in exchange for access to a safe investment vehicle that generated meaningful, reliable returns.

This kind of smart public-private financing can definitely facilitate the clean-energy transition in the broader economy, not just for buildings or power plants but for any climate-related project that required vast amounts of private capital. This is a strategy that Dan Reicher has long advocated. Reicher is an entrepreneur, a former Energy Department official in the Clinton administration, and the founding director of the Steyer-Taylor Center for Energy Policy and Finance at Stanford University.* A large feather in his cap is his having saved a forty-billion-dollar federal loan fund for clean-energy initiatives from being dismantled by a Republican-led Congress and the Trump administration. More recently, he brokered a major agreement between the U.S. hydropower industry and the environmental community that is now backed by significant federal infrastructure funding.

Reicher describes the two great passions of his life as "cleaning up the planet's energy supply and paddling down its great rivers." During boyhood canoeing trips, he began taking water samples and testing them for pollutants. A subsequent wilderness canoe trip in Ontario, Canada, and a four-month college expedition with Dartmouth classmates (including future Ohio senator Rob Portman) to canoe the entire length of the Rio Grande broadened his environ-

---

* In addition to these jobs, Reicher has also been director of climate and energy for Google, cofounder of the nation's first renewable energy project finance company, and a research scholar at the Stanford Woods Institute for the Environment.

mental concerns and led eventually to a biology degree. While in his twenties, he and a law school classmate took a collapsible kayak to China and paddled the Three Gorges of the Yangtze River before they were submerged by the Three Gorges Dam and its 410-mile reservoir. (Although large dams produce renewable power without burning hydrocarbons, they do pack an environmental wallop: the construction of Three Gorges, the world's largest power plant, uprooted 1.4 million people, flooding their villages and farmlands and forcing them to relocate. The dam also caused landslides and seismic risks.) In 1992, Reicher joined a kayaking expedition with three Kennedy brothers—Robert F. Kennedy Jr., Michael, and Max—on the wild Biobío River, in Chile, to arouse opposition to construction of the massive Pangue Dam. The protest failed, but Reicher's disappointment and his love of wild rivers helped set the stage for his future energy and climate work.

I first met Dan Reicher in 2018, at a talk in San Rafael, California, on how to finance the clean-energy transition; the event was hosted by the Environmental Forum of Marin, a local citizens' group. A year earlier, Reicher and colleagues had published a major new Stanford University study on green-energy investments. The report looked at the challenge of massively funding the clean-energy transition—and doing so as efficiently as possible.

The International Energy Agency was set up in 1974 by members of the Organization for Economic Co-operation and Development to ensure a steady supply of oil and to promote economic growth. Not long before Reicher's Stanford University report appeared, the IEA had issued a study estimating that if the world were to have a chance of staying below 2°C of global warming, it would need to invest about $2.3 trillion a year in public and private funds to combat climate change.* Reicher's study found, however, that the world is

---

* Coincidentally, that's about what the 2018 Trump administration tax cuts are expected to cost the nation over a ten-year period. See David Rogers, "Politico Analysis: At $2.3 Trillion Cost, Trump Tax Cuts Leave a Big Gap," Politico, February 28, 2018, https://www.politico.com/story/2018/02/28/tax-cuts-trump-gop-analysis-430781.

providing only about a third as much funding as the IEA believes is needed. The key questions Reicher's study sought to answer were why, and what to do about it. Reicher's research group pinpointed the bottleneck and offered a solution.

The problem was that the investment community saw clean-energy investments as "nontraditional"—i.e., they didn't fit the mold of the conservative, low-risk, "blue-chip" investment opportunities that financiers in New York, London, and Shanghai liked. However, these investors couldn't simply be written off: collectively, they wielded trillions of dollars, the bulk of the planet's investable capital, on behalf of massive financial institutions. Despite their swashbuckling reputation (as depicted in popular movies), these sophisticated professional investors are cautious. They don't want to make mistakes that might cost their clients—pension funds, sovereign wealth funds, insurance companies, endowments—billions of dollars and cause them to lose their jobs. Instead, they like to invest in established, financially stable companies and projects.

To get these wary institutional actors to invest on the scale required for a massive global mobilization on climate, Reicher and colleagues saw that clean-energy investments had to be "de-risked." Once that was done, climate investment would surge. If current global investment in clean energy were tripled—and fast—Reicher's study showed that the United States could be both a major participant in and a big beneficiary of "the single greatest economic opportunity of the 21st century." The bulk of the required capital was not going to come from governments, even wealthy ones; private institutions, such as pension funds, needed to step up. At the time, though, less than 1 percent of the trillions of dollars in assets under management on behalf of such institutions was directed toward nontraditional investments. Neither were multinational development banks filling the investment gap in the developing world, where so much new energy investment is needed but is relatively risky.

The IEA has determined that the world would need a formidable sum—$58 trillion in total by 2040—to fund the clean-energy transition and have a shot at staying below 2°C of global warming.* Reicher's group proposed a solution—one that, if adopted, could really matter to ordinary Americans because of the new jobs and economic activity it would stimulate. Currently, most of the trillions the world spends on energy still supports fossil fuels produced by multinational energy companies, while creating relatively few jobs per investment dollar. But were that capital channeled instead into building new clean-energy infrastructure and improving energy efficiency, the study found, far more jobs would be created and more economic activity and revenue would be generated. Reicher's research team demonstrated that if the United States participated robustly in this investment pool, the nation would be amply rewarded financially. Those dollars would create additional local, well-paying jobs on top of the three million jobs already in clean energy—more, in fact, than the number of bankers, farmers, real estate agents, or elementary and middle school teachers.** The new energy-investment dollars would be spent on efficiency upgrades for commercial and residential buildings; on manufacturing electric, hybrid, and hydrogen-powered vehicles; on constructing solar and wind farms; on residential solar sales and installations; on energy storage and charging stations; and on modernizing the electric utility grid.

After carefully analyzing the problem—and the multiple market, policy, development, and other major risk categories that investors face—Reicher's Stanford researchers identified the key factors preventing clean-energy investment risks from being reduced to tolerable levels. Moreover, their study suggested that once those risks were adequately mitigated for investors, the floodgates to massive private-sector investment in the clean-energy transition would open. New jobs would be created, and the larger economy would benefit as

---

* The Intergovernmental Panel on Climate Change (IPCC), in its *Special Report: Global Warming of 1.5°C*, 2018 (https://www.ipcc.ch/sr15/), has come up with a broadly similar estimate: $1.6 to $3.8 trillion would need to be invested annually for the world to reach carbon neutrality by 2050.
** Notably, three times as many people are employed in the clean-energy sector as in the oil, gas, and coal industries combined, where jobs are often dirty and dangerous.

demand for raw materials and finished goods swelled to supply the needs of the industry and those of the new clean-energy workers as their paychecks were spent. Reicher's group found that sometimes all it takes is a relatively small federal, state, or municipal investment in a clean-energy project to trigger private investors to jump on board alongside that government capital. In fact, each federal dollar spent on providing loan guarantees to promote decarbonization investments would leverage about ten dollars in private capital.

The lessons of the 2017 Stanford study were still top of mind for Reicher at the beginning of 2018, when he found himself on Capitol Hill defending a little-known but vital clean-energy investment program. Funding for the Department of Energy's Loan Programs Office—which supports clean-energy investment efforts with loan guarantees and loans to finance large-scale energy infrastructure*— had been authorized by Congress in 2009, but in 2018, $41 billion lay unused in DOE hands. This pile of cash did not escape the attention of the Trump administration: hostile to clean energy, it had threatened to eliminate the program in 2017 and, in 2018, still seemed determined to do so.

This threat was especially potent given how little funding the U.S. government devotes in total to energy efficiency, renewable energy, clean transportation, and the clean-energy transition. For example, the DOE's enacted budget for these objectives in 2021 was only $2.86 billion, a tiny sum for a leading industrialized nation. (By contrast, the Chinese government in 2017 announced plans to spend $360 billion on renewable energy by 2020, thereby creating 13 million new jobs.) To make matters even worse, the Trump administration had requested only $695 million in 2019 for clean-energy funding—a nearly 70 percent cut from 2018—while also proposing to eliminate

---

* The program has provided some $35 billion in loans to energy projects. As of 2020, it had $40 billion available for new projects in four program areas: $4.5 billion for accelerating the growth of wind and solar energy production, almost $18 billion for "advanced technology vehicle manufacturing," nearly $11 billion for advanced nuclear projects, $8.5 billion for fossil fuel combustion projects using advanced technology, and $2 billion for tribal energy projects.

the budget for the DOE's Advanced Research Projects Agency—Energy. Not only would this drastic underfunding have undermined the country's international advantage in clean-energy technology—much of which was originally funded by U.S. taxpayers—but it would also have made it impossible to fast-track technology that would protect the climate and, hence, U.S. national security, while also generating jobs and economic benefits.*

Reicher had been involved in boosting federal financial support for clean energy under President Clinton, notably by supporting the DOE's Innovative Clean Energy Loan Guarantee Program.** After leaving government, Reicher had helped organize an informal high-level group in the capital that worked for years to preserve the program. The coalition comprised people representing companies with projects focused on renewable energy, nuclear power, energy efficiency, energy storage, and carbon capture—all of whom potentially stood to benefit from DOE loans or loan guarantees.

Soon after President Trump's 2016 election, the coalition became energized when they heard from both the Senate and the House that the Trump administration was planning to end the loan program. Reicher was asked to testify before Congress on the program in January 2018, and he came primed with facts and figures to remind the committee of the geographic scope of the projects funded by the program and how they could favorably impact various committee members' congressional districts. Toward the end of a fairly humdrum subcommittee hearing, his best chance to defend the program came when a committee member posed a hypothetical about what each member of the panel would do if Congress appropriated fifty billion dollars for energy. Reicher shot back that more than forty billion was *already* authorized for just

---

* Robust federal funding for clean-energy technology would save taxpayers money, create new jobs, protect the environment, reduce climate disruption, improve productivity and energy reliability, and maintain or improve the quality of energy services enjoyed by consumers and businesses.

** The program provides financial assistance for renewable-energy deployment and innovative, advanced fossil fuel and nuclear energy projects, as well as supporting fuel-efficient and advanced vehicle manufacturing.

that purpose, and all Congress needed to do was resist the Trump administration's efforts to rescind it.

The case Reicher made for the program was so convincing that a bipartisan letter was sent the next day by committee members to leaders of the House and Senate Appropriations Committees. It informed the committees that the very program on the chopping block actually had already created more than $50 billion in total project investments, had saved or created 56,000 American jobs, and had prevented 35 million tons of CO_2 emissions—"the equivalent of removing 7.3 million cars from the road." The letter's clincher was a statement that the billions remaining in the program "could be used as a substantial down payment on the trillion-dollar infrastructure program" that both parties had discussed and that even President Trump had spoken of in a February 2017 joint address to Congress.*

A few days later, another letter went to the leadership of the House Appropriations Committee, pressing them to continue the program, followed by another from a bipartisan group of senators to the Senate Appropriations Committee. This correspondence highlighted the breadth and intensity of the influential support the program had in Congress and helped stimulate broader congressional approval. The Trump administration thereby got the message from both houses of Congress that the program would remain in effect, with full funding, and needed to move forward once again. Reicher's careful groundwork and well-prepared plea spared the loan program and its funding from extinction.

An Unconventional Alliance

Reicher was also busy on behalf of clean energy and the environment on a related front that same year. For years, he had kept his

---

* It was not until the summer of 2021, however, during the Biden administration, that a $1.2 trillion bipartisan infrastructure bill was finally adopted.

passions for rivers and clean energy separate, but in 2018 the two fused in a $63 billion-plus proposal he helped launch. Sensing that the time was ripe to try to resolve the multi-decade conflict between the hydropower industry and the environmental community, he tackled the problem with the help of a process he had codeveloped at Stanford known as the Uncommon Dialogue on Hydropower, River Restoration, and Public Safety. It now brought together a broad coalition of industry and environmentalist stakeholders from across the political spectrum to collaborate and articulate what needed to be done on the contentious issue of hydropower and the nation's more than ninety thousand dams.

Industry wanted help repairing and repowering existing dams. More than six thousand U.S. dams are considered "high hazard"—that is, a failure or serious operator error at them would likely result in loss of life. Environmentalists and conservationists wanted more clean hydropower to fight climate change, but they also wanted dangerous and destructive dams removed. (Removing dams likewise makes U.S. rivers more resilient to climate change by facilitating the return of wetlands and riparian areas that can reduce flood intensity.)

What helped bring the two sides together was the realization that only 2,500 U.S. dams currently generate electricity. Some of the other 87,500 dams—which variously provide flood control, irrigation, navigation, water supply, and recreation—were ripe for the addition of up to 10,000 megawatts of new power turbines. The hydropower industry agreed to support the removal of dams that no longer served a useful energy purpose, something that environmentalists sought, in return for support from conservationists for repowering useful dams. Eventually, after two and a half years of negotiations, a major agreement was reached. Its focus was on "three Rs" for the nation's dams: *rehabilitating* some for safety, *retrofitting* some for power, and *removing* some for conservation. Industry, the climate, and the conservation movement all took a victory lap.

When major adversaries perform the unnatural act of uniting for a common good, as they did on hydropower, even Congress sometimes listens. Thus, when the U.S. Senate finally adopted a $1.2 trillion domestic infrastructure package on August 11, 2021, it included $2.3 billion to begin implementing the three *R*s. With an achievement on this scale to its credit, the Uncommon Dialogue process could serve as a model for further coalition-building efforts to secure much-needed public funding for other contentious aspects of the clean-energy transition.*

Reicher's success in bringing together parties who don't usually communicate with each other is a model for how we can build creative solutions to all kinds of seemingly intractable problems in the clean-energy transition. We won't succeed in restabilizing the climate without breaking out of rigid ways of thinking and committing ourselves to opening channels for dialogue and information among all stakeholders—even those with whom we habitually disagree. In chapter 4, we'll see how a broad coalition of thinkers and doers came together to successfully create something never before seen: the world's first commercial "living building."

---

* As of this writing, another important measure that Reicher was deeply involved with remains pending in the House and Senate, with bipartisan support. Introduced in June 2021, the Twenty-First Century [*sic*] Dams Act would provide $21.1 billion to implement the three *R*s, multiplying ninefold the earlier $2.3 billion investment. If passed, it would supercharge the national effort to improve dam safety, modernize and upgrade hydroelectric capacity, and remove unnecessary dams to restore rivers.

PART II

Clean Technology, Inexhaustible Resources

4.

"The World's Biggest Energy 'Source'"

Energy efficiency is the pixie dust of climate remediation. Sprinkle it liberally on the tangles of pipes, valves, gears, wheels, walls, and wires that bring us energy services, and—lo and behold! Less energy is needed, but the same or better energy goods and services are delivered. With significant gains in efficiency, there's no need for belt-tightening or rationing. Plus, the energy savings usually mean less pollution, less environmental damage, and better health and comfort for users, all at a lower net cost. So, the logical first step in climate protection is to see where we can increase energy efficiency, regardless of whether the energy is produced from clean, renewable technologies or from fossil fuel. (Even clean technologies cause some greenhouse gas emissions when they're built.)

The cantilevered solar roof of the Bullitt Center, which bills itself as the world's greenest building, is like the flight deck of an aircraft carrier or a giant mortarboard. Some call it a baseball diamond in the sky.* Although metropolitan Seattle is the cloudiest major city

---

* For more about the Bullitt Center, see Mary Adam Thomas's book *The Greenest Building: How the Bullitt Center Changes the Urban Landscape*, Living Building Challenge Series (Seattle, WA: Ecotone Publishing, 2016), a sterling account of a platinum building. I've relied heavily on her work for some of the history and technical descriptions of the building.

in the United States, this award-winning six-story commercial office building manages to produce all its own energy via solar power and to supply its own water through rainwater harvesting. In addition, it treats all its own kitchen and bathroom waste, uses no toxic materials, requires little artificial day lighting, and is built to last two and a half centuries. Geothermal heat from the ground beneath the building radiates through the concrete floors to warm the interior. These exceptional features are seamlessly blended into an affordable, ultramodern, energy-efficient structure that is as inviting and beautiful as it is revolutionary. Its very existence is a powerful affirmation of clean energy's potential for urban transformation.

Using High Tech to Go Low Tech

Unlike conventional buildings, the Bullitt Center responds actively to its environment. As with the nerves of an organism, sensors detect and transmit the temperature, wind speed, sunlight, and precipitation outside the building's skin into its computerized brain. Then, custom software factors in the building's interior conditions and sends appropriate control impulses to adjust its heating, cooling, and water systems. Yet this super-high-tech, high-performance building actually uses only 17 percent as much energy as a comparable commercial office building. Over its lifetime, it will generate more than $13 million in carbon-reduction benefits. The building gets energy from the sun, water from the sky, heat from the ground, and daylight and ventilation through its windows. It thus models how people can prosper in harmony with nature. The building designers, in effect, used high tech to go low tech.

To find out how the Bullitt Center came to be, I spoke with the man most responsible for it: Denis Hayes, president of the nonprofit Bullitt Foundation. For a half century, he's been one of the nation's leading environmental thinkers—and doers. With an MBA and a law degree from Stanford already under his belt, Hayes saw his career really take off when former Wisconsin governor and senator Gaylord Nelson selected him as national coordinator for Earth Day 1970. Hayes made the event a worldwide success that helped put the modern environmental movement on the map. A post as head of the Illinois state energy office followed, and soon after, he was named director of the Solar Energy Research Institute (now the National Renewable Energy Laboratory). In 1992, he became president of the Seattle-based Bullitt Foundation, whose mission is to protect the environment and help build sustainable communities in the Pacific Northwest. The Bullitt Center headquarters building might be his crowning achievement.

The Birth of a Living Building

The inspiration for Hayes's extraordinary career and the Bullitt Center originated in a 1964 epiphany in the desolate Etosha Pan region of northern Namibia, a vast and usually dry lakebed that, after heavy rains, is sometimes flooded by the ephemeral Ekuma and Oshigambo Rivers. Ten thousand miles away from Etosha Pan, Hayes grew up in a working-class family in the tiny town of Camas, Washington, on the Columbia River. After high school, he enrolled in a local community college but was acutely aware of, and pained by, world problems like the U.S.-Soviet balance of nuclear terror, the Vietnam War, and racism in the United States.

Depressed about the suffering and injustice he saw no way to ameliorate, he dropped out of college after two years and spent the next three years hitchhiking alone throughout Iraq, Iran, Syria, Lebanon, North Africa, the Soviet Union (including Siberia), and Eastern Europe. The odyssey was, he said, part vision quest—a search for new, life-affirming values—and part educational journey, to satisfy his curiosity. Idealistic African political leaders had glorified the liberation of their people from colonial oppression, and Hayes wanted to see for himself if, once the chains of colonialism were broken, people would actually have the freedom to develop their human potential. "In every case, not only had they not made any progress, but it really seemed to be devolving into things that were worse and worse," he recalled. From Siberia to Africa, the new, inspiring values he longed for were nowhere to be found.

Then things got worse. It was the rainy season in Namibia, and the rivers had flooded into Etosha Pan. The area teemed with a rich diversity of wild predators and prey that brought to mind Eugene Odum's seminal text, *Fundamentals of Ecology*, which Hayes had studied during a National Science Foundation summer ecology institute right after high school. The scene was like being in the midst of a National Geographic special, yet it was far from idyllic. "It was very hot during the day. It was freezing at night. I was starving. I'd been alone for a very long time," Hayes said. The previous evening, he had just had a scarring emotional experience when he stumbled across the first German concentration death camp ever, in the coastal town of Lüderitz, a site marked by an obscure little plaque. "It's not much known, but they did there at a smaller scale what they later did at Auschwitz," he said. More than two thousand people of the Herero and Nama tribes who had fought back against German colonization of their land died in the camp due to overwork, disease, starvation, exposure to the elements, and medical experimentation. The Germans disposed of the bodies by dumping them on the beach at low tide, at a place called Shark Island, where they were washed out into the sea.

After seeing the camp, Hayes spent the following day crossing the desert and then paused in the evening to look back toward the coast. "Suddenly, the combination of my desire for some kind of a unifying explanation of how things operate on earth and *should* operate on earth, and how we could redesign societies . . . led me to believe that there was something to be learned from nature and from the way that it had sustained itself so well for so long." He realized then that some principles of ecology—for example, predator-prey relationships—operated in ways that would be destructive in human society. He was looking for natural principles that could make good templates for humanity. "I stayed up basically that entire night trying to think through—without having the relevant vocabulary of urban ecology, industrial ecology—what all this might mean for civilization."

Hayes decided that he would return home and focus the rest of his life on trying to find ways to apply the principles of natural ecology to the human prospect—to figure out how we could prosper in harmony with nature, instead of imposing our will on it. Back in the States, he returned to school, eventually turning his talents to environmental causes and the advancement of solar energy.

Design with Nature

Now in his seventies, walking slowly and unpretentious in jeans and a gray sweatshirt, Hayes captains the Bullitt Foundation from an airy office atop the Bullitt Center. The building "exceeded my hopes, not just my expectations," he confided to me. Back in 2007, the foundation had needed a new home, and Hayes had seen this as an opportunity to create a commercial building that would reflect the foundation's commitment to clean energy and climate protection. He wanted nothing less than to demonstrate the best sustainable energy performance possible in an urban environment in a way that was affordable and replicable, a model to inspire the creation of sustainable and resilient cities.

As a foundation executive, Hayes was wary of creating a self-aggrandizing monument to wealth that offered only a symbolic nod to sustainability. This led him to make an early and very risky decision to seek full certification for the Bullitt Center as the world's first commercial "Living Building." To qualify for full certification, a Living Building has to meet a set of standards so rigorous that even attempting to achieve them was a chancy business, requiring a unique, radical design.

The standards for certification that the Bullitt Center had to meet were developed by the International Living Future Institute's Net Zero Energy Buildings program. According to program director Brad Liljequist, the average net-zero building uses only 60 to 65 percent as much energy as a conventional building. A fully certified Living Building, however, has to do more. For full Living Building certification, a building must pass an independent audit to confirm that, on an annual basis, it uses net-zero energy, consumes net-zero water, and meets the world's highest energy efficiency and green building standards—that is, twenty stringent "imperatives," each with its own formal criteria, for performance areas that include water use, energy, health, materials, equity, and beauty.* Achieving the health standard, for example, means the structure must be toxin-free and supportive of its occupants' health; hundreds of common construction materials must be avoided just to clear this bar. All energy used must be produced on-site, without combustion, which means it must come from renewable energy sources. Buildings thus must simultaneously meet high aesthetic, efficiency, equity, construction, and durability standards. The standards range beyond the nuts and bolts of the building to include imperatives like "biophilia," or the nurturing of a sense of connection with nature.

Of four hundred buildings currently registered for the "Living Building Challenge," seventy-five have met some of the standards, but only a few have achieved them all. "The Bullitt Center is a world champ in its class," says Liljequist—one of only ten fully certified living buildings on the planet.

Defying the Odds, Together

Throughout his career, Denis Hayes had shaped public policy through the political process and public education. This challenge was vastly

---

* For more information about the Living Building Challenge, see "Living Building Challenge," International Living Future Institute, https://living-future.org/lbc/.

different: it was tangible and, he felt, high-risk. A belly flop, he feared, "would be visible to the world," and, worse, could well set back the green building movement: "If a foundation with an endowment was trying to do it and failed," he said, "then why would anyone in a profit-making company try?"

Hayes sought expert advice, and what he learned at first was not reassuring. No one doubted that it was possible to build a one- or two-story net-zero building in Seattle, but the commercial developers he spoke with all told him that Seattle was far too cloudy to build a six-story net-zero-energy, solar-powered building: not enough sunshine would fall on the roof to meet even tenants' plug loads, let alone the building's required operating energy. "None of the developers thought we had a snowflake's chance in hell," Hayes recalled. But he didn't take no for an answer and, instead, sought additional advice from solar-energy experts who studied and modeled Seattle's climate and insolation conditions and concluded that, although it would be difficult, it just might be possible, provided that all the contractors scrupulously met all design specifications.

This led Hayes to realize that creating a precedent-setting, convention-defying masterpiece of energy and resource efficiency would require a broad and intensive collaborative effort as remarkable as the building's physical engineering and construction. He therefore handpicked a team of architects and engineers, contractors, subcontractors, university faculty, expert consultants, and government officials. Still not satisfied, he also formed an advisory council. Early design ideas came from University of Washington students, organized into design studio teams.

Two-day, fifty-person design charettes (collaborative workshops) allowed participants, students, advisers, and the project's developer all to weigh in, capturing fresh ideas and anticipating problems. Because the project was so innovative and complex, unforeseen pitfalls abounded. The building's water-handling, waste-handling, and energy-generation systems all impacted one another. "The windows, for example, were an integral part of the lighting system as well as the cooling system," Hayes said, "not just, you know, *windows*." So no spontaneous major changes could be made to the building's design without causing a system failure. That made it supremely important to ensure that there would be no significant changes while the Bullitt Center was under construction.

To achieve this goal, the team had to create the most careful, integrated design possible. State-of-the art digital modeling was needed to assess the daylighting, energy, and thermal implications of various design concepts. "We modeled everything repeatedly," Hayes said. To keep costs down, the team identified the vendors for every major material and obtained cost commitments before construction to prevent cost overruns. Broad, interactive consultations among all participants continued intensively throughout the design, planning, and construction processes. Rather than being confronted with finished construction drawings, contractors were invited to contribute design ideas during the charettes and thus make early design changes as economically as possible, providing feedback on how to keep costs down.

As design modifications occurred, the building came to exceed even some of its creators' most exacting performance goals. Once built, the Bullitt Center was elegant, modern, and comfortable, yet it produced 60 percent more electricity than it used, a surplus that offset about 60 metric tons of CO_2 emissions annually. In 2015, the center sold back more than 90,000 unneeded kilowatt-hours to the local utility. Amazingly, the Bullitt Center didn't cost much more to build than other high-performance, Class A buildings.

Slashing Building Energy Use

How much energy could be saved across the nation if net-zero-energy buildings were the norm rather than the exception? "About 40 percent of the energy [in the United States] is currently being used in buildings," Brad Liljequist reminded me. "So, we're talking about taking that down toward zero percent."

Of course, no one's going to wave a magic wand and suddenly transform the entire building stock of the United States to operate on net-zero energy. But if zero-energy buildings became the norm in new construction through economies of scale, they likely wouldn't be any more expensive than current buildings, except for the installation of on-site solar power or other clean-energy generation systems. Zero-energy buildings are already starting to catch on: in 2021, the United States and Canada had seven hundred net-zero-energy commercial building projects and thousands of net-zero-energy and energy-positive single-family and multifamily home projects. Now that net-zero-energy and ultra-low-energy buildings are feasible virtually everywhere, federal, state, and local governments could establish ambitious annual goals for retrofitting existing buildings and requiring all new buildings to employ cost-effective efficiency measures and use 100 percent renewable energy. Whereas it isn't generally possible or cost-effective for high-rise office buildings to generate all their own power on-site, they could acquire 100 percent of their energy from off-site clean-energy power plants.

The United States has about 6 million commercial and industrial buildings and 105 million residential buildings. Could the nation set an ambitious deadline for ensuring that they all meet high energy efficiency standards? The U.S. Environmental Protection Agency's voluntary energy efficiency program, Energy Star, established in 1992, has to date certified only thirty thousand or so commercial and industrial buildings as Energy Star compliant. At that rate, it will take almost six thousand years to get all U.S. buildings certified.

A Trillion Saved, a Trillion Earned

Roughly $4 trillion is spent every decade to provide energy in U.S. commercial and industrial facilities where a third or more of that energy ($1.3 trillion worth) is wasted! So, not to have an effective national program in place to certify all U.S. buildings for energy efficiency is a notable policy failure.*

I discussed this problem with Steven Nadel, director of the American Council for an Energy-Efficient Economy (ACEEE). Not many people can say that their organization was instrumental in saving the U.S. public a *trillion* dollars, but the ACEEE did so through the impacts of the national appliance efficiency standards it helped establish. Nadel would like to see states and localities adopt the French system of promoting energy conservation with building energy ratings. "In France, they have these A-to-G labels," and a ban on selling or renting G-rated buildings took effect in 2020, to be followed by others: "You can't sell or rent an F[-rated building] after 2025," Nadel said. "People had many years to prepare. . . . I use that as an example of what may be needed to help generate impetus for the [energy] savings we need and the jobs." To retrofit all existing U.S. commercial and industrial buildings by 2045, about 5 percent of them would need to be fixed annually, starting in 2025. That would mean retrofitting nearly 300,000 buildings a year.**

Absent the widespread use of a mandatory building energy rating system to drive retrofitting, federal loan guarantees and tax incentives could be mustered. Retrofits are labor-intensive, Nadel reminded me, and that means an ambitious national building energy efficiency program could be a big source of new, well-paying domestic jobs that could not be offshored.

Although states and cities can provide building codes, their ability to provide financing is limited. The federal government, however, could provide leadership and ample, predictable financing for a major national building retrofitting campaign to address the climate

---

* If pursued throughout our economy rather than just in buildings, energy efficiency could be our nation's largest electricity-providing resource by 2030, allowing us to avoid spending billions on new power plants.

** Retrofitting rates for these buildings in the global north, however, are currently barely 1 percent. See Gayle Kantro, "Retrofitting buildings essential to reduce energy costs and combat the global energy crisis," *JLL*, November 10, 2022, https://www.us.jll.com/en/newsroom/retrofitting-buildings-essential-to-reduce-energy-costs-and-combat-the-global-energy-crisis.

crisis. According to ACEEE, not only do such efficiency investments repay themselves more than three times over, dollar for dollar, but they also boost GDP, create new jobs and, as an added bonus, often make buildings more comfortable and healthier for occupants while saving those occupants money on utility bills. Meanwhile, the reduction in energy use curtails pollution, improves public health, and protects the climate. Energy efficiency even improves social equity in a nation with vast disparities in income and wealth: greater efficiency means that low-income households that spend disproportionately on energy are relieved of some burdensome costs.

The Opposition

On my final afternoon in Seattle, I sat down for an in-depth conversation with Washington governor Jay Inslee. Inslee regards climate change as a global emergency and an existential threat; he also sees it as a tremendous economic opportunity. Our conversation and its aftermath illustrate just how challenging it can be, even for a charismatic clean-energy champion, to get major climate legislation approved in the United States today—even when the legislation would dramatically advance both efficiency and renewables while producing a cascade of public benefits.

Since his first election in 2012, Inslee has tried to make Washington a proving ground for the clean-energy vision laid out in his 2007 book, *Apollo's Fire*. Indeed, he has managed to take steps in that direction—by executive action, because throughout his time in office, state legislative support for decisive climate action has repeatedly failed. In 2015, for example, the Washington state legislature refused to consider a moderate cap-and-trade climate bill Inslee supported. A carbon-tax bill would also have been dead on arrival, and a 2016 carbon-tax initiative (I-732) failed to get voter approval.

Although the state house of representatives had been under Democratic control since 2002, when we talked in 2017, the governor was hopeful that his party, by taking back the senate, could pass a modest carbon tax of twenty dollars per metric ton of CO_2 and a pending bill (HB 2005) that would set the state on a path to 100 percent clean power by 2045, phasing out coal by 2030. In a special election in November 2017, Democrats did take back the state senate. Yet, to Inslee's chagrin, the Democratic legislature let the clean-power bill die on the last day of the session and also failed to pass the 100 percent clean-power legislation. Environmentalists blamed opposition from electric utilities. Supporters of the bills say they will try again.

In 2018, citizen activists from more than 250 supporting organizations put a carbon-fee initiative, I-1631, on the state ballot. Six thousand volunteers campaigned for "Yes on 1631." The initiative would have charged big oil companies and utilities a modest fee for their carbon emissions. Seventy percent of the revenue would have been invested in constructing new, clean renewable-energy infrastructure, clean-transportation options, and energy efficiency upgrades to lower people's utility bills. Another 25 percent would have been invested in clean-water protection, water supply, and water cleanup projects and in healthy forests; the final 5 percent of the money would have been spent on helping local communities prepare for and blunt the impacts of climate change. The initiative also would have provided assistance to fossil fuel industry workers and low-income communities hard hit by pollution and climate change.

Initial voter support for I-1631 was reversed by a $31 million flood of oil industry money, and I-1631 was defeated. "Despite the results, the Yes on 1631 Election Night bash felt a lot like a victory party—an immense, high-energy gathering of a movement on the rise," said KC Golden of Climate Solutions, a Washington nonprofit. "Of course, there was disappointment, but not an ounce of resignation." I-1631 supporters say they, too, will be back with another measure.

Putting Waste on a Diet

KC Golden once directed Washington's Energy Office and has served as a clean-energy policy adviser to the governor. When I met him, he was working for a small but influential nonprofit called Climate Solutions, in downtown Seattle. We were alone in the office and meeting to discuss energy efficiency. Not coincidentally, although it was starting to get dark, the office lights weren't turned on yet.

Golden told me that, based on his experience in the Pacific Northwest, if the United States took full advantage of all that energy efficiency has to offer, we could save more than ten trillion dollars nationwide—an amount that, in one-dollar bills, would stretch almost from Earth to the sun—and create an abundance of new jobs. "If you're dubious about the capacity of clean energy to meet the needs of a modern economy," Golden told me, "it's a little less of a stretch when you realize that we could cost-effectively cut waste in half. So, if we're building a clean-energy bridge [to the future], it only needs to be half as long."

But what if cutting waste in half were merely a start, and we could actually cut all energy *use* in half? Back in 2012, leaders and senior researchers with the authoritative ACEEE did some trailblazing work on the country's enormous potential for energy savings and pollution reduction. Their study showed how we could save nearly 60 percent of our energy use while saving $12 to $16 trillion (net 2009 U.S. dollars) on energy bills to 2050, by which time we also would have added a cumulative $3.6 trillion to the nation's GDP. Yet, as the Cat in the Hat famously said, "That's not all, that's not all!" Agriculture and land use are responsible for about 20 percent of total U.S. greenhouse gas emissions. Thus, conservatively speaking, merely cutting in half the other 80 percent of all emissions would slash our total GHG emissions by 40 percent. If we also cut our agricultural and land-use emissions in half, we would eliminate half the nation's entire GHG emissions.

According to the Northwest Power and Conservation Council (NPCC), the Pacific Northwest has saved 7,530 average megawatts of power cumulatively since 1978.* As the council points out, that's equivalent to almost three times the average annual output of Grand Coulee Dam and enough to avoid the need for more than thirty new natural-gas-fueled power plants. Efficiency has enabled the Northwest to avoid seven times the current power consumption of Seattle, its largest city. By 2016, efficiency had unobtrusively saved power consumers in the region more than five billion dollars—about a thousand dollars per household.

A vast unused ocean of energy surrounds us—in energy-inefficient buildings, kludgy machinery, outdated equipment, and dumb electrical control systems. Exploiting it is a lot easier, cleaner, and cheaper than mining or drilling for untapped energy. "The energy we've ceased to waste," Golden declared, "has been by far the biggest *new* energy resource [in Washington State]" since the 1930s. (That's when a fleet of large hydropower plants was built on the mighty Columbia River and its major tributaries.) Energy efficiency is the region's second largest power source, after hydropower, and, according to the NPCC, it will take care of all additional electricity demand in the region for the next twenty years. That will mean additional billions of dollars in avoided costs to consumers and businesses.

Doubling Down on Energy Productivity

KC Golden is far from alone among energy efficiency experts in extolling efficiency as a vastly underused energy resource. Less than two weeks before my meeting with Golden in Seattle, I interviewed Dr. Howard Geller in Boulder, Colorado, where he heads the Southwest Energy Efficiency Project (SWEEP), which he founded in 2001. Geller is a physicist and engineer by training and an interna-

---

* An average megawatt is one *million watts* delivered continuously, twenty-four hours a day, for a year.

tionally recognized energy policy expert. Before founding SWEEP, he directed the ACEEE in Washington, DC, for twenty years.

The amount of energy we use today in the United States for each dollar of economic output is already only half of what it was in 1980, Geller told me. So, we've doubled the productivity of each energy unit in forty years. How did we perform this magic trick? Turns out we did it bit by bit, by commercializing incremental advancements in research and development that gradually improved our energy technologies and thereby increased our energy efficiency. Three quarters of the historical increase in energy productivity since 1980 has come from making more efficient appliances, lights, vehicles, buildings, and industrial processes, Geller said. The rest has come from structural changes in the economy. (Because today we're importing more goods, rather than manufacturing them ourselves, we are, in effect, offshoring some of our energy waste along with our greenhouse gas emissions.)

What Geller said next really pricked up my ears: if we got serious about energy efficiency, we could cut our energy use per unit of output in half *again* in as little as twenty years. This would result in huge economic savings and would put even more money into the pockets of consumers, businesses, and government agencies. Even in a growing economy, we could still halve our greenhouse emissions if we boosted clean, zero-emissions energy production at the same time that we doubled efficiency. Fortunately, given recent technological advances and price declines in clean-energy technology and its storage, that's quite feasible.

Efficiency—The Gift that Keeps on Giving

Despite the doubling of U.S. energy productivity since 1980, most energy efficiency experts would agree that vast efficiency opportunities still abound today. Technologies that didn't exist five or ten years ago—like LED (light-emitting diode) bulbs, which provide at least double efficiency and sometimes more—offer "tremendous potential," Geller said. Increasing energy productivity will also produce a lot of new employment, as energy efficiency is generally much more labor-intensive than supplying energy. "The nation now has over two million energy efficiency jobs, according to E2 and E4The-Future—more than any other kind of energy employment," Geller said. Red-state voters, listen up: another million or so jobs could be added in the South, where energy efficiency gains lag those in the rest of the country. Even in the Southwest, a region known for fossil fuel production and, increasingly, for solar power generation, energy efficiency jobs greatly outnumber all other jobs in energy supply.

Geller was clear that we're not doing enough to increase U.S. energy efficiency. "ACEEE has an annual Energy Efficiency Scorecard. California, Massachusetts, Vermont, and Rhode Island are the leading states [in order], and have been year after year." (In 2022, however, New York took third place, displacing Vermont, which slipped to fourth place.) Geller would like to see us raise the efficiency of lagging states to the average of the top ten states. "We could do it," he said, "if we just made sure that our utility efficiency programs, building codes, and transportation policies matched the best practices already demonstrated. . . . I think it would put us on a trajectory [to double] energy productivity in twenty years."

Although energy-efficient new buildings may cost an extra 5 to 10 percent to build, they pay that investment back many times over during their lifetime. Meanwhile, efficiency technologies continue to improve by leaps and bounds. Today, energy-efficient LED fixtures and advanced lighting controls with low-cost sensors, for example, can produce 75 to 80 percent energy savings. "That's a critical opportunity for advancing energy efficiency—just better controls in our heating and cooling systems, our lighting, and our factories. That's part of the opportunity," Geller said. Every major piece of equipment in a business or home can now be interconnected and monitored; sophisticated energy analytics software then allows usage data to be quickly sifted to highlight valuable savings opportunities. Specialized contractors stand ready to deliver these savings, which can also be packaged and traded on energy markets to motivate utility participation (as shown by MEETS in chapter 3).

The opportunities don't end there. "Distributed energy generation"—putting solar panels on rooftops or microgrids on institutional sites—makes it possible to reduce transmission and distribution losses that otherwise sap electrical current and waste energy. Customer power demand can also be managed through peak-load pricing, which encourages customers to shift usage to off-peak hours, making more efficient use of power-generating assets. This reduces or even eliminates the need for new power plants to meet those peaks, and it saves utilities and consumers money. And the world as a whole benefits, whether we realize it or not, due to the emissions reductions that come along for the ride with the efficiency gains.

Thanks to work by Geller and SWEEP, utilities in the Southwest have upped their investment in energy efficiency programs for their customers from $20 million to over $400 million. This led to a 10 percent reduction in energy use by 2016 and, thus, now slashes CO_2 emissions in the Southwest by 15 million metric tons per year. "We estimate that energy efficiency utility programs implemented over the past decade will generate over seven billion dollars of net benefits for households and businesses in the region and will avoid the need for eight large [300 megawatt] baseload power plants," Geller said. Those seven billion dollars in net benefits, by the way, don't include public health and environmental benefits, just savings to utility customers minus the costs of the efficiency upgrades. The net benefits would rise 50 to 100 percent, Geller asserts, if we also valued the environmental and health impacts.

Rigorous pursuit of energy efficiency has enormous benefits and no downside. In the next chapter, we'll see how by tapping this vast resource and coupling it with renewable energy from wind, water, and solar power, all our nation's energy needs can be met while we continue to grow our economy and save the climate.

5.

100 Percent Clean
Energy for All

Stanford University scientist Mark Z. Jacobson was sitting in his office early one morning in 2017 when the email arrived: the editorial manager at the *Proceedings of the National Academy of Sciences* (*PNAS*), one of the world's most prestigious journals, was writing to advise Jacobson that one of his most important research papers was under attack. Ignoring it was not an option.

After developing breakthrough climate and energy computer models as part of his doctoral and postdoctoral research, Dr. Jacobson had begun publishing scholarly papers and popular articles showing how the world could reach 100 percent clean energy without the need for nuclear power, carbon capture and storage, or biofuel plantations. His work also pointed to an enormous bonanza—millions of jobs globally—that a well-planned clean-energy transition would generate. But his conclusions were anathema to the nuclear power and fossil fuel industries and to their political and academic allies, who contend that legions of new nuclear power plants are needed to fight climate change, along with systems that would capture carbon from natural gas and coal plants—thereby extending the reign of fossil fuels. Renewables, Jacobson's critics pointedly claimed, could neither scale up sufficiently nor furnish the many forms of energy people needed to live comfortably and conveniently.* It hadn't taken them long to lash out at Jacobson and his new research.

In 2015, Jacobson had gained worldwide attention for the prize-winning *PNAS* paper now under attack. Using a new, highly original computer model, he showed that most of the world could secure 100 percent of its energy needs using only power from the wind, sun, and flowing water. The paper detailed how a power grid could be stable using 100 percent renewable energy, an idea that countered the conventional wisdom that large amounts of clean power would make the grid unstable. Jacobson's paper had won the coveted Cozzarelli Prize, given to only six papers every year out of six thousand published by *PNAS*.

Now, however, the empire was striking back. A group of twenty-one scholars—including some with formidable academic reputations—were claiming in a new paper, which *PNAS* was preparing to publish, that Jacobson's seminal 2015 grid stability paper had two major modeling errors that "invalidated the study and its results." In a direct and caustic rebuke to Jacobson's findings, the authors concluded that a portfolio of energy options, including nuclear power, bioenergy, and carbon capture and storage, would be a more reliable and affordable path to a clean-energy system.

Any researcher receiving such withering criticism would have been dismayed, to say the least, and Jacobson was taken aback. When he read the critical paper, he quickly noticed that it was to be published as a full-length "research report," yet it contained no new research. Furthermore, although the authors claimed Jacobson's paper had two major modeling errors, none had actually requested the model or its output data for examination. "You can't write a paper saying somebody has a model error without actually looking at the model and the actual output!" Jacobson asserted. From notes on past correspondence with one of the authors of the critique, he came to the

---

* "Renewable energy" refers to all energy from the sun, wind, the earth's heat, flowing water, plants, and oceans. In the interest of brevity, however, the focus in this chapter is on solar, wind, and water.

conclusion that the main allegations were intentional misrepresentations of his work.

He quickly sent off a stiff protest to *PNAS*, charging that the authors had failed to disclose conflicts of interest and that multiple "authors" who hadn't had any part in the writing had been added just to give the paper credibility, another ethical departure. Only three authors acknowledged actually working on the paper. Three of the paper's signatories were being paid hundreds of dollars an hour by the Trump administration as expert legal witnesses in *Juliana v. United States* (aka the youth climate lawsuit), in an attempt to block a ruling that could force decisive action on climate change.

Over Jacobson's strenuous objections, the critical paper appeared in *PNAS* in early 2017. "[The authors] just wanted to publish this and then immediately issue press releases," he said. Indeed, scathing articles critical of his work soon did appear in the media, including a major story in the *New York Times*. Jacobson responded with a point-by-point rebuttal letter in *PNAS* and went on to make his case online and in published interviews.

Virtually all the co-signers to the Jacobson critique were proponents of one energy technology or another that Jacobson, in his research, had found unnecessary, such as nuclear power. But much to the chagrin of nuclear advocates and fossil fuel defenders, the promise of renewable energy sources is basically irresistible: Unlike finite fossil fuels, solar, wind, and tidal power are for all practical purposes eternal sources of energy; our civilization is not going to run out of them.* And the prices of wind and solar power have steadily plummeted for decades. With each doubling in the volume of solar cells produced since 1976, for example, the price has fallen by 22.5 percent, and annual volume increased over one thousand times from 2000 to 2022.

---

* Not all renewables are entirely inexhaustible. Whereas hydropower from dams is generally a renewable resource, if a reservoir is filled primarily by glacier-fed rivers, that hydropower will be lost if the glacier melts because of climate change. Geothermal reservoirs of steam and hot water can also be depleted over time, especially by overextraction. Generally, however, they will eventually be naturally replenished.

Unlike oil or gas, renewable power sources are also not subject to price volatility or energy embargoes; nor are they plagued by huge cost overruns, as nuclear power plants are. They also don't emit ionizing radiation or produce radioactive waste. And because renewables don't disrupt the climate, create noxious by-products, or make people sick from air and water pollution, they are better for society than fossil fuels and vastly preferable to current nuclear power plants and to energy from burning wood and other plant materials. They are instead clean, affordable, efficient, reliable, and versatile—able to meet all our energy needs across all sectors of the economy. However, until Jacobson's 2015 paper, nobody had actually *proven* that wind, solar, and water power alone could meet 100 percent of those needs with current technology.

A Clean-Energy Visionary

Jacobson's atmospheric science and climate studies began with a boyhood love of tennis. Playing matches as a teenager on "bad-air days" in Los Angeles was what first got him interested in the problem of air pollution. That interest eventually led to a PhD in atmospheric science from UCLA, where he specialized in building powerful air pollution models that also provided climate insights. Now in his late fifties, tall and still athletic, Jacobson is soft-spoken and unpretentious, a trifle boyish, relaxed and confident. He approaches the building and coding of his complex air pollution, energy, and climate models with the passion of a teenager hooked on an addictive new video game; solving problems that vex and stymie other people is fun for him.

Jacobson currently is a professor of civil and environmental engineering and director of Stanford's Atmosphere/Energy program; in his "spare time," he also cofounded two clean-energy advocacy nonprofits. His hard work—more than 150 peer-reviewed papers, plus hundreds of addresses and deliveries of expert testimony—

has had a major impact on the public discourse on, and adoption of, clean energy. Notably, he was one of the earliest energy experts to carefully analyze the broad questions raised by renewable-energy skeptics about the adequacy, affordability, and reliability of renewables. At a time when many research scientists were still cautious about discussing the dangers of climate change, Jacobson was boldly publicizing his conclusions in an effort to get his academic research applied in the policy realm to solve climate problems.

That research has so far helped convince more than two hundred U.S. cities and towns, including Denver, Minneapolis, and San Francisco, to commit, by law or executive action, to obtaining 100 percent of their electricity from renewables by 2050. Sixteen U.S. states (including California, Colorado, and New Jersey), districts, and territories are also committed to getting 100 percent of their power exclusively from wind, water, and solar, some as early as 2030. Hundreds of cities and some countries have followed suit internationally. A few cities and states, and ten nations, have also made commitments of various kinds to getting 100 percent of all their *energy* from renewable sources.*

Although he had made waves in atmospheric and climate science since the 1990s and published important scientific papers about renewable energy in the early 2000s, Jacobson burst onto the international stage as a clean-energy proponent in 2009, when he and co-researcher Dr. Mark Delucchi, from the UC Davis Institute of Transportation Studies, published "A Path to Sustainable Energy by 2030" in *Scientific American*. This groundbreaking article focused on the feasibility of powering the entire world with renewable energy and dispelled the notion that insurmountable technological and economic obstacles would make a global renewable-energy economy impossible.

---

* Although often confused, the terms *energy* and *electricity* are not synonymous. Primary energy is comprised of fossil and nuclear fuels and renewable energy; electricity is a secondary source produced from primary sources. *Energy*, as used in this book, means all energy used in society for manufacturing, transportation, heating, cooling, and producing electricity. Electricity is thus a subset of energy, and its production in the United States currently consumes about 38 percent of all primary energy.

Using his own climate and energy models, Jacobson and his colleagues subsequently developed clean-energy road maps for several large states and countries. This meant taking a rigorous, systematic look at the adequacy of renewable energy resources, their costs, their construction times, and the materials required for their construction and operation. He was also able to use his atmospheric and air pollution models to estimate the reduction in air pollution and air pollution–related deaths that each clean-energy road map would avoid. Gradually, over a period of years, with extraordinary diligence and the assistance of his students at Stanford, Jacobson created templates that enabled him, more or less, to automate the creation of state-level and national clean-energy road maps. He used these templates to produce road maps for 139 nations for which the International Energy Agency had data, all fifty U.S. states, and fifty-three cities.

Jacobson's research makes clear that the challenge of renewable energy is not economic or technological, but now largely cultural and political: cultural because of the necessity of rapidly educating a critical mass of humanity about the realities of climate change and the clean-energy transition and political because major economic interests want to slow-walk the clean transition and because issues of social justice are at stake. Fortunately, Jacobson's foundational work is being corroborated and confirmed by other prestigious new studies.

Disarming Clean-Energy Foes

Jacobson's modeling work has shown that a 100 percent clean-energy transition could avoid millions of deaths from air pollution each year, along with trillions in global-warming costs. His energy road maps show that a rapid energy transition could create more than 28 million net new jobs globally by 2050 and reduce energy costs and cut projected power demands by more than 42 percent.

His road maps for each U.S. state have given political leaders and policymakers confidence that setting ambitious decarbonization goals will cause neither blackouts and brownouts nor energy price surges and economic shocks. His disagreement with the idea that a broad, "all the above" menu of energy choices is necessary, however, galls prominent proponents of nuclear power and carbon capture and storage technologies, like the academics who lit into his prize-winning *PNAS* paper.

The fossil fuel and nuclear power industries have long disparaged renewable energy as too diffuse, too intermittent, and, hence, too unreliable. They have also variously claimed that it is too costly, would occupy too much land, and would take more energy to build than it would render, or that shortages in raw materials would stifle its growth and prevent its being scaled up to truly meet our energy needs. Yet Jacobson, armed with in-depth research and powerful computer models, took on all those claims and has arguably done more than any other scientist to convincingly put all those bogeymen to rest, prevailing over most of his critics.

Jacobson and others have shown that the once-daunting intermittency issue can now be successfully overcome in various ways, so that renewable-energy grids can be just as reliable as conventional grids. By studying continuous energy demand for specific states and nations on an hourly basis, Jacobson was able to show that wind, water, and solar could reliably match that demand. While intermittent power generation may be an issue when power is produced by a single local solar or wind facility, once these facilities are integrated into a renewable energy system and treated as an ensemble, it becomes a whole different ball game. Diverse renewables can pinch-hit for one another, and even similar kinds of renewables can backstop one another from different locales. By interconnecting complementary renewable-energy resources, the newly created system gains flexibility and reliability: When the sun isn't shining, the wind may be blowing. When the wind abates at one wind farm, a stiff breeze may be blowing at another. And when the sun isn't shining and the wind is slack, hydropower or geo-thermal power can be used.

Energy storage is another way to stabilize the output of inter-mittent energy sources. Once a major hurdle, storage costs have plunged in recent years; electricity, heat, and cold increasingly can now be stored economically for many applications. Solar thermal power plants have long been able to store excess heat when the sun is shining and retrieve it after sunset or during cloudy conditions;* photovoltaic (PV) power plants—which use those familiar solar panels—can now store surplus power affordably in industrial-scale batteries.** At night, when winds are strong, wind power can be gen-erated on large wind farms and stored for daytime use.

The dispersal of power-generating assets to sites at or near where the power is used is known as "distributed generation," and it offers certain benefits for utility customers. Instead of power being pro-duced in a large, centrally located power plant, distributed generators make an energy grid more reliable by reducing the likelihood of a system-wide power outage affecting a large number of customers at once. They also reduce power line losses that occur during trans-mission and can reduce transmission costs themselves by reducing demand for transmission capacity. In the case of renewable distrib-uted generators, such as solar panels, vast amounts of solar power can be produced on otherwise unused residential, commercial, and industrial rooftops to meet the power needs of those buildings and even others nearby. Solar panels on U.S. rooftops alone could produce nearly 40 percent of all the power sold in the United States! These systems don't require new land use, and when your solar power plant is directly overhead, transmission and distribution costs, along with transmission line losses, are reduced to zero. (Freestanding arrays over parking areas can produce even more power as well as create

---

* Solar thermal plants use mirrored surfaces to collect solar heat and then convert it into electricity, whereas photovoltaic power systems convert sunlight directly into electricity.
** Jacobson's research has shown that large, expensive, long-duration batteries are not necessary to the stability of a renewable grid.

desirable covered parking in hot climates.) Solar systems coupled with ever more powerful and cost-effective batteries* to store their power can now economically power schools and businesses. Different kinds of renewable energy sources can also be interconnected to form microgrid systems that could supply whole communities; homes or subdivisions can be powered by even smaller renewable "nanogrids," with backup power provided by the local utility.

Are renewable resources really adequate for our energy needs, as Jacobson contends? Well, in 2016, for example, the United States had sufficient total wind resources to generate roughly eight times its power consumption, according to one authoritative estimate, and more than sixteen times according to another. Looking at solar power, if we include photovoltaic power plants along with distributed rooftop generation, then total potential PV generation in the continental United States would be eighty times U.S. electricity sales. Of course, we don't need to rely exclusively on any single renewable-energy source for power, no more than we need to fuel our bodies with a single kind of food. We have a cornucopia of renewable energy sources available: The sun generously delivers more energy in an hour than all of humanity uses in an entire year. Hydropower and geothermal energy, the latter of which draws on subterranean reservoirs of steam and hot water, also provide economical power; both have been in commercial use for more than a century. And if we were able to capture even 1 percent of the heat energy found two to six miles underground, in hot rock, it would be enough to produce a thousand times the energy needed to meet all human energy needs.** Biofuels can serve the relatively few, hard-to-electrify energy demands that currently still require liquid fuels, such as long-distance air travel. (Short and medium-length flights can be electrified; see chapter 7.) Clean power can also electrolyze water to

---

* Lithium-ion batteries are falling in price at a rate similar to that of solar panels.
** Currently, this so-called "unconventional geothermal energy" appears to be relatively costly, and some technological and economic challenges must be surmounted in scaling it up, but we are already capable of drilling to these depths.

produce hydrogen, a fuel that can be burned to produce industrial heat, for example; or it can be converted back to electricity in a fuel cell to provide backup power; or it can be used as a building block for synfuels to power heavy industry.

Moreover, transformative new technology is now increasingly available to make renewable energy even more cost-effective and useful. Artificial intelligence protocols, big data, new inverters, and other power conditioning tools, plus advanced control and switching equipment, can dramatically stretch the capabilities of renewable-energy systems. Pricing incentives now possible through internet-enabled demand management technologies can persuade customers to shift their power use from one time to another, reducing peak demand. This reduces capital costs for the system and, thus, the average cost of power while enabling the grid to work more reliably and efficiently. It can also provide heating and cooling services (as described in chapter 11).

A Diffusion of Responsibility

Despite the abundance and versatility of renewable energy, if we rely on the fifty states to each pass its own clean-energy policy without adequate federal policy guidance and legislation, we may not achieve the comprehensive national clean-energy transition we need in time. Relying on individual regions and cities is a critically important and useful, but fragmentary, approach; more coherent national and international efforts are needed.

"Having a federal law would help," Jacobson told me when we met at his home in 2019. "But we should also have state laws. The more laws [at all levels], the better." (At this writing, the Biden administration has put various clean-energy proposals before Congress, and some have been included and enacted in the Infrastructure Investment and Jobs Act and the Inflation Reduction Act.) "Fortunately," Jacobson continued, "the costs have come down so much that transitions are going on without laws. . . . Iowa is 43 to 45 percent wind now [without statutes]. Nine of the top ten wind states in the United States are all Republican states without many laws favoring wind. You don't need to convince people when they're making money off something. It doesn't have to be a political issue. But if you want to get the thing sped up in all sectors, you do need enforcement of laws, and you have to push on all levels of government."

What specific steps are needed to turn your energy road maps into reality, I asked him. "Education," he answered. "Getting information out to more people about what's possible, what the benefits are." His own three-thousand-square-foot home showcases some of them. "This house is all electric—there's no gas." On the average, it annually produces 20 percent more electricity than it consumes. It's equipped with heat pumps for air and water heating, which use only a quarter of the energy of a gas or electric heater. The lights are LEDs, the house is super-insulated with triple-paned windows, and Jacobson cooks on an induction cooktop that boils water twice as fast as on a gas stove. "All these new technologies hardly use any energy," he said. By adopting them when remodeling or building a new home, homeowners can enjoy big savings. "I saved six thousand dollars just by not hooking up gas to the property," he told me, "and I saved another five thousand or six thousand dollars on pipes—I didn't need any gas pipes."

For clean personal transportation, Jacobson and his son use electric vehicles they can charge at home. Most people would envy his energy bills: "Last year," he told me, "I paid no gas bill, no gasoline bill, and no electric bill, and I was paid five hundred thirty dollars by my utility for the extra electricity. . . . With the subsidies that exist, [the payback time is] five to six years, at most. Without [a] subsidy, it would be nine to ten years."*

Jacobson's atmospheric research also has important implications

---

* Even homeowners and renters not enjoying a professor's salary at a top university can often access affordable financing for energy efficiency projects. Numerous federal and state programs exist to help with installation costs of solar and other clean-energy systems.

for transportation and biofuels. Through his atmospheric climate models, he has made a crucial scientific discovery: absorption of radiant energy by black carbon in soot, not methane, is actually the world's second most powerful cause of global warming. Thus, contrary to previous belief, biomass burning contributes to global warming on a life cycle basis; the dark particles produced by its combustion trap radiant energy. Jacobson has found, for example, that to many people's surprise, ethanol from corn actually causes *more* air pollution than gasoline, producing more ozone from tailpipe emissions in 80 percent of American cities, including Los Angeles. (The magnitude of the impact depends on what other pollutants are present in the air where the emissions occur.) By contrast, Jacobson said, "electric cars will eliminate 100 percent of the twenty thousand deaths caused each year by gasoline and diesel vehicle tailpipe air pollution."

Mainstreaming Clean Energy

Thanks to Jacobson's modeling, state and national policy support for a clean-energy transition have become less politically risky. The migration of his scientific work into the political arena first began almost a decade ago, when Marco Krapels, a businessman with a deep interest in clean energy and sustainability, invited Jacobson to a 2011 meeting with actor and activist Mark Ruffalo, film producer Josh Fox (widely noted for his 2010 documentary, *Gasland*), and others from New York State and California opposed to fracking. At the time, fracking was illegal in New York, but the governor was under pressure to legalize it. Krapels knew about Jacobson's research on the feasibility of a global renewable-energy transition; at the meeting, the activists asked him to create a clean-energy plan for New York.

Knowing the effort that would be involved, Jacobson replied, "I don't have time, but I'll write a paragraph. Then maybe you can hire a consulting company to write a plan starting with that paragraph." A couple of months later, after yet another meeting, Jacobson at last began writing the promised paragraph, "but I got inspired," he said, "and so, the next morning, I sent them a fourteen-page, single-spaced manuscript." Overnight, he had produced an entire conceptual energy plan for the state of New York. The document, he said, was greeted with "shock and awe." He, Ruffalo, Fox, and Krapels soon became close friends, and forty drafts later, in 2013, New York governor Andrew Cuomo was informed of the plan in a joint letter prepared by two nonprofits, the Natural Resources Defense Council and the Solutions Project, which urged Cuomo to begin implementing aspects of it.*

The plan's existence had another important impact. By 2012, fracking had become a controversial, high-profile political issue in New York State. The petroleum industry was eager to see it legalized; citizens who knew about the environmental problems it was causing in nearby Pennsylvania were vehemently against it. In 2013, thousands in the state protested fracking, and Governor Cuomo finally banned it in December 2014, clearing the way for adoption of clean-energy measures. The ban occurred, Jacobson said, "due to all this activism and the actual health and environmental concerns about fracking, but also because there was this alternative plan." The state proceeded to install lots of charging stations, established a "green bank,"** and a solar rooftop policy, and began laying the groundwork for large investments in offshore wind.

Continuing the momentum, Governor Cuomo in 2016 proposed that the state be required to meet a 50 percent clean-electricity stan-

---

* The Solutions Project was formed in 2011 by Ruffalo and Fox—who had extensive contacts with policymakers, celebrities, and nonprofits—along with Jacobson and Krapels, to educate the public and policymakers about the possibility of a transition to 100 percent renewable energy based on wind, water, and solar power and about Jacobson's road maps for individual states. In addition to scientists like Jacobson and business leaders like Elon Musk, the Solutions Project's advisory council included prominent celebrities like Leonardo DiCaprio, Woody Harrelson, Scarlett Johansson, and Julianne Moore.
** A "green bank" is a bank, usually publicly funded in whole or part (or structured as a nonprofit corporation) intended to provide low-cost capital to support clean-energy projects and an array of climate-friendly and environmentally friendly investments, for example, in climate-resilient infrastructure, energy efficiency, low-carbon energy sources, and water conservation. See "What Is a Green Bank?" Green Bank Network, accessed May 8, 2023, https://greenbanknetwork.org/what-is-a-green-bank-2/.

dard by 2030. The New York Public Service Commission adopted the standard in 2016, along with a requirement that half the power come from onshore and offshore wind, solar, and hydropower. The governor went even farther in 2018, proposing that the standard be raised to 70 percent by 2030 and "up to" 100 percent by 2040, and he included a comprehensive Green New Deal plan in the state's 2019 budget proposal. Finally, in a historic decision in June 2019, the New York State legislature adopted the Climate Leadership and Community Protection Act, codifying the governor's 2018 electricity targets into law and setting the goal of rendering the state's entire economy carbon neutral by 2050.*

An even more ambitious vision consistent with Jacobson's scientific research, New York Senate Bill S2878B, was introduced in 2019. It was designed to create a broad and powerful task force to plan for a statewide Green New Deal and to draft comprehensive legislation to implement it. The unprecedented governmental, industrial, and economic mobilization contemplated in the bill is also intended "to virtually eliminate poverty in New York state and to make prosperity, wealth and economic security available to everyone participating in the transformation." At this writing, however, the bill has not reached the governor's desk.

In the middle of advising (and pushing) New York State along the renewable-energy path in 2014, Jacobson thought, "Why not just do the same thing for California?" So, he began working on a California energy plan with students at Stanford and University of California researcher Mark Delucchi. This became a road map to 100 percent clean energy for California—a state with the world's fourth largest economy and a political climate favorable to clean energy—by 2050. After seeing the road map, the governor's senior energy and environmental adviser invited Jacobson to brief the Brown administration

---

* The state's plan, which includes environmental justice features and fair labor practice requirements, also calls for large expansions of solar and wind, including $1.5 billion in competitive awards for twenty large renewable-energy generation and storage projects and $200 million in port infrastructure investment to support a robust new offshore wind energy industry in the state.

on it shortly before the 2014 gubernatorial election in which Brown was seeking a second term. Governor Brown was reelected that November, and in his January 2015 inauguration speech, he proposed several laws that followed logically from the California road map, including a 50 percent renewable electricity goal for 2030 and a call to halve the energy use of existing buildings by that same year. These policies subsequently became law as part of the state's Clean Energy and Pollution Reduction Act.

Before leaving office in 2018, Brown also signed a nonbinding executive order (Executive Order B-55-18) that commits the state to carbon neutrality by 2045 and net-negative emissions thereafter; in addition, he signed a law (Senate Bill 100) that set a goal of 100 percent renewable power (from wind, water, and solar) for the state by 2045. And, starting in 2020, the state began requiring all new homes to have solar panels on their rooftops wherever appropriate. As for transportation, the California Air Resources Board (CARB) has the authority eventually to reduce transportation emissions to zero, and in 2022, it ruled that by 2035, all new cars sold would have to be electric or plug-in hybrids, in effect codifying an executive order previously made by California's current governor, Gavin Newsom, in 2020. "We [still] need legislation to address transportation, buildings, heat, and industrial heat," Jacobson said. "[T]here is no technical or economic barrier to actually doing this. The low-hanging fruit is in buildings and transportation. It's even easier than in the electricity sector."

As if in response, in late 2022, CARB issued its 2022 Scoping Plan, a new climate action framework that could double as a wish list for climate-protection advocates, though some might prefer an even more ambitious timetable and more stringent goals. The plan is designed to cut the state's emissions by 85 percent below 1990 levels and to achieve carbon neutrality by 2045. The really good news is that it will reduce petroleum use in the state by 94 percent, cut air pollution by 71 percent, create four million new jobs, and save the state two hundred billion dollars in health care costs. The plan also imposes a moratorium on new gas-fired power plants, supports mass transit, and establishes a multi-agency process to ensure equitable implementation.

Meanwhile, climate activists have also been working closer to the grass roots. In partnership with the Solutions Project, the Sierra Club mounted a campaign that ended in 2022 in which it went city by city to get communities to commit to 100 percent renewables. "They've been really successful with that, converting now, I think, one hundred ten cities and counties around the U.S.," Jacobson said, by getting city council resolutions or other commitments passed. Through the Solutions Project, Jacobson's clean-energy road maps also now have the support of more than one hundred other nonprofit organizations, in addition to the Sierra Club. "[The organizations] have been really instrumental in engaging the public and talking to policymakers," he said. Now, in addition to the governors of California and New York, those of Connecticut, Colorado, Illinois, Maine, Michigan, Nevada, and Wisconsin have also committed to 100 percent renewable energy.

Jacobson also has had an influence on on the federal level as well in important states. "I actually got a call from [Senator Bernie Sanders] before the presidential election in 2016," he told me. "[Sanders] said, 'I want to bring forward your hundred-percent-renewable plans for the fifty states to the Senate.'" (True to his word, Sanders cosponsored Senate Bill 987, known as the "100 by '50 Act," in the 115th Congress in 2017. It was explicitly designed to transition the economy away from fossil fuels to 100 percent renewable energy by 2050, but it did not become law.) "As a result of all the organizing, Sanders adopted our plans as part of his presidential platform in 2016 and . . . committed to 100 percent. [Former Maryland governor] O'Malley actually committed to 100 percent first. . . . Hillary Clinton said we have to go as quickly as possible to 100 percent renewable energy, and nothing should get in our way. Then the Democratic National Committee adopted 100 percent as part of their platform. We were small but pretty effective," Jacobson concluded. And despite all the efforts to discredit his work, that might be an understatement.

A Vindication, of Sorts

After the 2017 assault on his *PNAS* paper and modeling results, Jacobson felt his reputation and scientific credibility were on the line. He made vigorous efforts to convince *PNAS* not to publish the critique of his work; after it did, he sought a retraction. When all his efforts to settle the issue were rebuffed, he finally filed a ten-million-dollar lawsuit for defamation of character—but offered to settle the case without payment in exchange for a retraction. His critics and *PNAS* refused to budge, and as the case dragged through court and his legal fees mounted, Jacobson, facing another year of litigation and another million dollars in legal costs, decided to drop the suit. The vindication he could not achieve in court, however, has come in the realm of policy and law, as state after state has raised its renewable-energy targets in line with his research on the potential for 100 percent clean energy from wind, water, and solar energy. A recent federal study by the National Renewable Energy Laboratory has also confirmed that a net-zero-carbon electrical grid is possible by 2035 without any carbon capture, bioenergy, or net new nuclear power construction. (Moreover, of the four scenarios the NREL modeled, the one without these technologies was shown to produce the highest net social benefit.)

Jacobson's research first rocked the academic world, then had a transformative impact on the energy and climate plans and policies of nations, states, cities, utilities, and hundreds of corporations, including some of the world's largest. The Climate Group and CDP (formerly the Carbon Disclosure Project), through their "RE100" project, have now registered 360 companies that have pledged to get 100 percent of their power from clean renewables by various deadlines; they include the likes of Amazon, Google, JPMorgan Chase, and Microsoft. Jacobson has probably had a greater impact on legitimizing reliance on 100 percent clean energy from the sun, wind, and water than any other scientist in the twenty-first century. And that—despite the academic backbiting he has had to endure—is something his detractors can neither match nor ever take away.

6.

Beyond Fumes: 100 Percent Clean Transportation

It's only a matter of time before we are able to get everywhere we want cheaply and with clean-transit options. Cars, trucks, buses, and trains will eventually run almost entirely on clean electricity or hydrogen. That's very good news, given that transportation is the single largest source of U.S. greenhouse gas emissions. In this chapter, we will meet people who are working on ways to satisfy all our transportation needs with 100 percent renewable energy sources.

We're On the Road to Big Savings

Where is the United States on the path to clean travel, and what are the stakes? Turns out that by revamping our inefficient transportation systems along clean, sustainable lines, we could actually save seventy to one hundred trillion dollars globally by 2050. Most of the technology to do so already exists, and smart public policies could drive it swiftly into the market. We'd have cleaner air, a safer climate, and major fuel-cost savings—and, most important, nearly a quarter of the world's greenhouse gases would be eliminated.

Huge challenges still need to be faced, however. Enormous legacy fleets of long-lived rolling stock of all kinds will have to be replaced. For example, the United States currently has close to 25,000 diesel-powered locomotives with an expected lifetime of twenty-five to thirty years. The nation also has 2 million diesel-powered semi-tractors. Then there's the minor problem of the increasingly long-lasting 1.5 *billion* cars and light trucks now on the world's roads, roughly 98 percent of them operating on gasoline and diesel. Together, these vehicles—whose number is growing rapidly—produce 15 percent of all fossil fuel and industrial CO_2 emissions per year: nearly 6 billion metric tons of CO_2 globally on an annual basis.

Fortunately, clean transportation is bigger than just the increasing number of hybrid and electric cars we see on the road today—and it's coming sooner than you might expect. For roadway use, Daimler, the world's largest truck maker, recently announced that it would switch to 100 percent zero-emission vehicles (ZEVs) by 2026 or sooner. For long-haul eighteen-wheelers, it is testing, in collaboration with Volvo, a $GenH_2$ fuel cell truck that will go six hundred miles without refueling; these trucks will be able to refuel at a network of hydrogen fueling stations in Northern Europe that Daimler is building in partnership with Shell. Daimler also plans to bring battery-operated short- and medium-haul trucks to market by 2026 that will be cheaper than diesel models. They'll be chargeable at a network of high-voltage charging stations in the United States and Europe that Daimler is going to build in partnership with Siemens.

As you might imagine, Tesla created quite a splash, in 2017, when it announced that it was jumping into the pool of companies intent on fielding two models of a heavy-duty, long-haul Class 8 electric semi-tractor, the Tesla Semi. The shorter-range model, with an expected three-hundred-mile range, was originally to cost $150,000, while its sibling, with five hundred miles of range, was supposed to sell for $180,000. However, by mid-2023, when some of the less expensive trucks finally became available for delivery, their range had stretched to four hundred miles, but their price had swollen to $250,000, about twice the cost of a conventional Class 8 diesel. Time will tell whether Tesla will make big waves in the heavy-truck market with the Tesla Semi, as it has in the luxury-car field. Tesla also announced a partnership in 2020 with a Chinese company to produce advanced batteries for between $80 and $100 per kWh, about half the cost of today's best batteries.

Another indication that the writing is on the wall for new gasoline and diesel vehicles was the decision in 2022 by CARB to ban their sale in California by 2035. New York and Washington State now have both followed suit. The governors of twelve states have petitioned the Biden administration to make the ban national, and many other states may follow their lead. Abroad, Austria has proposed banning new gasoline and diesel vehicle sales by 2030, and Norway by 2025; for Germany, Ireland, the Netherlands, Sweden, and India, the date is 2030. A total of sixty nations have adopted bans by various dates, and the twenty-seven-nation EU bloc as a whole is banning them by 2035. Even China will ban them after 2040. It's no exaggeration to say that the sun is setting on gasoline and diesel vehicles.

These bans are possible because, as the percentage of clean power on the grid ramps up, truly emission-free clean-electric transportation is becoming ever more feasible. Electric vehicle price tags are shrinking along with the rapidly falling cost of their batteries. CARB has projected that new EVs will reach price parity with gasoline and diesel vehicles by 2030; some experts believe it could happen much sooner.

Driving an EV is also becoming ever more practical: the United States already has more than seventy thousand charging stations, and President Biden has announced plans to build five hundred thousand more. If funded, that build-out will make it easy for drivers to find charging spots. At the same time, more powerful batteries are making long trips on a single charge possible, and EVs are becoming hip and stylish; U.S. drivers have forty-four different EV models from which to choose. Volvo is phasing out gasoline and diesel car sales entirely in 2030; GM, by 2035.

Major automakers like Honda, Toyota, and Hyundai are also producing zero-emission fuel cell cars. While more expensive now, they offer more than three hundred miles of range and can be refueled in minutes, without long waits at charging stations. So, it's no surprise that electric and fuel cell vehicle sales are generally trending steeply upward. Worldwide, an estimated thirty million electric cars, buses, and delivery vans are on the road. Almost 60 percent of all new-vehicle car sales are expected to be electric by 2040.

Fueling Revolution

Dan Sperling, relaxed and congenial, with a playful sense of humor, is a man who wears many hats. The day we met, at the Institute of Transportation Studies that he founded at UC Davis, he was inside his office wearing a dapper brown Italian fedora that matched his brown slacks and loafers. He owns at least thirty hats.

His trademark fedora and well-trimmed beard gave him a slightly rabbinical air, an unpretentious, whimsical look for a fellow who, more than anyone else, is the go-to person for all transportation issues in the state of California. Sperling not only serves on the state's Air Resources Board but also holds twin professorships at UC Davis. He is a winner of the esteemed Blue Planet Prize and many other high professional distinctions.

A big-picture guy and a man of strong opinions, Sperling believes EVs "will eventually sweep gasoline and diesel cars from the market" and that driverless cars are inevitable. One of his foundational beliefs is that the easiest way to begin reducing GHG emissions in transportation without any technological revolution is simply to make vehicles more efficient. However, he's well aware that instead of trying to squeeze a few extra droplets of efficiency out of mature combustion engine technology, much larger energy efficiency benefits can be harvested through the design of new transportation systems and through the adoption of the many new transport technologies that are the subject of this chapter.

Sperling wants to revolutionize transportation and has proposed melding electric propulsion, driverless automated vehicles, and ride-sharing to maximize their benefits—a suburbs-without-garages-and-driveways scenario. He is, in effect, an ambassador and advocate for safer, cleaner new mobility systems. The car-free-suburbs scenario is detailed in his latest book, *Three Revolutions: Steering Automated, Shared, and Electric Vehicles to a Better Future.*

Although today he works on urban problems, Sperling grew up in the countryside, on a chicken farm in rural New York that reeked of manure dumped into open lagoons not far from the family's well. Friends refused to ride in the family's smelly car. Understandably, Dan was not drawn to a career in farming, and instead of getting engaged in farmwork, he read voraciously. At Cornell University, he got a degree in environmental engineering and urban planning and later earned a PhD in transportation engineering from UC Berkeley. In 1973, during his senior year at Cornell, a VW van full of Peace Corps recruiters picked him up while he was hitchhiking to visit his girlfriend and convinced him to join the corps. Soon after graduation, he did so, and found himself in Honduras, working for the Ministry of Public Works as an urban planner, trying to create a better society for ordinary people—a mission he's been on ever since.

In California, half the state's GHG emissions and 80 percent of the nitrogen oxides that create ozone pollution are produced by the transportation sector, so to eventually reach net-zero carbon emissions, California energy policymakers have known for some time that deep reductions in transportation sector emissions would be necessary. The state took comprehensive action on climate change in 2006, with its Global Warming Solutions Act, also known as Assembly Bill 32. The law's aim was to reduce California's overall GHG emissions to 1990 levels by 2020. To help achieve this goal, CARB in 2009 took up a mechanism that Sperling and Professor Alex Farrell, from Berkeley, came up with known as the Low Carbon Fuel Standards (LCFS). Sperling regards it as his most important work.

The essence of the LCFS program is to create performance standards for *fuels*, in contrast to the national fuel efficiency standards for *vehicles.* The idea is not only to reduce GHGs and air pollution, but also to diversify the state's fuel supply and reduce reliance on petroleum; the program works in conjunction with other AB 32 provisions to reduce GHGs. Adopted in an amended form by California in 2011, the LCFS now requires oil companies to reduce the carbon content of their fuels 20 percent below 2011 levels by 2030. Other states and provinces, such as Oregon and British Columbia, have adopted the standard with variations; still others are considering doing so. Sperling is hoping the LCFS will eventually be adopted nationwide and in Canada.

The program requires fuel providers either to meet the standard or to purchase carbon credits from those who have reduced their GHGs. It is expensive for oil companies to make refinery changes to lower the carbon content of their fuels, so after they've made the least costly fixes, it is cheaper for them to buy carbon credits under the LCFS; the program, in effect, obliges the oil industry to subsidize lower-carbon alternatives by buying credits from suppliers of low-carbon fuels.

The idea sounds like a win-win, but it produces mixed results. It provides subsidies to renewable biogas and other renewable-energy producers, as well as electricity suppliers, and it funds rebates to buyers of electric and plug-in hybrid vehicles. However, it also subsidizes ethanol, which is not a good idea, as explained in chapter 5. Moreover, it hinges on the sale of carbon offsets, which allow polluters to keep on polluting (often to the detriment of low-income neighbors), provided that the polluter buys carbon credits from those who reduce their emissions. It's a little like the sale of indulgences in the medieval Christian church, in which sinners could be partially or fully pardoned for their sins by making payments to the church.

Still, the LCFS "pressures oil companies to transition to a lower-carbon future," Sperling says. And LCFS carbon credits in California are now hovering at their peak cost of two hundred dollars per ton of CO_2, plus inflation. In contrast, "a fuel tax or a carbon tax would have to be huge to motivate the same level of innovation," Sperling maintains.

Of course, it's not just personal vehicles that will need to go green. According to the United Nations, by 2050, more than two thirds of the world's population of nearly ten billion will live in urban areas. Riders in densely populated megacities will need affordable, pollution-free public transit. That's where electric buses, light rail, and trains must come in. Even freight and passenger trains will need to be fully electrified with batteries, hydrogen fuel cells, or overhead wires.* Currently, U.S. freight locomotives together burn 3.5 million gallons of diesel fuel a year in their 4,400-horsepower engines, polluting the air with black-carbon soot and GHGs. (Advanced diesel locomotives can greatly reduce emissions across the board, but relatively few are currently in service.) BNSF, North America's largest railroad, however, is already testing a battery-electric locomotive, as is Union Pacific.** Stadler Rail, a Swiss company, has contracted to deliver a hydrogen fuel cell passenger train to the San Bernardino County Transit Authority by 2024. Some half-dozen companies in Japan, Germany, the United Kingdom, and the United States are developing battery-electric or fuel cell trains.

A Chariot Without Fire

Electric public transit offers the public a double whammy cost savings and environmental benefits—and there's no better example than the transit bus. Because their routes are easily altered, buses offer transit planners greater flexibility than fixed-rail systems. A typical full-size bus, though, gets four to five miles per gallon on diesel fuel and annually emits two hundred thousand pounds of

---

* Though technologically simpler, overhead wires energized with clean power may be the option railroads prefer least. The cost of these lines can be prohibitive—four billion dollars per one hundred miles—and can interfere with a railroad's freight operations.
** The 7-megawatt FLXdrive battery-electric locomotive that BNSF is testing, made by Wabtec of Pittsburgh, is coupled to two diesel locomotives, reducing diesel fuel consumption by 11 percent. Diesel savings are expected to reach almost 30 percent in two years, though, and the company plans to build a FLXdrive locomotive powered by hydrogen fuel cells to bring diesel use to zero.

GHGs, along with harmful conventional air pollutants. An electric bus powered by renewably generated electricity has *zero* tailpipe emissions and—depending on diesel fuel costs and routes—can save bus fleet owners over $300,000 in fuel costs (based on the $5.25-per-gallon national average U.S. price of diesel fuel in November 2022) and at least $100,000 in maintenance over the lifetime of a bus. That's enough savings to more than offset the electric bus's higher purchase price, which may be as much as $300,000 more—though, this should decline steeply as battery costs continue plunging.* For the New York City Transit Authority, which operates thousands of buses, the savings offered by electric buses would amount to billions of dollars. Since this translates into a return on investment of about 10 percent, and because cities can float low-interest, tax-free municipal bonds, it makes financial sense for transit districts to retire their entire diesel bus fleets.

Electric transit bus (e-bus) sales are booming around the world and are projected to grow 43 percent annually from 2022 to 2027. EV transit bus sales in Europe—where zero-emission and fuel-cell buses combined hold 30 percent of the market—grew to 26 percent, while in the United States, sales growth reached 66 percent but still only comprises about 2 percent of transit bus sales. Globally, 400,000 e-buses (including hybrids and fuel cell vehicles) are already on the road, almost all in China. An additional 112,000 battery e-buses, plug-in hybrids, and fuel cell electrics were sold in 2022. Some U.S. transit districts are still slow to embrace them, however, doubtless deterred by the high initial costs, reputation for range limitations, or deficient cold-weather performance. Yet, despite these potential drawbacks, it is hard to understand why any city fleet operator today would not want to make their next bus electric.

Proterra ("for the earth" in Greek) is the leading U.S. manufacturer

---

* The maintenance savings are possible because EV technology is much simpler than internal combustion technology. A battery-powered bus, for example, has only a handful of electrical parts—a motor, stator (the stationary part of an electromagnetic circuit), and two bearings. A diesel engine bus has 3,200 parts. Which bus is going into the shop more often?

of purpose-built electric buses and bus-charging systems.* Headquartered in Burlingame, California, the company has sold more than 1,300 buses to 135 transit agencies in 43 states and Canada—more than all its competitors combined, so it's the number one electric bus company in North America in sales and miles traveled. By leasing its bus batteries to its customers, the company has made sure that the initial cost of owning a Proterra bus is equivalent to that of a conventional diesel bus.

Unlike some competitors who retrofit existing bus chassis with electric propulsion, Proterra reimagined the electric bus from the wheels up. Its lightweight composite body weighs four thousand pounds less than a steel-bodied bus, and under ideal test conditions, the bus can travel 1,100 miles on a single charge. Under real-world conditions, a Proterra bus can go close to 200 miles on a charge. Because the average bus serves a route of only 130 miles daily, Proterra's range of 150 to 200 miles means the Proterra bus does not have to recharge its battery until it returns to its depot at night. Its electric drivetrain takes steep hills at a faster clip than a diesel bus. This puts performance concerns and battery range anxiety in the rearview mirror. In addition, the bus is so energy efficient that it can be loaded with ten thousand pounds of passengers and still get the same urban mileage as a conventional gasoline SUV or a pickup truck. Pound for pound, its new batteries are as powerful as those in a Tesla, yet also affordable.

Because most e-buses and delivery fleet EVs of the future will be charged at night, they will provide a market for utilities just when power demand is lowest. Conversely, during the day, when demand peaks, parked EVs (and banks of used EV batteries) connected to the grid could, for a fee, help utilities meet their peak loads without invest-

---

* Chinese competitor BYD Auto, however—with major subsidies from the Chinese government and millions in investment by billionaire Warren Buffett—is nipping at the company's heels. BYD has a manufacturing plant in California and is offering a solution to the high initial cost of electric buses by leasing them to transit agencies. See Hannah Norman, "San Francisco Investor Teams Up with Buffett-Fueled Chinese Manufacturer to Lease Electric Buses," *San Francisco Business Times*, July 11, 2018; and "BYD Receives Largest Battery-Electric Bus Order in U.S. History," press release, BYD, November 13, 2019, https://en.byd.com/news/byd-receives-largest-battery-electric-bus-order-in-u-s-history/.

ment in new power plants. Battery buses and trucks will thus help solve a major problem utilities face: declining demand for service as some customers cut their demand through energy efficiency or reduce their power purchases by installing their own rooftop solar arrays.

Proterra's founder, Dale Hill, is a charismatic, self-described "hillbilly engineer" with a disarming old-boy air that belies a shrewd business sense and relentless work ethic. With his practiced monologue, white goatee, and flashy tie, he could be the Colonel Sanders of electric buses.

Hill graduated from LeTourneau College in Longview, Texas, having studied mechanical engineering and welding engineering. After years of broad industrial experience, he got a contract to build the first hybrid-electric buses for Denver's Regional Transportation District. When a major investor backed out of his company, Hill mortgaged his house and took out a federal Small Business Administration loan.

"We wound up manufacturing and designing just about everything on the bus from scratch," he told me. His new company built a novel right-hand-drive hybrid-electric-drive bus with a vacuum-formed plastic body. One of the country's first low-floor buses, it had four doors on one side and a Ford Pinto engine that ran on natural gas to charge the bus's batteries, which were nothing fancier than ordinary car batteries. Archaic as they seem today, those buses operated for seventeen years and logged over seven million miles.

Hill's next major project became Proterra, his ninth start-up. The Federal Transportation Administration (FTA) contacted him in 2004 and told him they wanted someone to design "the bus of tomorrow" and that Hill's group was the only one they knew that had ever designed a "new technology bus" from the ground up. Hill was awarded a multimillion-dollar contract to produce a prototype that was delivered in 2008 to the American Public Transportation Association's annual meeting.

At the meeting, Hill heard the FTA's head of R&D deliver a white paper entitled, "The Bus of 2030: The Electric Bus." After the presentation, Hill went up to him, put his arm around his shoulder, and said,

"Hey, Walt, how would you like to have your 2030 bus twenty-one years early?"

"What are you talking about?" the FTA administrator asked.

"Well," said Hill, "if you remember, four years ago, I told you I had no idea how we were going to make a bus run all day on battery power. We figured that out, and it's sitting down on the [convention] floor."

Two weeks later, Hill was in Washington, giving FTA leaders an invited presentation. In response, the FTA created the three-year Transit Investments for Greenhouse Gas and Energy Reduction (TIGGER) program, which provided forty million dollars to transit agencies to buy Proterra battery-electric buses. "That's how we got the funding to attract customers and to be a success," Hill told me. He predicts that "by 2040, 90 percent of the buses in the U.S. will be battery-electric." In California, starting in 2023, public transit agencies have to begin buying electric buses. Starting in 2029, they must buy only battery-electric buses.

While I was doing my research on Proterra, I spoke to then-president and CEO Ryan Popple, a Harvard MBA and former partner in the venture capital firm of Kleiner Perkins Caufield and Byers. Whereas Dale Hill knows how to build instant rapport with clients in a folksy way, Popple is a cautious, expert manager. Yet his rhetoric on electric buses was far from conservative: "The goal of the company," he said, "is to *eliminate* diesel buses." When I pressed him on that, he said, "What we need to do is make that electric bus so much better than a diesel bus that a diesel bus is in a museum. Just like the horse and buggy is an interesting historical relic. . . . I actually think the company will maximize shareholder returns if it succeeds in its larger purpose, which is to take a dirtier product completely out of the market. I don't think we should stop innovating until there is no reason to buy a diesel bus."

Tomorrow's Trucks

At this point, you may be wondering why I'm spending so much ink on buses and trucks—hardly the most glamorous of vehicles to most people. The answer is that whereas they're only 10 percent of all vehicles on U.S. roads, they account for a hefty chunk of all air pollution from the transportation sector: 45 percent of all nitrogen oxides and 60 percent of all particulates. As we'll see, a lot more could be done to clean up these vehicles.

Battery technology is not just revolutionizing the bus industry; it's invading trucking as well. Among the world's heavier trucks, those restricted to a one-hundred-mile range on a well-defined route are likely to be electrified first. These include the drayage vehicles that shuttle goods from ports to warehouses or distribution centers; the largest trucks are used for very heavy loads, up to eighty thousand pounds. Because regional haul (under three hundred miles) and "last-mile" trucks usually return to a depot at night, charging them overnight should be relatively easy, even today. And with low operating costs and relatively small batteries, these medium-duty e-trucks should cost no more than diesels on a life cycle basis—and maybe less as innovation continues.

Mike Moynahan is the fleet manager for HEB, a large grocery store chain in San Antonio, Texas. HEB has more than 100,000 employees and thousands of trailers that serve more than 150 communities in Texas and Mexico. When we spoke, he had just ordered a dozen powerful electric trucks he calls "yard mules." Unlike real mules, these vehicles work tirelessly, twenty-two hours a day, to move trailers around HEB's terminals; Moynahan can charge their batteries at rock-bottom off-peak-power prices. (One day, that power will come from clean renewables. Today in Texas, about three quarters of it still come from fossil fuels and nuclear power.)

At Moynahan's suggestion, I spoke with the maker of the e-truck he bought, Lonestar SV Electric. Company manager Blake Yazel explained to me that although these terminal trucks cost $115,000 more than an equivalent diesel, once Lonestar has taken out the diesel engine and refurbished the truck with a sophisticated battery-electric drive, it repays the extra cost within two years in moderate use; in 2020, the Lonestar truck saved buyers up to $60,000. Those savings go straight to a company's bottom line—the kind of investment that should grab any manager's attention. Even when diesel fuel prices were relatively low, as during the COVID-19 recession in 2020, the e-truck still provided huge savings in fuel and in avoided diesel engine maintenance of up to $25,000 a year. Today, diesel fuel is more than 50 percent more expensive, so the savings are much greater.

Faster Charges, Longer Hauls

Unlike Mike Moynahan's yard mules, the long-haul, heavy-duty truck will probably be among the last to be electrified. Batteries powerful enough for a long-hauler are expensive and add thousands of pounds to the tractor's weight, reducing its maximal payload by up to a quarter. Because haulers are paid by weight, this costs them money. Even for a fleet operator like Moynahan, whose trucks serve mostly short or medium-length routes, battery-powered trucks would work for only a quarter of his fleet. The rest need to carry their maximal payloads.

The other big problem for battery-electrics, especially long-haulers, is the lack of widely available roadside fast-charging facilities. Even some fleets that, in principle, could charge electric tractors overnight in their "hub-and-spoke" depots have sometimes found their local utilities unprepared to supply the large amount of power needed. Yet, with dramatic drops in battery costs, improvements in the trucks' power-to-weight ratio, and continuing progress in fuel cell–electric technology, major diesel truck manufacturers are eyeing the future. They see the potential for a large, heavy-duty e-truck in a market in which more than 280,000 conventional heavy trucks are sold each year in the United States alone.

As batteries continue improving in range performance and dropping in price, electric trucks will start to be phased in on different timetables and rates in different market segments. At least until 2050, the truck industry will have a mixture of diesel, natural-gas, hybrid, fuel cell, and electric vehicles on the road.* Yet, in a recent survey of three hundred commercial fleet managers, 90 percent said that EVs are "the inevitable future of commercial fleets." Whereas diesels are a mature technology that has been around since the 1890s, e-trucks are a young, dynamic technology that is improving in performance and falling in cost at rates diesels will be unable to match.

The displacement of diesels will start most quickly in those markets where the route can be most easily served and the demands on the vehicle most easily met with electric technology. The burgeoning of e-commerce and online shopping is creating greater demand for "last-mile" service to homes and businesses. These kinds of deliveries tend to be bulky and to "cube out" (i.e., take up all available onboard space) long before the vehicle's load weight limit is reached. This leads to lighter loads that electric trucks with heavy batteries can easily handle. Shippers like Amazon, FedEx, and UPS rely mainly on light- and medium-duty trucks for local deliveries. Amazon recently placed an order for one hundred thousand electric delivery vans to be made by Rivian, an EV manufacturer.

Smaller light trucks—half-ton pickups, minivans, and small cargo vans—can use electric drive about as easily as a passenger car. Light- or medium-duty delivery trucks will "go electric" next, especially those operating in stop-and-go urban traffic conditions that activate their regenerative braking systems, recharging their batteries.

Mike Roeth directs the North American Council for Freight

---

* Daimler Truck AG, Volvo Group, Honda, Hyundai, and Isuzu are all engaged in various collaborations to advance hydrogen-electric fuel cell technology and to develop heavy-duty fuel cell trucks because of the long driving range possible. While fuel cell vehicles will travel long distances with the least extra weight, they are going to face stiff competition from battery-electric trucks.

Efficiency (NACFE) and has worked in the commercial vehicle industry for thirty years. He sees the dearth of truck charging stations as a stumbling block for e-trucks. Even if utilities handle the costs and can provide sufficient electricity, fleet operators will still have to pay for chargers, each of which can cost $30,000 to $50,000. Federal investment tax credits or loan guarantees for these investments could help accelerate EV truck adoption, but installing the infrastructure will take not only money but also time. In a textbook case of how sound government policy can drive private corporate investment, the European Union's goal of a carbon-neutral freight transportation sector by 2050 spurred Daimler Truck AG, Volvo Group, and Volkswagen AG's heavy-truck subsidiary (Traton Group) to announce plans to collaborate in building a $593 million network of 1,700 charging stations for heavy trucks in Europe.*

The Future of Trucking

Nationally, oil-fueled transportation is the largest source of greenhouse gases, according to the U.S. EPA. Internal combustion vehicles release 28 percent of U.S. greenhouse gases, nudging out power generation and industry for first place. (If you include all the refinery and other oil industry emissions, it could be as high as 35 percent or more, even without considering the energy and pollution needed to fabricate all those vehicles and the two hundred million tires that passenger vehicles require annually.)

Battery-electric or fuel cell trucks will change the game. They and e-buses have zero noxious emissions, zero noise, and zero vibration. They offer the excellent torque of an electric motor and, overall, are

---

* The European Union, however, could need up to 50,000 charging stations by 2030 to meet its goal, and the United States could need nearly 600,000 ordinary charging stations to service the then-expected 15 million electric vehicles. See Abby Brown et al., *Electric Vehicle Charging Infrastructure Trends from the Alternative Fueling Station Locator: Second Quarter 2022*, Technical Report NREL/ TP-5400-84263 (Golden, CO: National Renewable Energy Laboratory, December 2022), https://afdc.energy.gov/files/u/publication/electric_vehicle_charging_infrastructure_trends_second_quarter_2022.pdf.

much more efficient and economical to operate than a combustion engine vehicle. Battery-electric trucks completely eliminate a truck's engine, fuel system, crankcase oil, emissions-control system, and engine coolant. All major truck manufacturers are introducing them; Toyota Motor Company doesn't even plan to sell any more fossil fuel–powered vehicles after 2050.

Here are some basic "rules of the road" for understanding what clean trucking will look like. Trucks come in nine different weight classes—from under 6,000 pounds to over 33,000 pounds—and in myriad styles, ranging from pickups, vans, SUVs, semi-tractor trailers, and local delivery vans to tankers filled with fuel, grain, milk, or chemicals. Then there are the trucks manufactured to perform specialized functions, like hauling waste or concrete, trimming trees, refrigerating perishables, carrying mined ore, or transporting shipping containers. Even huge ore-hauling trucks used on mine sites, however—the ones that require a long ladder to reach the cab—will one day be clean. Fortescue Future Industries, an Australian company, is planning to have a hydrogen fuel cell truck like this in commercial operation sometime in 2023.

Charging a large e-truck fleet requires power on a scale comparable to that of a large manufacturing plant—as much as 5 to 30 megawatts for the largest operations. Providing that much power means engineering work—perhaps the undergrounding of new power lines—and, not infrequently, substation work. Some small utilities just can't provide more power without blacking out their other customers until they can add new generating capacity. Another problem is that charging an e-truck takes longer than refueling a diesel, a disparity that NACFE expects will persist until after 2030. In order for e-trucks to be efficient and economical for long-distance use, extremely fast charging systems will be needed to impart up to a megawatt of power to a large-truck battery pack within thirty minutes. These powerful, fast-charging systems would also have to be integrated with the electric grid, energy storage, and renewable-energy sources.

"We're talking about a new paradigm, with super-sized grid connections, new power lines, transformers, and on-site switch gear to feed high-power chargers," said Paul Stith of engineering consultancy Black and Veatch. Stith is a specialist in the high-power charging infrastructure that will need to be installed at ports, rail yards, warehouses, and distribution hubs, as well as truck stops. The planning for, and building of, new utility substations and overhead or underground power distribution lines can take years, he says. For large-scale electric truck deployment, Stith advises, utilities and fleets need to begin serious conversations about these issues *now*. In fact, major modernization of the utility system is going to be necessary for any massive adoption of EVs of any kind.

Alex Voets of Daimler Truck North America is another expert on accelerating electric truck adoption. He sees a need for "pervasive incentives all over the country for publicly available commercial charging infrastructure" to make coast-to-coast electric truck trips feasible. A few small steps in that direction are already being taken. The Federal Highway Administration is currently planning an alternative-fuel corridor along Interstate 45, from Dallas to Houston, to provide hydrogen and EV charging for medium- and heavy-duty trucks and buses.

Some states, however, are already leading the way toward a clean trucking industry. The California Air Resources Board, for example, has adopted an "Advanced Clean Trucks" rule, with the intent of increasing sales of clean trucks under new standards that would ultimately require truck manufacturers to sell only zero-emission new medium- and heavy-duty trucks in the state by 2045. Such policies do not emerge spontaneously, in a vacuum; they would need to be the fruits of vigorous organizing and lobbying by a coalition of environmental, labor, social justice, health, and community groups.

Living More "Carlessly"

There is, of course, one very rapid route to lowering GHG emissions from personal transportation that this chapter has yet to examine: What if more of us simply chose to live without *any* kind of car? Cleanly powered, shared transportation could take millions of vehicles off the roads, eliminating not only the pollutants they emit in operation but also those they generate through their manufacture and shipment to market. That's the future Dan Sperling, of UC Davis, would like to see: One day, instead of ubiquitous gasoline and diesel vehicles, quiet, safe, and efficient electric vehicles will pick us up at our homes and take us to our destinations or to transportation centers, where we'll be able to board renewably powered electric trains, trolleys, buses, or air taxis. All these transit modes could be integrated into one smart, efficient transit system, accessed with a smart card. Given the low operating costs of EVs, congestion-relieving public transit could increasingly be offered as a free public service, like most roads, to incentivize their greater use.*

Those who chose not to own a car could easily rent one as needed and summon it automatically to their home, pocketing thousands of dollars a year in savings. Private homes might therefore not need large driveways; perhaps just a delivery drop-off spot would suffice. Garages could become guest rooms, home offices, or gyms, or perhaps even be converted into studio apartments. Who couldn't use a little extra income these days? What town couldn't use a little extra affordable housing? And if concrete driveways were ripped out, new vegetable gardens and fruit trees could be planted. Growing food locally avoids the shipment of food from afar, which causes more emissions.

In a more carless world, unnecessary on-street parking spaces could be eliminated, making room for larger sidewalks, bikeways, sidewalk cafés, and mini-parks and other green spaces. Buses, trucks,

---

* In addition to roads, some municipalities already offer free shuttle buses, airport trams, or moving sidewalks.

light rail, and, eventually, automated vehicles could have dedicated lanes on major arteries. Unneeded parking lots and garages on valuable urban land could be converted into much-needed multifamily housing or other uses.

Less space would be needed for car repair shops and parts providers. Millions of EVs with bidirectional charging could provide grid support for utilities, reducing their peak power-generation needs and providing revenue for vehicle owners and cost savings for utilities. Power companies could then afford to cut rates, which customers would greatly appreciate. Meanwhile, conventional vehicles could become much more energy efficient in response to tightened federal fuel efficiency standards designed to cut CO_2 emissions. Higher efficiency requirements for automakers would accelerate the transition to zero-emission vehicles.

Ironically, having studied cars during his whole career, Dan Sperling was "never a car guy." Today, he no longer owns one. To him, they were just a means of getting places—and he believes there are better, more economical ways to do so. "I don't pay [for] parking, I don't pay [for] insurance, I don't pay [for] registration," he says. "I don't pay for repairs. I don't pay for the car itself. I don't pay for gasoline. When you add up all those dollars, it's really a lot of money." When he does need to go beyond walking or bicycling range, or if it's raining or he's stranded, he calls a Lyft or an Uber. For longer trips, he will still rent a car.

Did Sperling have any regrets or anxieties about giving up his car? "I felt liberated," he replied. "I didn't have to deal with a car anymore." When taking a ride service on longer trips, "I've got all this time. I sit in the car and read or do work or nap. I *like* being chauffeured."

In his book *Three Revolutions* (2018), Sperling says that with all costs counted, cars on the average cost about $0.57 a mile to own and operate. (Doubtless the costs would be far greater today.) "From my research," he says, "the average cost of owning and operating a moderately new car is about $8,500 a year"—more than $100,000 before the average car is retired. He has calculated that travel by more intensively used automated commercial vehicles would be much cheaper—"about $0.20 per mile"—because the fixed costs would be spread over many more miles. The costs would be halved again if the rides were pooled, further spreading the costs among multiple passengers—cheaper, he contends, than subsidized public transit.

A more carless future is certainly a "policy good" worth pursuing. Because current land vehicles of all kinds will be in service for ten to twenty years or more, though, it is vital also to employ some policy accelerants to help speed us up the road to cleaner transportation, shared or otherwise. Here, for a start, are a dozen easy policies that can be implemented in most places in the United States in the short and medium term to cut the emissions of legacy vehicles until they can be replaced.*

1. Set dates for the phaseout of fossil fuel vehicle sales.

2. Adopt a steadily increasing carbon tax to accelerate the retirement of fossil fuel technologies.

3. Continue implementing zero-emission vehicle mandates, to require manufacturers to steadily increase their production.

4. Generously fund "cash for clunkers" programs that get the most-polluting vehicles off the road sooner.

5. Continue requiring vehicle manufacturers to improve fuel efficiency and lower the carbon content of their fuels.

6. Provide enticing incentives for the purchase of fuel cell

---

* Some of these measures are already selectively in effect in some jurisdictions. Additional policies are discussed in later chapters of this book.

and battery-electric vehicles, including cars, buses, trucks, motorcycles, motor scooters, electric bicycles; agricultural, mining, and construction equipment; and even the humble lawnmower.

7. Provide ample funding for mass transit.

8. Support R&D on the most cost-effective ways of retrofitting an array of existing vehicles with electric propulsion.

9. Through zoning ordinances, locate jobs and services where people live, to help reduce travel demand.

10. Resist efforts by developers to breach urban boundaries with urban sprawl.

11. Generously provide financial assistance to speed the deployment of an ample national EV charging network for cars, trucks, and buses.

Even without the forceful, definitive clean-transport policies needed, our economy is gradually moving toward pervasive electrification, and the cost not only of batteries but also of clean hydrogen fuel is falling as large (gigawatt-scale) hydrogen production facilities are being built around the world. How long it will take to complete the transition to green transport is now a matter of political will and national policy. The technologies needed are either close to deployable or already available—ready to be scaled up whenever enough votes are cast for game-changing policies and the needed funding is made available. If we want a cleaner, greener transportation future tomorrow, however, coalition building and political organizing will be necessary today, to mobilize public support so that appropriate laws and policies are promptly passed and fully funded.

7.

Soaring the Skies, Sailing the Waves—Cleanly

Transportation is undergoing a technological revolution not just on the ground; change is also taking to the air and sea.

Remember the Jetsons and their flying car? Well, dozens of companies are now developing small EVs that take off and land vertically but are lighter and cheaper than helicopters. Think of air taxis—Ubers and Lyfts of the air—soaring over urban traffic jams to drop you off at your doorstep or on your rooftop. And in Norwegian coastal waters, the world's first fully electric, zero-emission high-speed ferry has gone into service, while a number of luxury electric hydrofoil speedboats are already skimming the waves elsewhere. The new ferry can cruise at 23 knots while carrying 143 passengers, not to mention 20 bicycles!* For the open ocean, Maersk Line, the world's largest shipping company, plans to introduce a container ship fueled with bio-methanol and e-methanol in 2023.** Other oceangoing

---

* The world's first "ordinary" electric ferry, the MV *Ampere*, was built in 2015 in Norway, and today the country already has seventy electric ferries in operation. See Naida Hakirević Prevljak, "World's First Electric Fast Ferry Is Here," *Offshore Energy*, July 14, 2022, https://www.offshore-energy.biz/worlds-first-electric-fast-ferry-is-here/.

** E-methanol is methanol produced electrically from natural gas, a fossil fuel. Yet methanol burns far more cleanly than either diesel or bunker oil fuel (i.e., any fuel used aboard water vessels) and is one of the least-polluting marine fuels; upon combustion, it releases only CO_2 and water. Bio-methanol is produced from carbon-containing plant material or solid waste. If plant material or plant-derived refuse (paper, for example) is gasified and burned, then—on balance and in principle, at least—no net CO_2 should be produced, because the complete combustion of organic matter releases the exact same amount of CO_2 as the plant matter from which it was derived itself removed from the atmosphere by photosynthesis during its growth.

cargo vessels are expected to follow, powered by hydrogen and ammonia.

But, first, let's check out what's starting to happen overhead.

Come Fly with Me—Electrically

Electric aviation is at an earlier stage of development than e-trucking. The same promise and barriers to adoption exist for electric airplanes as for the e-truck, but they are even more technically challenging: in contrast to electric trucks, electric aircraft are designed to leave the ground.

The environmental case for electrifying aviation is easy to grasp. Before the COVID-19 pandemic, people were taking three billion flights per year, and air travel was growing rapidly worldwide. In 2018, aviation produced almost a billion tons of CO_2. Under business-as-usual assumptions, those GHG emissions would triple by 2050, cumulatively releasing more than forty billion tons of CO_2 into the atmosphere by then.

If the aviation industry were a nation, its 2019 emissions would make it the seventh largest CO_2 emitter in the world—and much of that CO_2 is discharged at high altitudes, where it has a more powerful warming effect. Airplanes also emit nitrogen oxides (NOx), which are potent greenhouse gases. In addition, small piston-engine light planes still burn leaded gasoline. Ampaire, a start-up electric plane manufacturer in Hawthorne, California, estimates that in California alone, "small aircraft emit 70,000 pounds of lead, 1,700 tons of NOx, and 90,000 pounds of particulate matter annually."

Although the U.S. EPA and international organizations recognize that GHG emissions from aircraft may endanger public health and welfare, targets for their reduction in aviation were not set in the Paris Agreement, and the EPA has yet to adopt aviation emissions standards. Totally eliminating aircraft GHG emissions in a cost-effective manner is certainly a difficult technological and policy challenge. But *greatly reducing* emissions is quite feasible. Unfortunately, aviation emissions were excluded from international regulation on the unpersuasive excuse that these emissions "are difficult to allocate to a particular country."

No one-size-fits-all solution will eliminate all aircraft emissions. The solutions will involve a combination of improving airplane engines' efficiency, making airplanes lighter and more aerodynamically efficient, substituting sustainable biofuels for fossil kerosene, and deploying new electric, hybrid-electric, and hydrogen fuel cell technologies.

By using more energy-efficient aircraft, airlines have already reduced the fuel burned per passenger by half since 1990, but they could reduce it by half again with yet-more-efficient planes. In addition, sustainable biofuels could cut emissions by 70 to 100 percent compared with conventional jet fuel emissions. Biofuels, however, present cost, supply, and environmental issues that electric propulsion and fuel cells do not. (See chapter 9.)

Beyond improving fuel efficiency, airlines are partial to schemes like carbon credits, which often cost only a few dollars per ton of CO_2. But instead of reducing or eliminating airline emissions, these carbon offsets allow the airline to continue its emissions on the presumption that the credits purchased will result in reduced emissions elsewhere. Unfortunately, the road to climate hell is often paved with good intentions. Thus, an emissions reduction that should or would have been undertaken anyway, or that might turn out to be transitory, can nonetheless sometimes be credited as if it actually offset the airline's carbon emissions. (The carbon credit issue is treated in chapter 14.)

Debut Electric Airplanes

The most environmentally responsible solution for preventing aviation emissions is the electric or hydrogen-fueled (aka fuel cell) airplane, technologies still in their infancy for aviation. Electric planes are currently capable of carrying only a small number of passengers, on relatively short flights. Yet Boeing, Airbus, and the entire U.S. aerospace industry are looking at opportunities in electric aviation, as are 170 companies worldwide. It's not hard to understand why: Electric planes have zero exhaust emissions. They're much quieter than conventional aircraft. They're also more energy efficient and cheaper to operate,* and because they are much simpler—for example, they have no fueling or exhaust systems—they require less maintenance. And because they can take off faster, they also don't need large airports with long runways and can, instead, use any of thousands of small and regional airports.

Electric aviation is making progress, and the industry is bristling with new entrants. Even as I was writing this chapter, a small company called AeroTEC sent a nine-seat Cessna Caravan with a magniX electric motor into the air for a thirty-minute maiden voyage. AeroTEC claims the craft is the largest all-electric airplane to fly to date. In addition to AeroTEC, a few other small companies already have their own small electric planes aloft. Eviation, an Israeli company, has flown a nine-passenger all-electric commercial plane for the regional air travel market with a 250-nautical-mile range and a top speed of 300 miles per hour. The company already has more than $2 billion in orders and expects to have the plane in commercial service by 2027.

In Sweden, Heart Aerospace plans to have its four-propeller hybrid-electric nineteen-passenger ES-19 airliner certified and ready to fly by 2026, with a 250-mile hybrid range—a game-changer because of the number of passengers it serves. Now the company has upped the ante and is developing the thirty-seat ES-30, which is to have a 500-mile range in hybrid mode and a 125-mile range in all-electric mode and is to be ready for commercial delivery by

---

* Heart Aerospace, a Swedish manufacturer of electric airplanes, estimates that electric planes reduce maintenance costs by 90 percent compared to turboprop planes. Ampaire, a small California company, estimates that fuel-equivalent costs are 90 percent less than with conventional aircraft.

the end of 2028. Major investors include Air Canada, Breakthrough Energy Ventures, Saab Aerospace, and United Airlines Ventures.

Harbour Air, of Seattle, is now flying a classic workhorse seaplane called a de Havilland Canada DHC-2 Beaver that has been retrofitted with a magniX electric propulsion system, the same propulsion system used by Eviation. The Beaver is being used for short flights—only a fifth the distance of its fossil fuel–powered cousin—but it offers impressive fuel savings. Whereas the original Beaver engine costs $300 to $450 an hour to operate, largely for fuel, flying the magniX costs just $12 an hour for electricity.

Some market entrants will have relatively niche applications. For example, Pipistrel, a small Slovenian aircraft company, is selling two-seat electric training airplanes, an electrified glider, and an electric propulsion kit for retrofitting suitable light planes.

As mentioned, even flying cars are on the horizon—literally. Billions of dollars are about to be invested in small electric, propeller-driven vertical-takeoff and -landing vehicles, known as VTOLs or eVTOLs, that would provide affordable air taxi service. A VTOL—initially with a pilot, but someday robotically controlled—would pick you up at a midtown urban "vertiport" on a multistory building or parking structure. It would then whisk you along at two hundred miles per hour for a quick local commute over traffic congestion to the suburbs or to a medium-range destination. Prototypes are already taking to the air as piloted and autonomous test vehicles.*

Joby Aviation has plans to offer an airborne urban ride-hailing service; it is partnering with Toyota to build the vehicles and has acquired an Uber spinoff known as Uber Elevate to provide ride-hailing expertise that Joby says "will combine the ease of conventional ridesharing with the power of flight." It claims the service will be "a green alternative to driving." Joby envisions getting costs down to $0.50 a mile, similar to the costs of driving a private auto, and is

---

* Of course, we have yet to confront the future problem of providing air traffic control for thousands of such small aircraft, so they don't collide with one another, killing people and raining dangerous debris on crowded urban areas.

in the process of taking the company public, in a deal to be valued at $6.6 billion. Advocates contend that the costs eventually would be similar to those for a metropolitan taxi ride. The firm has set its sights on delivering ride-hailing services—like an airborne Uber or Lyft—as early as 2024.* Joby Aviation's four-passenger eVTOL takes off and lands vertically, like a helicopter, but is twice as fast and "a hundred times quieter" once airborne. It recently flew a successful 150-mile test flight, including takeoff and landing.

Beta Technologies of Burlington, Vermont, makes an eVTOL called the Alia with which it is uniquely targeting the short-haul freight delivery market now served by box trucks. The Alia requires no runway and is designed to be flown also without a pilot. Currently, the company has 350 employees, $400 million in investment, and 10 orders from UPS, which plans to order 120 more.

Going Mainstream (or Slipstream)

How long will it be before electric aviation becomes common? What are the obstacles to carrying a planeload of a hundred passengers on a long, electrically powered flight? Black and Veatch's Paul Stith estimates that by the late 2020s or early 2030s electric planes will be capable of serving regional markets within a distance of 500 miles and will have the technical capability of going the roughly 1,100 miles from New York to New Orleans. Three such flights together could take you from New York to San Francisco, but the passenger capacity and economics of these flights are still unknown, and planning for airplane charging infrastructure is still in its infancy.

---

* The Federal Aviation Administration would first have to approve the plan, and initially prices would likely be high if not prohibitive for most people. Uber has asserted that eVTOL flights could cost about what an Uber Black service costs now before dropping to UberX pricing levels; autonomous eVTOLs would cost far less than those with pilots (see Paul Stith, "Charging Tech Key to Bringing Urban Air Mobility to the Masses," *Metro*, September 16, 2019, https://www.metro-magazine.com/10007188/charging-tech-key-to-bringing-urban-air-mobility-to-the-masses). But it might initially be more difficult for them to get FAA approval.

The big stumbling block for electric aviation is the long-haul, fully loaded flight. Here the problem is similar to that for long-haul trucks: the energy density of batteries (the energy per unit of mass) is low compared with the energy density of aviation fuel. Today, the best lithium-ion batteries hold up to 250 watt-hours of energy per kilogram. But a kilogram of jet fuel kerosene has nearly 13 *kilo*-watt-hours, fifty times as much energy. A Boeing 747 uses about 17,000 gallons of fuel in flying from New York to London. To provide the same amount of energy would take almost 3,000 metric tons of lithium batteries!

Thus, absent some very remarkable advances in battery technology, the first large airplane to fly fully loaded across the Atlantic is more likely to be propelled by hybrid-electric drive—and, later, by hybrid-electric fuel cells—than by batteries alone. Yet it took only sixteen years from the time of the Wright brothers' first twelve-second flight near Kitty Hawk for John Alcock and Arthur Brown to make the world's first nonstop transatlantic flight. It doesn't take a great stretch of the imagination to see that, in a decade or two, medium-size, all-electric planes will routinely be serving regional markets. Norway is so confident of this that it is planning to require that all domestic flights be electric by 2040.

In addition, scientists are hard at work improving the energy density and lowering the cost of electric batteries. Until these new advanced batteries are available commercially, governments could speed the transition to electric flight and cleaner fuels by raising taxes on aviation fuel and requiring an increasing reliance on sustainable biofuels, such as the kerosene made from waste CO_2 (which will be discussed in chapter 8). They could also require that a steadily increasing proportion of short-hop and regional flights be made electrically.

The Global Shipping Challenge

If the challenge of greening aviation is formidable, however, consider trying to get a large container ship across an ocean without fossil fuel. It's a gnarly problem—or, more precisely, a yucky, oily one. There are no financial incentives to clean up shipping, no business case to do so, and no regulations to necessitate it. You could say that shipping is the neglected orphan of global CO_2-mitigation efforts.

Ocean shipping is the most energy-efficient way of moving goods, and it currently moves 90 percent of the world's freight. However, shipping produces about 3 percent of global carbon emissions and perhaps 13 percent of the world's conventional pollutants. Just as with aviation, if shipping were a country, it, too, would be the world's sixth or seventh largest GHG-emitting nation. If international commerce swells under business-as-usual assumptions, and if other, more easily controlled emissions sources are simultaneously curtailed, shipping could be responsible for as much as 17 percent of the world's GHGs by 2050.

The largest container ships weigh over two hundred thousand tons *before* taking on their hundreds of thousands of tons of cargo. It takes powerful engines indeed to force these "large floating shoeboxes" through the water; if they could be driven electrically, it would take an onboard power plant capable of delivering up to 90 megawatts of power. Batteries are not practical for these huge ships, because of their low energy-to-volume ratio and high cost. Fuel cells could do it, but the fuel would be relatively bulky and expensive compared with diesel fuel. Ammonia is another alternative fuel that could be burned without producing CO_2, but 90 percent of it is currently made from natural gas, a fossil fuel (which thus generates carbon emissions as a by-product of its production), and it has less than half the energy density of bunker fuels.*

---

* Bunker fuels are generally the heavy, "residual" fuel oils (left after distillation) used to fuel a ship engine.

And here's yet another obstacle: If the maritime industry is to go beyond improving the hydrodynamic efficiency of hulls and propellers and the thermodynamic efficiency with which ships burn oil or gas, it would need to adopt low- or zero-emissions technologies. Because oceangoing vessels travel long distances to different ports, they would not only have to substitute relatively expensive cleaner fuels for today's cheaper and dirty fuels before leaving their home ports—something few companies would voluntarily do—but would also need access to that alternative fueling infrastructure at their destinations. For the most part, such infrastructure doesn't currently exist, and no one has stepped up with an offer to pay for it because—in a classic chicken-or-egg situation—there is currently no market for it.

Another problem is the world's legacy maritime fleet: even if all new ships used zero-emission propulsion, it would be years before maritime emissions dropped significantly, because most ships are in service for twenty-five or thirty years, and sometimes much longer. Unless government regulators come up with an "early retirement plan" to buy out and scrap legacy fleets, it will therefore be decades before the maritime sector could be carbon neutral. "If I put perfectly imaginary zero-emission vessels into the fleet now, the fleet would barely see a fifty-percent reduction by 2050, just because the legacy fleet would still be high-carbon," said James Corbett, a transportation expert from the School of Marine Science and Policy at the University of Delaware. (One potentially very effective solution would be to retrofit the existing fleet with zero-emission propulsion, something that, at least conceptually, could be done using clean—aka "green"—ammonia as a fuel or feedstock for hydrogen fuel cell propulsion systems. I'll explain more about ammonia shortly.)

These constraints imply that the problem of greenhouse gas emissions from international shipping is not going to be quickly or cheaply solved; nor can a single nation do it alone. Unfortunately, the International Maritime Organization, a powerful but secretive UN organization that in theory could expedite the process by firmly regulating international shipping, has not done so. Instead of treating the climate crisis as a global emergency, the IMO has allowed the shipping industry to police itself, leaving its GHG emissions exempt from regulation. Sadly, the IMO appears to have been "captured" by the industry it should be regulating and has merely set an "aspirational" target of making half the fleet "carbon neutral" by 2050. According to reporting in the *New York Times*, "the organization has repeatedly delayed and watered down climate regulations, even as emissions from commercial shipping continue to rise."

"Mandatory targets or regulations are definitely needed to encourage the shipping companies to do more, but the problem is that international maritime lobbying is so strong," Professor Zheng Wan, of Shanghai Maritime University, told me. In a gaping regulatory loophole, international shipping, like aviation, was excused from regulations and specific targets under the 2015 Paris Agreement. Signatories to the agreement, including Brazil, India, and China, have opposed emissions caps, and Chile has opposed fuel-saving speed limits.

Early Breakthroughs

Without strong regulation and vigorous introduction of scalable new technologies, the reduction of maritime carbon emissions will continue to be slow. It's much cheaper for the industry to continue burning the heavy fuel oil or bunker fuel it has used for generations than to switch to clean fuels and to a new worldwide fueling infrastructure. Even just burning cleaner fuels and using particulate filters, however, would reduce damaging emissions of black carbon and help reduce Arctic warming.

Absent a sound business case for cleaner oceangoing vessels, some progressive companies have nonetheless stepped up and are voluntarily starting to decarbonize. Maersk, which has voluntarily promised carbon neutrality by 2050, introduced a cargo vessel in 2021 that can be fueled by "carbon-neutral methanol or bio-methanol."

(Those fuels, however, result in only very modest emissions reductions, and the combustion of methanol results in the release of CO_2.)

One industry leader, however, Hurtigruten, a Norwegian cruise ship company, has begun launching a battery-hybrid-powered fleet, and Compagnie Maritime Belge thinks that a hydrogen-powered container ship may be possible within ten years. CMB has already launched the HydroTug 1, a tugboat powered with two engines that burn hydrogen in combination with a small amount of conventional diesel fuel. The company expects the tug to be fully operational in 2023 and capable of replacing the emissions of 350 cars. The company says it's fairly easy to convert a conventional medium-speed engine to dual fuel, and it has its eye on the global tugboat market. Of course, if the world prioritized the rapid development of a hydrogen engine for large vessels and invested heavily in the required R&D, as it has done with aerospace technology, there wouldn't be any fundamental scientific obstacles.

Not surprisingly, eliminating pollution from smaller marine engines is less challenging than eliminating the emissions of ocean-going vessels. Smaller vessels, typically used for shorter journeys, don't require as much onboard energy, and they often ply a fixed route or return to their home ports daily. A few pioneering companies have thus already begun using hydrogen or electricity to power small boats for local or short trips on predictable routes where recharging or alternative fueling is relatively easy. "There's a fair number of hydrogen ferry, electric ferry, or biofuel ferry demonstration projects operating short distances around the world," Professor James Winebrake, of Rochester Institute of Technology, told me. But this is just the beginning of the coming revolution in marine propulsion.

Accessible but Partial Solutions

When we look at oceangoing freight as part of a larger goods-movement *system*, Winebrake explains, opportunities then begin appearing for cost-effective early decarbonization. Ports, for example, can use electric gantry cranes to load and unload ships. Cargo haulers could move goods cleanly inland from ports using electric trucks and trains. Ships could also be required to connect to clean electric power while in port, so their diesel generators aren't firing away just to power shipboard equipment. There's no one solution, no single easy way to eliminate *all* shipping industry emissions, however. "You can't simply get there with a new toy," says Professor Corbett, at the University of Delaware. The problem is far too complex. Just as there is no monolithic trucking industry, "there is no one shipping industry." Different industry segments will require different solutions.

Transoceanic vessels, Corbett believes, will be last to adopt really clean technology, although his research has convinced him that hydrogen technology would work well. His assessments probably count more than most: This specialist in sustainable freight transportation not only has a PhD but is also a professional engineer. And unlike most other academics, he actually spent years at sea, in the engine rooms of oil tankers, right after graduating from the California State University Maritime Academy. "The truth is, I wanted to be a writer and read Joseph Conrad, so I went to sea," he told me.

Some relatively easy ways of mitigating ship pollution exist, but ship owners are reluctant to adopt them. Ships can save fuel and avoid emissions easily just by reducing speed ("slow-steaming"), which can cut emissions 20 to 30 percent, says Corbett. But to motivate shipping companies to do so (and possibly lose business to a competitor), fuel prices have to be high for a sustained period of time, or new regulations have to force ship owners to slow down. Those who believe that a carbon tax can solve all climate ills, take heed: Corbett's research has shown that merely to change industry practice and motivate slow-steaming (let alone to induce the embrace of new propulsion technologies), a carbon tax or fee of at least $125 to $300 a ton would be needed. For comparison, carbon credits in the EU Emissions Trading System were trading at close to $100 a ton in early 2023.

In the past, even when oil cost $150 per barrel, ships did not switch to zero- or low-carbon fuels, and even if a $300-per-ton carbon tax were tacked on, it would raise $20-a-barrel oil only to $147 a barrel. Moreover, the potential economy-wide impacts would likely preclude such a high carbon tax. Alternatively, adding high fuel taxes to the cost of conventional shipping fuels could put many companies out of business. Other short-term options for cutting emissions include problematic measures like fueling engines with natural gas. But studies have shown that so much natural gas leaks during its production and transport that its global-warming impacts are at least as bad as conventional bunker and heavy oil fuels.

Biofuels are no panacea, either. Farming oil seeds and other energy crops on a large scale to produce biofuel would compete for land and other agricultural resources with food production. Moreover, biogas and other biofuels offer only carbon neutrality at best; at a time when a climate crisis is already upon us, no avoidable additional carbon should be added to the atmosphere. Also, as shippers would be quick to point out, biofuels are more expensive than conventional fuel, and sufficient quantities are not widely available.

One tried-and-true formula for reducing emissions from ships is requiring them to reduce drag, adopt better propellers, and observe lower speed limits unless cleanly powered. Shippers have gradually been adopting some of these efficiency improvements in new ships anyway. Wind power assists or solar assists could also be required on certain kinds of vessels, to shave a few percentages off their fuel consumption; collectively, even these marginal reductions in fuel consumption would add up. In addition, financial incentives could also be offered to induce companies to scrap older ships earlier and replace them with newer, more efficient, cleaner ones.

Government incentives could also be used to make it profitable for companies to create hydrogen fueling infrastructure at ports around the world, and the U.S. Department of Transportation or the Department of Energy could invest heavily in demonstrating clean-fuel vessel technology to speed its adoption. Finally—especially if governments coupled incentives with regulation—shippers could be required to increase the percentage of clean-fuel ships in their fleets every year (once they are available) or to buy high-cost carbon credits if they prefer.

Thus, much could be done about maritime emissions. But until we mobilize the political will to motivate the federal government to assertively regulate shipping in cooperation with other nations to create an international schedule for decarbonizing the industry, the cleanup of transoceanic shipping is likely dead in the water. Self-regulation by shippers and efforts by lone nations clearly cannot work.

Green Alchemy: Fuel from Air and Water

Before despairing about the difficulties of cleaning up cross-ocean shipping, let's not forget ammonia, the stuff we all know in solution form as a pungent household cleanser. It turns out that, when concentrated, ammonia has extraordinary and surprising potential as a clean fuel and for storing and distributing clean energy around the world, provided that regulations and technologies are developed to eliminate the production of harmful nitrogen oxide emissions. (These include nitrous oxide, a powerful greenhouse gas.*) With the use of renewable energy as a power source, however, ammonia can be made from air and water without any carbon emissions. And because it is a hydrogen-rich compound, it's suitable as a fuel for hydrogen fuel cells. Thus, ammonia fundamentally has what it takes to be a CO_2-free combustion fuel for an engine *and* a clean input for a highly efficient hydrogen-electric fuel cell. In addition, ammonia has other important assets to recommend it as an industrial fuel.

---

* Emissions can occur during the production, transport, and combustion of green ammonia and must be "almost completely eliminated" for ammonia to be a clean fuel. Most of the emissions occur during combustion; these can potentially be abated by removal with catalysts combined with combustion gas aftertreatment systems. See Martin Cames, Nora Wissner, and Jürgen Sutter, "Ammonia as a Marine Fuel: Risks and Perspectives," Öko-Institute, e.V., Berlin, June, 2021, https://safety4sea.com/wp-content/uploads/2021/06/NABU-Ammonia-as-a-marine-fuel-2021_06.pdf.

Given all its potential, it's surprising that ammonia is not better known or more touted as a promising way of decarbonizing global shipping. As you'll see in a moment, however, the ammonia story is a complicated one.

The ammonia molecule comprises three atoms of hydrogen and one of nitrogen. Currently, this clean-fuel wannabe is generally derived from natural gas in a highly energy-intensive, high-temperature process known as "steam reforming." The reaction separates the carbon from the hydrogen in natural gas, resulting in the production of hydrogen, with CO_2 as a by-product. (The steam itself is generally produced by the burning of fossil fuel, an additional source of CO_2.) Then nitrogen is removed from the atmosphere (which is 78 percent nitrogen) and combined with the hydrogen to create ammonia.

Because of this intimate association with fossil fuel, ammonia made using natural gas, coal, or another fossil fuel is commonly known as "brown" ammonia, even though ammonia itself is a colorless gas. Most of the world's ammonia is made this way; that's why ammonia production is actually the largest industrial source of CO_2. (Globally, nearly 2 percent of all CO_2 is created by ammonia production.) Fortunately, though, now that the cost of solar and wind power has fallen so dramatically, this need not be the case.

Hydrogen for ammonia synthesis can be readily produced by the electrolysis of water. It can then be converted into ammonia by combining it with nitrogen from the air. If the electrolysis is powered by clean, renewable energy, and the nitrogen is removed from the air with power produced from renewable energy, then the resulting hydrogen and ammonia create *no* carbon emissions. This zero-carbon "green" ammonia is the stuff we want for its many beneficial uses: propulsion, heat, power, and energy storage. (Propulsion means that, as noted, ammonia can be used directly to fuel an internal combustion engine or in a fuel cell that produces electricity that, in turn, can drive a vehicle.) Green ammonia could not only help decarbonize the global shipping industry but could also cleanly power trains, heavy trucks, and other large industrial equipment—all without producing any CO_2. The one downside is that ammonia is not that easy to ignite and, therefore, has to be combusted at high temperatures. This produces hazardous nitrogen oxides (NOx) and the dangerously powerful GHG, nitrous oxide (N_2O). Research on how to combust ammonia while keeping production of NOx* and N_2O to a minimum is ongoing and shows promise.

Green ammonia, according to the prestigious Royal Society, could nonetheless be the basis for an entirely new, integrated worldwide renewable-energy storage and distribution solution. That's because ammonia can be easily and safely stored and very inexpensively transported by truck, train, ship, and pipeline. It also has several other advantages: It has about ten times as much energy as lithium-ion batteries and, so, is preferable to batteries for large trucks, trains, and any large vehicle that must operate over long distances and periods, demanding large amounts of energy. In addition, a well-developed shipping and storage system exists for ammonia, as the gas is already in widespread use as the main ingredient in the production of the inorganic fertilizer that helps grow food for half the earth's population and is widely used as a refrigerant and to make many other industrial chemicals.** The United States already has more than ten thousand ammonia storage sites and thousands of miles of ammonia pipeline.

Nations with large solar and wind resources could use their renewable-power resources to cleanly produce green ammonia at major ports around the world. Where renewable power is cheap and abundant, green ammonia is either cost-competitive with conventional brown ammonia or close to it. When the cost of carbon capture and storage is added to the cost of making brown ammonia, or when a carbon fee is imposed, the economic case for green ammonia gets even stronger.

---

* For example, nitrogen dioxide, NO_2, is a precursor to the formation of nitric acid (HNO_3), which leads to acid rain.

** Ammonia is commonly and relatively inexpensively shipped in bulk today in international commerce, so it is currently stored at many ports and is readily available around the world. In addition, it is even cheaper to store and ship in bulk than liquid hydrogen, which has to be pressurized to 300 to 700 atm (atmospheres) and stored in high-pressure tanks or refrigerated to -253°C for storage.

The Challenges Ahead

Despite these advantages, many obstacles stand in the way of ammonia's becoming a common energy source in global shipping. First, to have a real impact on conventional maritime fuel use, a new green ammonia industry would need to be created almost from scratch. To date, virtually no industrial quantities of green ammonia are being made. For large ships to one day operate on ammonia fuel, new ammonia-burning engines and powerful fuel cells would also need to be developed and commercialized. (Because of its low flammability, mentioned earlier, ammonia combustion still presents certain technical challenges, apart from its production of NOx, and is the subject of ongoing research.)

Another formidable challenge to mainstreaming ammonia as a shipping fuel would be having to ultimately replace the three hundred million tons of petroleum-based marine fuels now in use. More than five hundred to six hundred million tons of green ammonia would be needed—which is more than three times the world's entire current production of brown ammonia today. (This large quantity of ammonia would be needed in part because ammonia has only about half the energy density of petroleum-based fuels.)

As ammonia production ramped up, vast fuel storage and delivery systems would also have to be built at ports around the world, along with a great many large, dedicated wind and solar power plants capable of generating the hundreds of thousands of gigawatt-hours needed to power sufficient green ammonia production for the world's shipping industry. If past is prologue, we know that such pervasive changes would take decades to accomplish.

Despite all these challenges, green ammonia still has many important advantages over conventional fuels. It can be used to fuel an internal combustion or gas turbine engine directly, or it can be fed into a fuel cell in which it reacts with oxygen to produce electricity without producing any NOx, as there's no combustion in a fuel cell.*

---

* Hydrogen and oxygen combine electrochemically to produce electricity without combustion.

(Today, a proton-exchange membrane fuel cell would likely be used, but PEM cells aren't powerful enough to propel a large ship.) In the future, ammonia could be used in more advanced, high-temperature solid oxide fuel cells. These could be efficient and powerful enough to fuel a large ship or an aircraft, thereby one day potentially reducing their emissions.* The first order of business, however, is to design and build virtually emission-free ammonia-burning marine engines.

Not only can ammonia be used directly to fuel engines, but it could also be decomposed into hydrogen and nitrogen so that the hydrogen could fuel the engine. When ammonia itself is burned, however, its combustion, as noted earlier, releases both nitrogen dioxide, the precursor to smog and acid rain, and nitrous oxide, the GHG gas. The good news is that these emissions could be captured and that doing so is a relatively mature technology, some of its elements—such as the ubiquitous automotive catalytic converter—having been available since the 1970s.

Instead of concentrating only on building new marine engines, we could also modify existing liquid natural-gas marine engines to burn ammonia instead, thus offering a possible shortcut to decarbonizing large oceangoing vessels. For all the reasons previously mentioned, though, this will not be done quickly, cheaply, or speedily. Decarbonization of the entire world's shipping fleet by conversion to ammonia would cost, by one estimate, $1.9 trillion.

Carbon-Free Shipping: The Wave of the Future

The challenges notwithstanding, bold first steps are currently being taken. Equinor, a Norwegian energy company, hopes to retrofit an oceangoing supply ship with an ammonia fuel cell by 2024, while MAN Energy Solutions, a Volkswagen subsidiary that builds marine

---

* While ammonia holds less energy in a given volume than hydrocarbon fuel, ammonia-powered fuel cells operate at much higher efficiency than internal combustion engines, so ammonia's performance as a fuel is similar to that of liquid propane.

engines, plans to have an ammonia engine commercially available by 2024 and is working with other companies on ammonia engines capable of powering large container ships. Together with Samsung Heavy Industries, MAN is also part of an initiative to launch an ammonia-powered oil tanker by 2024. (The irony of developing green technology and then using it to transport oil is hard to miss.)

In addition to its applications in transportation, ammonia combustion also offers opportunities for the zero-carbon production of industrial process heat. Hydrogen derived from green ammonia, for example, could reduce the steel industry's carbon emissions by substituting for fossil fuel in providing process heat for steelmaking. As we'll see in the next chapter, heavy industries like steel are exploring some truly revolutionary new energy technologies apart from hydrogen to reduce emissions in what has long been thought of as nearly impossible-to-decarbonize industries.

8.

Putting Captured CO2 to Work

American industry today produces more than a fifth of all U.S. energy-related greenhouse gas emissions. So, if the United States is ever going to have a truly clean-energy economy, industry will have to rise to the challenge. Arresting the climate crisis requires transforming every sector of the economy to clean-energy sources for heat and power and to plant-based hydrocarbon feedstocks. Thus, in industry, we need a whole new production strategy for synthesizing the basic chemicals used to manufacture finished goods. An individual company that makes cosmetics, polyester clothing, or dentures, for example, might strive to develop a new production system that uses natural, non-fossil hydrocarbons as a raw material, reducing demand for fossil fuel–derived inputs. Naturally, the less fossil fuel used in any production process, the fewer the possibilities for creating GHG emissions somewhere in the cycle of extracting and processing that fuel.

Another firm that relies on fossil fuel as a chemical input or for fuel might strive to avoid producing waste—this, again, would reduce the throughput of material and would likely reduce emissions. Yet another firm might focus on turning waste from one process into a useful production input or profitable end product; scraps from used tires, for example, can be repurposed to make rubberized flooring for sports arenas, and used plastic bags can be recycled into new carpet.

All these approaches would help reduce fossil fuel dependency and, so, potentially, might reduce emissions. This sort of industrial "alchemy" is now occurring as part of a larger evolutionary effort in industry toward zero-waste production,* efficient resource use, and "biomimicry"—a process of borrowing design ideas from the natural world (where everything is in one way or another reused or recycled)—to thereby create a circular industrial economy, the ultimate form of waste reduction.

How would this change the life of an ordinary American? Unless you work in the factories where these goods are made or live near them and are exposed to their emissions, you probably won't notice any difference. Manufactured products will be indistinguishable physically and chemically. But the foundations of the chemical industry will have been transformed. The fundamental building blocks from which at least one hundred thousand of our industrial chemicals are synthesized will come from plant materials instead of fossil fuels. All the goods then created from these chemicals—paints, dyes, coatings, adhesives, etc.—will be fossil fuel free. The name for this new industrial revolution is "green chemistry." If it is adopted forthwith, our children and grandchildren might have a chance of escaping the worst-case climate catastrophe now brewing.

How Green Chemistry Works

From a green chemistry and sustainability perspective, one doesn't look just at the immediate functions of a synthetic chemical, especially one produced through fossil fuel petrochemistry. As Professor Julie Zimmerman at Yale and her colleagues point out in a brilliant overview of green chemistry, creating green chemistry solutions to a problem means looking more deeply and broadly, from a systems perspective,

---

* Zero waste is the hallmark of the "circular economy." In this case, it also implies closing the carbon cycle, so that when CO_2 is generated, it is treated as a resource rather than as waste and, unless permanently and safely stored, is then fed back into a production process as an ingredient.

at the molecular nature of each compound made. This generates new insights about the compound's good and bad properties and encourages consideration of how the new product will impact human and planetary welfare. This means looking holistically, far beyond how well a product does a narrowly defined job, such as killing an agricultural pest or a bacterium. It means considering how toxic and persistent the product is in the environment and the extent to which it bioaccumulates in our fatty tissue. It means factoring in how much toxic waste is created in producing it, how much greenhouse gas is released, how much energy and water are consumed, and how recyclable the product is. Only then can a comprehensive "balance sheet" be created that accurately reflects the product's toxicity, environmental persistence, climate impact, and the degree to which it depletes finite resources. Yes, consumers want products that will do what they're designed to do, but with minimal hazard and impact. An implication of green chemistry is that if, on balance, a product, however profitable, is harmful, we need to be prepared to forgo producing it, even if powerful chemical companies clamor to sell it.

Green chemistry tends to use more benign reactants than petrochemistry, places fewer demands on the earth's valuable rare metals,* and tamps down the production of toxins, reducing their release into the environment (and, ultimately, our bodies).** When it works as intended, green chemistry also produces fewer nonrecyclable by-products. (To see how well our industrial chemistry is doing on this score nowadays, just check your local beach or roadside!) To be clear: this isn't just about the climate or public health or altruistically saving nature. As you'll see later in this chapter, a green chemistry approach in a single nascent industry alone could, by 2030, become a trillion-dollar global business.

---

* These rare metals are now required as industrial catalysts, but a green-chemical industry would use different, more common materials.
** We all, to varying degrees, carry traces of petrotoxins in our bloodstreams, whose epidemiological effects are not adequately known. It is probably correct to state, however, that, on average, industrial petrochemistry is quietly poisoning us.

Despite its well-recognized essential features, green chemistry today still means different things to different people. To some chemists, it simply means making things less wastefully, with fewer dangerous chemicals and fewer toxic by-products, whether the carbon inputs come from fossil fuels or from a benign natural source. However, I advocate—and this chapter will focus on—a more specialized, ambitious, even revolutionary kind of green chemistry, one that operates without fossil fuels and that can even safely turn legacy fossil fuels into useful products.

An important caveat is in order here. Even this latter kind of green chemistry does not absolve society from carefully scrutinizing each compound produced. Green chemistry *will* enable us to make fossil fuel free some industrial chemicals that are now made from fossil fuels, but the resulting products will still be chemically identical. Hence, if the fossil fuel–based product had toxic properties or was hard to recycle, the "green" substitute will be no better. Polyvinyl chloride is a good example. Millions of tons of PVC are sold each year for more than forty billion dollars. Commonly made from ethylene, a fossil fuel, and chlorine, PVC can be made without fossil fuel. The latter version, however, will still be difficult to recycle (only 5 to 6 percent of plastic ever is recycled anyway), will create poisonous phosgene gas and dioxin when burned,* and will cause numerous other environmental problems. Making something without fossil fuels doesn't necessarily make it safe or desirable.

We're Surrounded

Look around you right now. If you're not out in the wilderness, chances are you're virtually surrounded by manufactured objects made, one way or another, from oil, coal, or natural gas. In your

---

* In World War I, phosgene was used as a chemical warfare agent to kill 85,000 people, and dioxin is a highly toxic carcinogen that can also cause serious diseases other than cancer.

medicine cabinet, the aspirin, toothpaste, shampoos, lipstick, combs, and Vaseline are all petroleum products. Your child's crayons are made from petroleum wax. If you've had a heart valve replacement, the valve was made from petroleum. The sunglasses you wear on the tennis court and the tennis racket you wield are both made with petrochemicals. Ditto for the basketball your kids play with, and your golf balls. Even the polyester slacks and sweaters or nylon leggings you might be wearing contain fossil fuel hydrocarbons.

Indeed, polymers and plastics of every conceivable kind—from films and lenses to calculators and the keyboard on which I'm typing—are made from chemicals extracted from fossil fuels. So, too, are detergents, lubricants, dyes, preservatives, fertilizers, and agricultural chemicals (including pesticides and herbicides). But, someday—I hope not too far off—we'll be looking for "Fossil Fuel Free" labels on the goods we buy, just as some products today are marked "Organic" or "Fair Trade." And at some more distant time, perhaps those labels won't be needed, because *all* products will be made from renewable materials.

How will this transition happen? Twelve, a small start-up "carbon tech" company has set out on a mission to develop and commercialize new technology that uses catalysts to convert waste CO_2, captured from industrial flue gases or other sources, into useful industrial chemicals. It does so by cleverly combining waste CO_2 with hydrogen, electricity, and proprietary nanocatalysts, thereby creating next-gen chemicals entirely without fresh fossil fuels. Twelve's future products could include large volumes of industrial and transportation fuels, along with commodity chemicals.

Twelve is far from alone in its potentially lucrative quest to produce useful products from waste CO_2 while protecting the climate. For example, Carbon Corp. of Calgary, Canada, turns waste CO_2 into valuable, lightweight, super-strong structural materials known as carbon nanotubes, which can serve as replacements for steel, aluminum, and other metals. Using high-yield electrolysis, Carbon Corp. makes these "diamonds from the sky" along with other carbon nanoproducts, such as graphene and carbon fibers, that can be woven into textiles. The CO_2 raw material for Carbon Corp.'s products is free, as it comes either from the air or from flue gases. But the carbon nanotubes the company produces are currently worth one hundred thousand dollars a ton, thanks to their remarkable properties. How's that for turning trash into treasure? The tubes are harder than diamonds, a hundred times stronger than steel but eight times lighter, and thinner than a human hair. Theoretically, they can replace almost any structural material. In addition, Carbon Corp. is also on the side of the angels in powering its production with solar instead of fossil energy.

Newlight Technologies, of Huntington Beach, California, is also among the companies turning CO_2 and methane into more valuable commodities. Newlight feeds bacteria with these gases to create polyhydroxyalkanoates, or PHAs, naturally occurring biopolymers produced by microorganisms and stored in their cytoplasm. The PHAs are then cast into pellets that can be melted and reformed into many products usually made from petroleum. PHA biopolymers, however, are biodegradable, sustainable, and "climate-friendly," as their synthesis removes GHGs from circulation. Not a bad example of green chemistry in action.

In addition to the firms just mentioned, many others are doing R&D or are in the early commercial stages of turning waste CO_2 into algae for fuels and fertilizer, soil amendments, construction materials, polycarbonates (for plastic packaging), plastic bottles, films, laminates, and coatings for food and beverage containers. Alchemists of old would be awestruck.

Reincarnating Carbon

As the example just given illustrates, waste carbon can be reformulated and reincarnated as soap, vodka, and jewelry—or as gasoline, diesel, and other fuels. One company has even flown a Boeing 747 on a transatlantic flight on a fuel blended with a next-gen aviation fuel made of biokerosene. One day, next-gen fuels will necessarily disrupt the petroleum industry's business model. Today, however, fossil fuels are still relatively cheap, and in the United States, no carbon taxes are assessed for the pollution and climate havoc they wreak. So, fossil fuel producers will likely continue to sell them until either steep carbon taxes are imposed or regulatory policies force these companies to forswear fossil fuel raw materials.*

In theory, the carbon compounds in reformulated products made by companies like Carbon Corp. and Twelve could be recycled indefinitely. If the production of these new products is powered using renewable energy, essentially no new carbon from fossil fuels would be needed to make new products. Chemical engineers could just extract and then recycle hydrocarbons from scrap manufactured goods as raw materials to turn into new products—a pretty heady prospect.

Whereas Twelve is still relatively small, the company nonetheless has about a hundred employees and is rapidly doubling its workforce, thanks to $157 million in capital raised in two recent funding rounds. And despite its size, it has big plans: it aspires "at a minimum" to reduce global CO_2 emissions by half a billion metric tons per year, which would be like reversing the total global emissions of all of Shell Oil's refineries. The firm's name, said CEO and cofounder Nicholas Flanders, signifies that Twelve is "creating a new story for carbon," whose most common isotope, not coincidentally, has an atomic mass of 12.

The company's vision, Flanders said, "is to be a profitable solution to carbon emissions." It plans to produce ethylene, which is used in industry to make plastics, like PVC piping; and also methane, which is both a fuel and an input that can be used to make ammonia fertilizers, alcohol, and other chemicals. By transforming captured waste CO_2 into the fundamental chemical building blocks for everyday goods, Twelve avoids burning any additional fossil fuel and hopes to

---

* Or until a global supply shock or global shortfall in production causes fossil fuel prices to spike.

keep CO_2 out of the atmosphere for each new product's lifetime. In climate policy jargon, these products are "carbon neutral," adding no net new carbon to the atmosphere.*

The holy grail of climate-protection technologies, however, is the development of carbon-*negative* chemistry, as carbon-neutral technologies have an intrinsically limited role to play in decarbonizing the atmosphere. That's because the multibillion-ton scale of our global carbon emissions far exceeds the mass of carbon that could be used in manufactured products. Carbon capture and reuse, therefore, can never be a comprehensive climate solution. (It will always remain an important way to reduce carbon emissions, however.)

Twelve has already delivered pilot-scale CO_2-conversion systems to flagship customers that include Mercedes-Benz (which will use the newly synthesized compounds in making car parts), Tide (laundry detergent), and Pangaia (polycarbonate-lens designer sunglasses). Twelve intends to start to fill customer orders for its shipping container–size, industrial-scale systems in 2023. The units are designed to integrate seamlessly with a customer's production line, providing chemicals that customers need for their products. The company thus enables the customer to use recycled CO_2 instead of petrochemicals and to reuse carbon emissions instead of simply offsetting them.

Very Like a Plant

Twelve is the brainchild of three Stanford graduate students. Chemical and mechanical engineering PhD students Kendra Kuhl and Etosha Cave were working side by side under Stanford professor Tom Jaramillo when they met Nicholas Flanders, a joint MBA and chemical and mechanical engineering student at Stanford.

---

* If the product produced is a synthetic fuel, its carbon will be released back into the atmosphere quickly as CO_2 once the fuel is burned, so that's just really recycling CO_2 before its release. But it's still better than releasing it as exhaust after only one use.

Much of the research in Dr. Jaramillo's lab concerns chemical reactions (powered by clean, renewable energy) that convert water and CO_2 into fuels and chemicals. (Petroleum products are hydrocarbons, chemicals made entirely of carbon and hydrogen.) The young scientists were looking for ways to take the hydrogen from water and combine it with water and CO_2 from power plant or industrial exhausts to create synthetic hydrocarbon compounds.

To accomplish this hat trick, they had to create new nanomaterials to act as catalysts for the reactions that disrupt the chemical bonds in water and CO_2.* To the extent that this could be done very cheaply, it would create clean products that could compete with fossil fuels and their industrial chemical offspring, thereby making it easier and less expensive for ordinary citizens to reduce their dependence on fossil fuels and petroleum products.

Twelve's three cofounders took some of the beaker-scale research in Professor Jaramillo's lab to see if it could be scaled up for commercial use. No one really knew if it could be done; many people advised them to do something else. "They had experience in developing new clean technologies, and they were saying, 'This seems too hard. Don't waste your time,'" Kuhl recalls. "It was hard to push back on that. You think, 'This person has more experience than me, so I should do what they say.'" But there was a breakthrough moment: "I know this technology better than they do," she realized, "so we're going to push forward, because we think this will work."

It was a bold move, because Twelve's young founders knew it would mean taking financial and professional risks. Failure was not only an option; it was a likelihood. The failure rate for start-up businesses of all kinds is 90 percent; the failure rate for people trying to commercialize cutting-edge science and engineering is likely to be even higher.

Flanders and Kuhl describe what they're doing as "industrial photosynthesis"—that is, using CO_2, water, and energy as a plant does.

---

* Catalysts increase the rate of a chemical reaction without being consumed in it.

But whereas a plant combines energy, water, and CO_2 from the air to make sugars and cellulose, Twelve uses a rectangular, washing machine–size electrolyzer in which waste CO_2 gas makes useful carbon-based industrial products, with oxygen as a by-product.

An ordinary water electrolyzer is used to separate water into hydrogen and oxygen. But Twelve redesigned a typical electrolyzer into a device that deconstructs CO_2. It's basically a chemical reactor outfitted with a stack of square flexible-plastic electrolyte membranes bolted together into a "membrane electrode assembly." Each plastic sheet is coated with a proprietary catalyst that separates incoming gas. Water and CO_2 enter separately on one side of the device; oxygen and carbon monoxide (CO) bubble out the other. Both gaseous products have industrial value and will be captured and used once the process is commercialized.*

The magic new catalyst that makes all this happen is a dark, ink-colored fluid that robots spray onto the proprietary polymer membranes. Twelve is relatively tight-lipped about this proprietary "secret sauce," and for good reason: the novel nanotech material is key to the entire enterprise. If brewing the special sauce were analogous to cooking, then Twelve's scientists would be some of the world's most highly qualified chefs, manipulating nanomaterials so tiny that several thousand could fit on a red blood cell.

The company's newest test reactor initially won't produce vast amounts of CO, but it will be enough for the commercial needs of certain smaller customers. After testing it thoroughly, Twelve will set up on-site pilot operations at prospective customers' commercial facilities—most likely food processors, electronics manufacturers, and pharmaceutical firms—where Twelve's oxygen and carbon monoxide will first be used.

---

* Carbon monoxide has many industrial uses, including in metal purification and production of chemicals such as acrylic acid, used in the production of diapers, textiles, and other products. With the addition of hydrogen, it can be converted to syngas, a useful fuel. Oxygen has many industrial uses, including in welding, steelmaking, and the production of common chemicals, such as methanol and ammonia.

Early Days

Twelve is situated in an industrial section of Berkeley, California. Whereas its one-story building might be prime real estate elsewhere in town, at this location, it's so close to the Southern Pacific railroad tracks that when a train passed, it sounded as if the locomotive were coming through the front door, its air horn obliterating my conversation with CEO Flanders.

He and Twelve's two other cofounders are idealistic thirtysomethings. Kendra Kuhl grew up in Montana, not far from Glacier National Park, where each year she saw the glaciers shrinking and the winters getting warmer. Because her father was a wilderness ranger, she spent her early years at remote wilderness outposts and developed a deep love of nature. Later, when her father, as president of the Montana Wilderness Association (now called Wild Montana), lobbied and testified before Congress, Kuhl saw nature pitted in Congress against mining and timber interests.

"I really wanted to do something for the environment," she told me—and because she had both a passion and aptitude for math and science, she decided to focus on developing new technologies that would support the transition to a more sustainable economy. This led to a chemistry degree from the University of Montana and a PhD in chemistry at Stanford.

Twelve's cofounder, Etosha Cave, grew up in Houston, near a neighborhood built on the site of three unlined and abandoned Gulf Oil storage tanks or pits (later acquired by Chevron when it merged with Gulf). In the 1960s, a developer bought the land, filled in the pits, and built homes on top of them. These were bought, as anticipated, mainly by low-income Black residents. Their neighborhood water lines broke frequently, allowing polyaromatic hydrocarbons to contaminate their drinking water.

Subsequently, these residents suffered an epidemic of cancers, lupus, pregnancy disorders, and headaches. Chevron, which now owned the oil pits, denied that the pits were responsible. The EPA found toxic chemicals in the soil but ruled that the hazards were low and declined to establish a Superfund site. A lawsuit on behalf of residents was unsuccessful. Ultimately, the claims of some 1,700 plaintiffs were resolved through mediation for a total of only $12 million. The size of the settlement prompted charges of racism and environmental injustice.

Cave heard this dramatic case discussed extensively by family and friends while she was in seventh grade, and it both angered and motivated her. Environmental laws and policies had not protected residents. But soil tests by scientists and engineers had provided crucial proof of the contamination, which had made some restitution possible. Already gravitating toward math and science and now wary of the oil industry, Cave attended an engineering magnet school and became keenly interested in renewable energy and clean technology.

Twelve's CEO, Nicholas Flanders, was the son of two research professors and grew up doing his homework sitting on the laboratory floor while his dad conducted experiments. His early fascination with engineering and large industrial machines first emerged when, as a small child, he built Lego rocket ships and elaborate space stations; later, he feasted on Tom Swift novels in which the hero invented new technologies to solve one crisis after another.

Today, Twelve is looking to the day not far off when it gets to build its first industrial-scale facility. With an operating budget of over ten million dollars and a recently completed first commercial-scale reactor, it now seems just a matter of time. Flanders, Cave, and Kuhl are no longer the trio of starry-eyed grad students they were only a few years ago.

Beating the Odds—With Uncle Sam's Help

The path of a start-up is never easy, especially when you're trying to do something very hard very quickly. At first, nothing came easily to Twelve: Initial grant proposals were rejected. Its first experiment didn't work. And for six months, it failed to produce even 1 percent of its desired end product: carbon monoxide.

"Nicholas and Etosha and I were talking all the time about whether this is going to work," Kuhl said. Then, one day, their yield suddenly jumped to a breakthrough 23 percent. Instead of celebrating, however, Flanders said, "We need to repeat this experiment, just in case it's an error." After the result was verified, the new, higher yield represented an important proof of concept and enabled the trio to secure a $125,000 NASA grant. At that point, although the group rarely allowed themselves a celebration, they did go out for French fries.

Today, their yield is 99 percent of useful product.*

The company got its real start in 2016, when it won a competitive federal fellowship program called Cyclotron Road. The prize was two years of free lab space at Lawrence Berkeley National Laboratory, to build a prototype reactor, and some initial funding for it. With experimental data generated there, the team then got a Small Business Innovation Research grant, which made it possible for them to make new discoveries and scale up their technology. "There is government involvement at each stage of the development of a technology like this," Flanders said. These investments pay off royally for taxpayers: "I've seen one report that you get a five-to-one return on every tax dollar spent on science and R&D."

Government research funding—not dollars from venture capitalists—generally is the first money in for "deep tech" companies like Twelve; private investors usually require either commercial revenue or a thumbs-up from a big company to join the party. So, when Twelve got funding for some feasibility work from Shell Oil's GameChanger program, it "really changed the investor conversa-

---

* This means that 99 percent of the electrolyzed CO_2 was converted to CO instead of to other chemical species.

tion," Flanders said. Backers today include famous investors like the Dolby family, which funds science-driven technology, and libertarian PayPal billionaire Peter Thiel. Not bad for a team that, seven years ago, had nothing more to show than a concept, some bench-scale lab results, and a PowerPoint presentation.

Twelve now plans to partner with large industrial companies and to convert a lot of CO_2 emissions into new products as it further refines its technology. It has already found some heavyweight partners, such as SoCalGas and a major electrolyzer company. With SoCalGas, Twelve hopes to create renewable natural gas from waste CO_2 produced along with methane by dairy digesters.

With another collaborator, LanzaTech, the simple, energy-rich molecules Twelve currently makes are fed to specially engineered bacteria that turn those molecules into complex chemicals used for jet fuel. Nowadays, in industry, most of the bacteria-making alcohols and other useful chemicals are metabolizing sugar from crops like grain or sugarcane. But growing the crops takes up land and other resources and creates environmental impacts. "Second-generation" feedstocks like those LanzaTech uses are instead comprised of cellulose, often from waste,* and waste industrial gases, which certain bacteria can ferment into desired compounds. In this "direct gas fermentation," the bacteria consume a mixture of CO, or hydrogen and CO_2, bubbled into an aqueous solution.

LanzaTech's operation is a bit like retrofitting a brewery onto a steel mill or a landfill site to create fuel with which to fly planes. Steel mill exhaust, high in CO, is piped directly into vats of specially bred microbes that happily metabolize it into sugars, which are in turn fermented into alcohol to make a range of basic industrial chemicals: LanzaTech takes the ethanol, for example, and performs a several-step chemical process that turns it first into ethylene and then into jet fuel. The company's demonstration refinery

---

* Sources of cellulose include agricultural residues, forestry waste, wood chips, municipal compost, and organic municipal waste.

at a Chinese steel mill has already produced over ten million gallons of jet fuel. Gallon for gallon, this alternative jet fuel, according to a Boeing executive, could reduce an airplane's net carbon emissions by 80 percent.

"Steel mill emissions today are tomorrow's running shoes, yoga pants, or containers!" LanzaTech proclaims. "This is the circular economy in action." The company makes another exciting claim: captured and recycled carbon from organic wastes or industrial off-gases could eliminate the need for 30 percent of the oil used today, while cutting global carbon emissions by 10 percent.

What's Next for Twelve?

The syngas (a mixture of CO and hydrogen) that Twelve can now produce is just one of its intended commercial products.* Kuhl and Cave's Stanford research showed that not just ethylene and eventually methane, but also a dozen different useful chemical building blocks for many everyday products, could be produced from CO_2 and water.** One day, Twelve plans to make as many as ten of them, including alcohols, formaldehyde, and even more complex molecules. In my first interview with Nicholas Flanders, he confidently declared, "We can make the building blocks for a jet fuel, for the foam in your running shoe, for the headlights of your car, and for the jug of milk that you buy at the store." These claims are increasingly credible.

To succeed long term, Twelve will have to make a powerful business case for the products it plans to create—which will have to compete with those of the entrenched petrochemical industry, which uses, among other fossil resources, cheap, plentiful shale gas

---

* Syngas can be fed into today's natural gas pipelines.
** Ethylene, for example, is used to make some laundry detergents and various plastics, including the PVC pipes commonly used in household plumbing.

from the Permian Basin, in the U.S. Southwest.* In the company's favor, though, the cost of renewable electricity, the cost of carbon capture, and the cost of electrolyzers have all gone down by 80 or 90 percent over the past ten to fifteen years. Inasmuch as electricity is by far the main expense in Twelve's chemical process, the fall in renewable-power costs means that the company's production costs have also plummeted. "Ten years ago," said Flanders, "there was no way to compete, because renewables were still too expensive."

Commercial products using bulk CO and ethylene from carbon tech companies like Twelve can be competitive with products made from fossil fuels at electricity prices of 3.5 to 6 cents a kilowatt-hour (in 2020 dollars), says Flanders. The cost of new renewable power from wind and solar is already within or below that range. If a new product is cost-competitive with a fossil fuel product and has the exact same properties, plus all the environmental benefits of a lower carbon footprint, that makes a great business case, he said. "You get instant market share." This bodes very well for Twelve and other carbon converters.

An Ambitious Time Line

Flanders expects the company to have small industrial plants for CO at customer sites by late 2023 and to be building a large, industrial-scale chemical plant by the mid-2020s. Whereas Twelve faces some modest competition from rival carbon tech companies, it appears to have a competitive edge in turning CO_2 into chemicals or fuels. Other firms use a two-step process. "Our edge," Flanders says, "is that our device *directly* converts the CO_2 into the [needed chemical] building blocks." Twelve's unique, proprietary catalysts also transform CO_2 into new products at low temperature, avoiding

---

* Shale gas and oil are so plentiful, in fact, that more oil and gas are now exported from the United States than from Saudi Arabia.

the need other companies have for costly high temperatures.

Because the oil industry today makes CO out of CO_2, it already has billions of dollars of infrastructure in place for turning CO obtained from fossil fuels into finished products. Twelve's product could be fed into that existing infrastructure to make all the legacy products now being made in petrochemical refineries. "Shell has been sponsoring some of our work around [converting] CO_2 to CO because they could use that as a way to make a carbon-neutral jet fuel," Flanders said.

Twelve's vision extends far beyond its current efforts to convert waste CO_2 into CO or other chemicals. The founders see themselves as being in the vanguard of a transition to fully closing the carbon cycle, creating a system in which every molecule of carbon used is either stored or reused. "We kind of give carbon a second life now," Flanders said.

In that future, "biogenic carbon"—carbon made from organic materials, such as food waste or forest residue—will also be available to companies like Twelve. "I think to start the transition," Flanders said, "we may take the CO_2 coming from a chemical process, for example, and give that a second or a third life—you have to start somewhere." He added: "Over time, if you imagine a far distant future ... if the cost of direct air capture goes down enough, we will be the first customers [for that CO_2]."

Ramping Up

As explained, Twelve and other companies in the field have gotten where they are today much more quickly and efficiently thanks to federal research and demonstration funding. Continuing ample federal support of these companies so that they can ramp up quickly to industrial-scale operation would further reduce the costs of carbon-conversion technologies like Twelve's and speed up their adoption. Because the carbon tech industry is new and small, costs today are higher than they would be in a fully mature industry. If government wanted to supercharge the industry and accelerate its growth, it could levy carbon taxes to make fossil fuel–based chemicals less competitive. Government could also set low-carbon procurement standards for the manufactured products it acquires or regulates, so as to favor products that incorporate waste CO_2 instead of fossil fuels. In addition, governments could provide carbon credits for companies that invest to avoid carbon emissions. Some already do.

Government could also greatly increase its investments in energy storage and clean-energy R&D and procurement for all routine operations. This would continue to drive down the costs of renewable energy and would increase the value of the energy generated, as it would be more easily available when needed, and clean-power surpluses could be stored more economically. Keeping the price of renewables down and improving clean-energy infrastructure is likely to benefit Twelve and other carbon-conversion companies.

A lesson to be had from the groundbreaking early commercial work being done in the carbon-conversion industry is that a great deal of the technology needed to replace refined products from coal, oil, and natural gas with more benign chemicals already exists. If government and private industry now resolve to act decisively, we can accelerate the move to a cleaner world in which the chemical building blocks used in industry are clean. This would make it possible for the many products we use in daily life to have a sparkling-clean environmental record.

✝ ✝ ✝

To continue our survey of green chemistry's potential to help protect the climate, we'll now hop across the Atlantic to Belgium, to see what a nascent "biobased" economy looks like. That's one in which plant-based molecules, instead of fossil fuel hydrocarbons, are the foundation of a new kind of economy—and a cleaner environment.

9.

A Plant-Based Economy

Having now discussed how to repurpose waste carbon, let's check out how the chemical industry could be revolutionized.

The key is the creation of new chemical processes that can produce what we need from plant material instead of petroleum. The technologies that will enable us to do this are broadly known as "biobased," meaning they use raw materials from plants or animals. (If you want to be a stickler, coal, oil, and natural gas also are derived from ancient plant material, but that's not what is meant by "biobased." Biobased stuff was not only alive, but alive fairly recently.) Biobased green chemistry offers society the opportunity of making the everyday goods and fuels we need out of things that grow (and that grow back), rather than from minerals that are mined or from chemicals made from petroleum or natural gas.

Biobased technologies hold the potential to slow the rate of climate change while enabling us to convert plants (or materials derived from them) into food, feed, and other products. They also can simultaneously protect the environment by recycling resources that otherwise might be wasted and by enhancing resource efficiency—by taking what may have been organic waste material or very-low-value commodities and transforming them into useful commodities. This not only reduces greenhouse gas emissions, but also often produces new jobs and revenue. In addition, it may enhance food security by enabling inedible material to be converted into feed or fertilizer. As an extra bonus, sometimes the new products even have superior properties to those made from petroleum: they may be nontoxic and recyclable, or better lubricants.

For the past decade and a half now, an innovative industrial research center in the Flanders region of Belgium has been pointing the way to this greener future. The center's founder-director, Professor Wim Soetaert, explained to me how the "biobased" companies he fostered in the nearby Port of Ghent are now profitably producing hundreds of thousands of tons of biofuels and hundreds of megawatts of green power every year while also developing new industrial technologies to make green consumer products that are creating new jobs and bringing new business to the port. But, for reasons we'll soon see, biobased technology is still a work in progress.

What Good Are Biomaterials?

Before we plunge in to explore the new bioeconomy, it would be useful to define a few basic terms. For example, "bioresources" can include forestry, agricultural, and livestock residues as well as algae, sewage sludge, and organic municipal solid waste. These resources—often referred to as "feedstocks" because they are "fed" into other processes as raw materials—are classified into first-, second-, third-, and fourth-generation raw materials.

First-generation biomass comes from edible crops or from the starches and oils derived directly from them. Second-generation materials are inedible, including wastes and so-called lignocellu-losic materials (such as woody material), which contain cellulose and lignin, two tough carbohydrates that give wood its stiffness and structure. First-generation biomass feedstocks—including crops like corn, sugarcane, soybeans, and palm oil—have serious liabilities, primarily because they are foodstuffs. By contrast, agricultural residues, tree prunings, and crop stubble, for example, are second-generation biomass raw materials. Agricultural residues do have valuable roles in nature, recycling nutrients back into the soil. Thus, their removal can, in the long run, deplete the soil. But while this isn't optimal, at least it doesn't divert food crops to industrial use and, so, doesn't encourage clearing of new land for crop plantings or pose a threat to global food security.

More advanced biotechnologies can avoid the soil depletion problem, however. Third-generation biomass materials, for example, are those that have been produced by bacteria from algae, sewage sludge, or municipal solid waste. Food crops are not involved, nor are soils depleted. Finally, fourth-generation biomaterials are produced by genetically engineered microorganisms acting on either genetically engineered algae or other plant material or directly on solutions of waste CO or CO_2.

Many common consumer products can be readily manufactured by the chemical, pharmaceutical, and plastics industries from biomaterials, including food, paper, fuels, textiles, various plastics, and other manufactured goods. Biobased products even include certain composites of natural resins reinforced with wood fibers or other natural fibers. Many people might be surprised to learn that there is a long list of 100 percent–biobased products already; the list includes natural dyes, enzymes, fatty acids, detergents, elastomers, surfactants, solvents, and cosmetics. You can buy laundry liquid and hand soap, shower gel, skin oil, and even a window de-icer that are 100 percent biobased. Other products mix biological inputs with fossil fuel, reducing the amount of fossil fuel needed. Some European companies do even more, producing sneakers from apple leather, cosmetics from sunflower oils, insulation from hemp (instead of fiberglass), lampshades from fungal mycelia (instead of plastic), and even synthetic breast milk sugars for baby formula.

How far can we push the biobased production model? Pretty far, it turns out. Biobased technologies can, in theory, provide us with all these goods, but without worsening climate change, just by directly converting plant materials into all that useful stuff without fossil fuels. I say "in theory," however, because there's a hitch here, and it's a big one: the energy that's used to make all this happen today still comes mostly from fossil fuels. Gas and diesel, for example, are used for fueling the machinery for growing and gathering plants in fields and forests and then transporting them to the factory. Coal or natural gas is often used to generate the power that operates the factory equipment needed for producing final goods, biobased or not. But all this need not be so. In the foreseeable future, those field and factory processes will be powered by renewable energy.

The expenditures of fossil fuel energy mentioned are considered as indirect production inputs, not inputs directly used in synthesizing the product itself. Biobased technologies, by definition, don't use fossil fuels as direct production inputs in the way that, for example, natural gas is directly used to make fertilizer today. Thus, despite their dependence on indirect fossil fuel inputs today, biobased goods can already offer major environmental and economic benefits now, with many more likely in the future.

Growing a Bioeconomy—Preferably Without Biofuels

In Ghent, Belgium, a nonprofit public-private partnership, the Flanders Biobased Valley (FBBV), and its key allies have spent the past sixteen years working to transform the Flanders region into the hub of a new European center for biobased economic activity. As of 2020, more than €750 million had already been invested in biobased development in the region. Traditional biobased companies, like producers of oils, paper, yeast, and gelatin, have long operated in the Ghent area, but FBBV has sought to add value to the regional economy by aggressively developing a suite of newer, more advanced biobased industries.

FBBV's initial goal was to set up a cluster of biorefineries. This has now been accomplished, and its capacity continues expanding. Yet, from an environmental standpoint, the biofuels industry may be the most problematic of all the biobased industries. Whereas bio-feedstocks create useful chemicals and products without petroleum, biofuels (even the advanced ones) are ultimately not an ideal solution, as almost all, when burned, produce air pollution similar to that from fossil fuels. So, whereas the use of advanced biofeedstocks to make industrial chemicals and products is an environmentally sound way of mitigating climate change, using renewably generated electricity or "green" hydrogen is a far more effective way to meet fuel needs.

FBBV's Soetaert is a university teacher of industrial biotechnology. He's both a professor and an investor, with ten years of experience in the sugar and starch industries. Soetaert is an enthusiast for bio-products as well as biofuels, despite their environmental impacts. He heads the Centre for Industrial Biotechnology and Biocatalysis,* a research group that strives to transform certain microbes "into cell factories" to make useful chemicals and hasten the transition to a biobased economy. "Why make it yourself," they ask, "if a bug can make it for you?"

The industrial transition they're working on is already clearly under way on an area of 1.5 square miles within the North Sea Port, where bioethanol, bio-methanol, and biodiesel plants are all sited close together. It's the largest integrated production site for bio-energy in Europe, according to Sofie Dobbelaere, FBBV's former managing director. Today, 200,000 tons of bioethanol, 400,000 tons of biodiesel, and 205 megawatts of bioelectricity are produced here every year. Dr. Dobbelaere estimates that this has led to the creation of about five hundred new jobs and millions of tons of new transport business for the port. All this sustainable new economic activity in the region has also avoided the production of millions of tons of CO_2, Soetaert says.

Soetaert sees biofuels as essential from a climate perspective. The situation, however, is not so clear-cut. The CO_2 contained in any bio-

---

* In biocatalysis, natural substances such as enzymes are used to speed up chemical reactions.

fuel has incontestably been removed from the atmosphere by plants. When burned, it returns an equal amount of CO_2 to the atmosphere. So, viewed narrowly, these two processes offset each other—and the CO_2 released when a biofuel is burned is therefore not a fossil fuel CO_2. "We want to get rid of fossil fuel," Soetaert notes.

But, as described earlier, fossil fuel energy today is generally used in growing, gathering, processing, and transporting energy crops for bio-fuel—for example, in tractors that till the fields or trucks that deliver the biofuel to service stations. (Fossil fuels are also used in the produc-tion and distribution of second-gen biomass today, but less so.) Thus, the fossil fuel consumed in getting a biofuel to its point of use must be included in the energy balance ledger in which its net greenhouse gas benefit or cost is tallied. In addition, common nitrogen fertilizers used on biofuel crops result in the release of nitrous oxide, a powerful GHG hundreds of times more potent than CO_2.

Furthermore, when edible crops are diverted from the dinner table to the fuel tank, it tends to drive up food prices, which most burdens those who can least afford it. Simultaneously, these higher prices also provide an incentive for clearing new land, which destroys existing vegetation, releases CO_2 stored in the soil, and eliminates the CO_2 that the cleared forest or other ecosystem would have stored over time. These problems, some specific just to "first-generation biofuels" (including trees from tree plantations), can all be major drawbacks to heavy reliance on biofuels, even without considering their resultant air pollution. To assess the carbon footprint of any given type of bio-product—namely, the CO_2 expended in its production and the fossil fuel CO_2 that its use avoids—a careful life cycle analysis thus has to be performed.

It is true that biofuels do tend to emit fewer GHGs when burned than most conventional fuels. Burning a gallon of biodiesel, for example, produces about six pounds of CO_2, while conventional diesel releases twenty-seven pounds. But, ultimately, the alternative to which biofuel needs to be compared for effectiveness in the long run is clean electricity or clean hydrogen.

Thinking Beyond Biofuels

Wim Soetaert is well aware of the environmental issues in first- and second-generation biofuels. "We have to go from first-generation to second-generation biofuels," he says unequivocally. "We're working on that. Currently, we're still using first-generation biofuels here. Moving to second-generation biofuels based on liquid cellulosic waste is going to take a significant amount of time, but it's happening." But going straight to second-generation biofuels "is not going to happen," he cautions. Second-generation biofuels are more expensive than first-generation ones, and the processing of waste materials into chemicals presents technological challenges. "We still need to develop these technologies," he notes, although pilot plants are operating, and demonstration plants are being built.

Ultimately, 100 percent of aviation fuel could be biofuel, Soetaert says. Being able to substitute truly carbon-neutral biofuel for commercial aviation fuel derived from petroleum would go a long way toward constraining the rapidly growing tonnage of GHGs blasted into the atmosphere every day by thousands of jet aircraft. No technology exists today capable of powering a large airliner a long distance on electricity or hydrogen, so, for the time being, biofuels look like the best technology we have for this application.

In addition to its biofuels work, FBBV has been active in promoting Flanders as an ideal site for new factories producing bioplastics and biodetergents. Although the money for all this has come from corporations and investors, FBBV's critical role was as a networking organization that generated ideas for new biobased products and supply chains, calling attention to new commercial opportunities. Networking was initially difficult to orchestrate, Soetaert said, because the companies were very competitive and leery of sharing trade secrets. Eventually, working as an unpaid volunteer—or, in his words, "a missionary"—he wrestled them to the table and demonstrated the benefits of collaboration.

A national production quota system for biofuel production in Belgium helped to create financial incentives for the companies. The quota system gave selected companies the right to fill local biofuel demand. This made it economically attractive for them to go into the biofuel business. Still, "a lot of people had to be convinced—investors, politicians, you name it," for a new factory complex to arise. "I know how to speak the language of investors and politicians," Soetaert said. As a university professor, he also had the advantage of appearing independent and impartial. "If I worked for one of those companies, it wouldn't have worked."

Nurturing New Industries

FBBV not only focused on obtaining financing for new factories and promoting the creation of biobased industry clusters, but also helped companies in the young industry solve common regulatory and other problems, including access to raw materials, technology, expertise, and publicity. In addition, FBBV worked to improve public awareness of biobased products' value, and it assisted its industrial partners in developing technological expertise and forming new partnerships. Finally, FBBV fostered industrial partnerships that, for example, allow one company to use another's intermediate or waste products as a raw material for its value chain, integrating their production cycles.

In addition to heading the biotechnology and biocatalysis center, Dr. Soetaert is a professor at Ghent University and director of the Bio Base Europe Pilot Plant, which for-profit companies can use to accelerate the development of industrial processes to produce biobased products. The pilot plant can take a customer's biobased laboratory-scale protocol and bring it to industrial scale or provide custom manufacturing services; it also provides training and other educational services. Since Soetaert founded it in 2012, the plant's workforce has expanded to more than 150 employees and is still growing rapidly, with €27 million in new investments expected in

2022 and 2023. It is the world's largest pilot plant for biobased products and processes.

In his scientific research, Soetaert is tackling important challenges that inhibit the world's transition to a biobased economy, like turning microbial strains into cell factories that manufacture new chemicals—for example, biosurfactants—that were traditionally made from fossil fuels. (Surfactants are the active ingredients in soaps and detergents.) He is also working on producing milk sugars similar to those in human breast milk (important for making infant formula) and aminoglycosides, which are highly potent broad-spectrum antibiotics. By 2050, he thinks, "around 50 percent of all chemicals produced will be biobased."

Here are some fundamental reasons that this astonishing forecast just might be correct. First, biofeedstocks have intrinsic efficiency, not to mention technical advantages over petroleum in producing industrial chemicals. These will eventually give biofeedstocks an economic edge, too. "Petroleum contains no oxygen," Soetaert said, "so you have to get the oxygen in, and I can tell you it's technically very challenging." By contrast, "if you start with sugars that already by nature contain a lot of oxygen, it simply makes sense. . . . So, from a cost perspective, this is the winner." Polylactic acid, from which bioplastics are produced, is already in commercial production. "It makes perfect sense," he said, "because if you produce polylactic acid, you have a very good yield starting from sugars. The efficiency is really very good." That means that the process is more likely to be cost-competitive and commercially competitive.

To continue the work of FBBV, which recently lost its corporate funding, Soetaert managed to reorganize it within the Bio Base Europe Pilot Plant, now known as the Bio Base Connect network. The network will perform services similar to those performed by FBBV—especially its networking function—but with a more secure funding source.

A Young Industry's Challenges

In a 2022 ceremony attended by the prime minister of Belgium, Alexander De Croo, and many other dignitaries, ArcelorMittal, the world's largest integrated steel and mining company, opened Europe's first commercial-scale, carbon capture and utilization bioethanol plant in the Flanders bioindustry cluster. By capturing and recycling as much as possible of its waste carbon and other wastes, "the plant is preparing for a future when green hydrogen will remove the need to use any fossil carbon," said the company's executive chairman, Lakshmi Mittal. ArcelorMittal plans to reduce the carbon intensity of its steel by 25 percent globally and by 35 percent across its European operations. It has already invested over $200 million in its XCarb Innovation Fund to accelerate the decarbonization of steelmaking.

The company currently uses LanzaTech's gas fermentation process, in which biocatalysts plus bacteria use CO and hydrogen from the steel mill gases and heat from the combustion of waste biomass to make ethanol for blending with gasoline or processing into jet fuel. At full capacity, the €200 million plant will produce 63 million gallons of bioethanol per year from its waste steelmaking gases and will reduce its carbon emissions by 125,000 metric tons per year. And, next year, ArcelorMittal will also open a €35 million reactor designed to process waste wood into biocarbon for its blast furnace. This will cut the plant's annual carbon emissions by 112,500 metric tons by reducing the amount of (coal-derived) coke needed in its blast furnace, and the company will double the savings over the following two years, when it plans to open a second reactor.

The huge steel mill currently produces five million tons of steel annually, resulting in ten million tons of waste CO_2. Because that CO_2 originally came from coal, the ethanol made from it is not a biofuel, but, Soetaert says, it brings "a touch of biotechnology." More important, however, the process recycles CO_2 that would otherwise be emitted into the atmosphere.

LanzaTech itself appears to be on a roll: in addition to the Arce-lorMittal plant, the company has a plant in Shougang, China, that converts ferroalloy factory emissions into ethanol; an Edmonton, Canada, plant that uses a new bioreactor design to convert waste biomass into ethanol; and another plant in Japan, in partnership with Sekisui Chemical, that turns municipal solid waste into ethanol. The global ethanol market is projected to be worth $170 billion a year by 2030. LanzaTech is excited by the commercial prospects of these new processes, but is also engaged in ventures to make clothing, perfume, propane, running shoes, a sustainable natural-rubber substitute, and tires from its recycled carbon. The company aspires to eventually impact the multitrillion-dollar bulk chemical market, and went public in 2022.

For the time being, though, despite FBBV's work, pure biobased industries, especially those using third-generation feedstocks (produced by bacteria or algae), are still very much in their infancy. Though progressing fast, this third generation of feedstocks is not yet a major part of the bioeconomy. Instead, most of the output of the North Sea Port's bioeconomy, for one, is still concentrated in biofuels, which are at the low end of the bioeconomy "value pyramid"—far below bioproducts like medicines, cosmetics, and food—and their environmental impacts place them outside the realm of long-term climate solutions.

Still, Dr. Soetaert and FBBV have much to be proud of. Today, everyone in Belgium fuels up with gasoline and diesel that contain either bioethanol or biodiesel, both of which are locally produced in the country. "That may not be exceptional, but we nearly missed the bioenergy boat completely," Soetaert says. "I guess no activity whatsoever would have happened in Ghent on biofuels if it wasn't for Flanders Biobased Valley." As a result, today the Port of Ghent has Europe's largest biorefinery cluster and ArcelorMittal's (and the world's) first "steelanol" plant. "We've been working on it [for] nearly ten years," Soetaert said. "We've been pioneers in sustainable production at a time when everybody considered this as 'nice to have,' but not real." So, the head start that FBBV gave Belgium is now paying off handsomely.

Carbon repurposing and the bioeconomy can change much about the way we make and consume products in a greener future. But we will not get to that future without addressing the substantially greater problem of reducing, and ultimately eliminating, the excess GHGs already in the atmosphere—pursuing a future that's carbon negative. In the next chapter, we'll look at what it would take to decarbonize some of the biggest industrial polluters on earth: the cement and steel industries.

10.

Green Cement and Steel

Some industries are fiendishly difficult to clean up. Among heavy industries, cement and steel have long been thought of as not just difficult but almost impossible to decarbonize. Most industries—say, electric power or bread baking—can be decarbonized just by energizing them with clean, renewable energy sources. But the cement and steel industries are actually *built* on carbon; it's a vital constituent of their products. So, eliminating the emissions of these two industries is especially challenging. If it can be done, however, little should stand in the way of decarbonizing all heavy industry. Now two new industrial processes are revolutionizing cement and steel-making, legacy industries that have been essentially unchanged for decades or far longer.

Green Cement

Given the ubiquity of cement in the built environment, it's unfortunate that its production is so environmentally damaging. Mining and transporting billions of tons of limestone, chalk, and clay-bearing rock every year to make cement creates a lot of emissions, as does the milling to prepare the ingredients for the kiln. Next, roasting them releases still more CO_2. However, those GHGs are dwarfed by the emissions from the chemical reactions *within* the kiln—about twice those created in firing it, which is invariably done with coal, the fossil fuel with the most GHG emissions per kilowatt-hour of any conventional fossil fuel. Ultimately, making concrete from the world's cement also consumes three billion gallons of water annually—a tenth of global annual industrial water use, mostly in water-scarce regions—and a great deal of the concrete is then used to cover much of the earth with a hard, dead, impermeable surface.

To produce cement, limestone, clay, and some secondary materials are pulverized and added to a rotating cement kiln. Coal—burning with a flame temperature of about 2,000°C outside the kiln—heats the kiln's interior to 1,500°C. Then, within the kiln, limestone breaks down into lime and CO_2, and the cement ingredients react. Thus, for every ton of Portland cement (the most common kind), about a ton of CO_2 on average is released in total, from the coal combustion and the chemical reactions inside the kiln.* (The exact quantity varies greatly depending on the efficiency of the kiln and how much fly ash and slag are added to the mix.)

Once the ingredients of cement are done reacting within the kiln, small, rocklike nodules of an intermediate product called "clinker" tumble out. When crushed and mixed with a bit of uncooked limestone and a touch of gypsum, the result is a dry powder: Portland cement.

Though often conflated, cement and concrete are not the same. Concrete is a mixture of cement, sand, gravel, and water; the cement

---

* According to Solidia's chief technology officer, Dr. Nicholas DeCristofaro (author communication, email, March 23, 2020), in the most modern cement plants with the most efficient kilns, about 0.8 tons of CO_2 are produced per ton of Type 1 Portland cement (the type widely used in the Western world). Older plants produce a ton or more of CO_2 per ton of cement. Thus, the average for Portland cement might be one ton of CO_2 per ton of cement. However, Portland cement can be blended with fly ash, blast furnace slag, and other materials whose CO_2 footprint is nominally regarded as zero (to avoid double counting) because, in the case of the ash and slag, CO_2 emitted in creating these additives would have been generated whether the ash or slag was used for cement production or not. In the United States and Europe, these non–Portland cement additives are 10 to 20 percent of the final product, but in China, they can be up to 50 percent or more. For a product that was half cement and half additives, the CO_2 emissions in the most efficient plant would fall to 0.4 tons per ton of cement.

acts like a pasty glue, holding all the ingredients in the slurry together. As conventional concrete cures, the cement chemically combines with the water, causing the mixture to harden in any desired shape as it dries. The basic concept has been around since Neolithic times, when builders in Jordan and Syria used a primitive form of cement. In more recent antiquity, the ancient Romans used concrete for roads, bridges, and buildings—the unreinforced concrete dome of their Pantheon still stands today after 1,900 years.

Concrete is so useful that, by mass, it's the world's most-consumed commodity after water, used virtually everywhere in the built environment for roads, sidewalks, buildings, bridges, and dams. But both from its environmental impacts and its GHG emissions, modern cement—and, by extension, the thirty billion metric tons of concrete made from it every year—has a large, dirty environmental footprint.

From the standpoint of cutting GHGs in a world where global cement demand is still growing, cement production emissions are a dastardly problem. Cement and concrete account for about 8 percent of global CO_2 emissions, which is almost as much as the entire global agricultural sector and more emissions than any industrial process except power generation and steel.* Were all the emissions released by the world's concrete production generated instead by a nation, that would make it the world's third largest emitting country, after only China and the United States. To make matters worse, demand for concrete is rising steeply. The total floor area of buildings worldwide, for example, is expected to double by 2060.

Although cement making has been spewing CO_2 for thousands of years—before climate change became a global crisis—people historically ignored its emissions. But now, if the world is to meet its climate goals as articulated at the Paris and Glasgow climate change

---

* Steel produces about 11 percent of global CO_2 emissions and 7 percent of all GHGs, according to Ali Hasanbeigi, "Steel Climate Impact: An International Benchmarking of Energy and CO_2 Intensities," *Global Efficiency Intelligence*, April 2022, available at https://www.globalefficiencyintel.com/steel-climate-impact-international-benchmarking-energy-co2-intensities. Others put the industry's emissions at 7 to 9 percent.

conferences, some means must be found to sharply reduce or eliminate these emissions.

From Lab to Slab

Enter Solidia Technologies. The scientists and engineers of this small but sophisticated start-up believe they have a way of decarbonizing the cement and concrete industries. They're convinced that once their revolutionary new technology is scaled up, it could radically transform this huge global emissions source into an enormous carbon sponge capable of absorbing billions of tons of CO_2 every year. Big bucks are involved: the global concrete industry is a trillion-dollar enterprise; the global cement industry is a three-hundred-billion-dollar market. So, innovators and entrepreneurs like those at Solidia, who transform these industries and reduce their environmental footprint, are hoping eventually to reap huge rewards.

When I visited Solidia, it seemed well positioned to achieve its mission of playing a major role in decarbonizing concrete. The company holds a hundred patents, with more than two hundred more pending. In addition to Kleiner Perkins, the multibillion-dollar venture capital firm, investors include oil companies BP and Total, plus industrial gas provider Air Liquide. These investors were reassured by the fact that the new concrete had already been verified for strength and durability and that its production technology had been demonstrated at scale in fifty commercial concrete plants in ten countries.

I was struck by the upbeat mood at the company's headquarters in Piscataway, New Jersey. "The excitement is not really about what we're doing, but what we're doing for the planet," said chief technology officer Nick DeCristofaro. The genial, intense metallurgist with an MIT doctorate in materials science and engineering came to Solidia after academic posts and stints at major corporations. Joining him was Dr. Vahit Atakan, formerly Solidia's chief scientist, originally from Ankara, Turkey. In a sense, the new cement recipe Atakan and his MIT PhD adviser, Professor Rik Riman, developed can be traced back to Atakan's childhood home in Ankara. There, Atakan did so many cooking experiments that his mom finally exiled him from the kitchen. From early childhood, he had always wanted to be an engineer. "I destroyed pretty much all my toys," he said. "If something is working, I like to find out why. If something is not working, I want to know why *not*." The complex and daunting chemical engineering challenges he had to surmount at Solidia were a perfect fit for him.

Nowadays, when wet concrete is poured at a construction site or to repave a sidewalk, the cement in it is a source of CO_2, because of all the cement-making emissions just described. How could they be reduced? What if the chemistry of cement could be altered, Atakan and Riman wondered, so that when it cured into concrete, it reacted with CO_2 instead of water? That would bind one molecule of CO_2 to one atom of calcium in the cement. Or, better yet, what if each calcium atom were bound to *several* carbon atoms and took even more CO_2 out of circulation? The whole concept of a new "green cement" intrigued Atakan and his professor.

While Atakan was doing postdoctoral work with Professor Riman, the two scientists tried to fuse ceramic materials at high temperatures before the materials decomposed. The technical term for what they were doing is *sintering*, that is, heating a material (usually a dry powder) so it forms a more compact solid without first liquefying—a kind of high-tech cooking. Their studies of the compounds they made led them to wonder if they could use CO_2 to produce calcium carbonate (found naturally as limestone), synthetic marble, and, ultimately, concrete.

Bill Joy, of Kleiner Perkins, already had an interest in climate change. Kleiner Perkins took an interest in Riman and Atakan's notion of applying their research to cooking up some new cement and concrete. The firm made an early investment. Other investors followed.

R&D—A Better Recipe

Dr. Atakan knew from his materials science training that if he found a mineral that would form a carbonate, the carbon footprint of the concrete might be reduced. The search was on. Atakan began testing candidate materials. One year stretched into two and then approached three. Riman and Atakan tested hundreds of minerals. "We lost count," Atakan said. The carbonation conditions for most minerals required either excessively high temperatures or high pressures for commercial purposes.

Finally, they hit upon a relatively rare mineral called wollastonite. It could be carbonated at near-ambient conditions—so far, so good. But another problem remained: there just wasn't enough wollastonite in nature to allow Atakan to scale up his new process commercially. He would have to be able to synthesize it, and no one knew for certain that this could be done. Finally, he and his colleagues succeeded in doing so, and that, said Atakan, "is how Solidia was born." Synthetic wollastonite became the principal ingredient in Solidia Cement 1.0.*

The new compound significantly reduced the required kiln temperature, cutting energy consumption and CO_2 emissions. A reduction in the calcium-to-silicon ratio in wollastonite compared to that of Portland cement meant that synthetic wollastonite would further cut emissions from inside the kiln, so overall cement production emissions dropped by a total of 30 to 35 percent.

Fortunately for Solidia, the synthetic product was superior to Portland cement in a number of ways. It cures in a small fraction of the time of normal cement—fewer than twenty-four *hours* instead of twenty-eight days. So, instead of accumulating inventory, concrete block producers could slash costs by switching to just-in-time manufacturing, making Solidia Concrete more cost-competitive

---

* Because of its chemical structure—one calcium atom for every silicon atom—wollastonite releases less CO_2 when heated than Portland cement, which on the average has two and a half to three calcium atoms for each silicon atom. Each calcium atom is bonded to a carbonate molecule, which releases CO_2 when heated, adding to the greenhouse effect.

(potentially even with asphalt). Globally, this could reduce concrete waste by one hundred million metric tons per year. Additionally, leftover Solidia Concrete in the "cement mixer" or residue in concrete block molds can be easily removed and recycled, because Solidia Concrete doesn't react with water to form a solid.

Compared with conventional cement, Solidia 1.0 has two major differences. Not only does it "cook" inside the kiln with less limestone and at a lower temperature than traditional Portland cement, saving energy and avoiding emissions, but it is cured into concrete in a warm breeze of concentrated CO_2 instead of with water, so it chemically bonds with some of the concentrated CO_2, taking it out of circulation.* This technology could not only significantly reduce ambient levels of CO_2 but also save billions of gallons of water globally—a trillion concrete blocks a year will be poured by 2027, and if they were cured with Solidia's process, imagine the savings.

Solidia 1.0 has one big limitation, however: currently, it can be used to make only *precast* concrete. That's because the concrete has to be cured in a controlled environment, where it can be treated with gaseous CO_2. This can't be done when it is poured outdoors.

Cleaner, Cheaper Concrete—Indoors *and* Out

Today, the precast concrete (blocks and slabs) made with Solidia 1.0 cuts the net CO_2 footprint per ton of cement by up to 30 percent from raw material to finished product. Solidia, meanwhile, is ambitiously working on Solidia 2.0, which will extend the benefits of this precast technology to on-site concrete pours, a much larger global market. Moreover, Solidia 2.0 could, in theory, store *multiples* of the amount of CO_2 released during its production. This would be the

---

* This CO_2 currently is provided by industrial gas suppliers. Ultimately, it will be captured at factories burning fossil fuel and transported to cement plants; capturing it directly from the atmosphere is technically feasible but much more costly and difficult because atmospheric CO_2 concentrations are much lower than in industrial exhaust streams.

closest thing there is to greenhouse gas Aikido, turning a wicked problem into a lovely solution. Moreover, the new concrete would safely store CO_2 for millennia—until the concrete was torn up and tossed into a kiln at 900°C or above. Within five or six years, Solidia expects to be making this "carbon-negative" cement commercially.

Solidia's current-generation cement is compatible with existing cement and concrete manufacturing equipment. In addition, Solidia's new cement recipe produces more durable concrete that takes less time to cure, creates a smaller environmental footprint, and generates less waste. According to Solidia's former chief executive Tom Schuler, a mechanical engineer whom I interviewed at Solidia, the company's "first-generation" new concrete is already delivering important economic benefits to its customers alongside the payload of double-barreled environmental benefits just mentioned.

Where do all those economic benefits come from? Well, an ordinary conventional cement kiln is an enormous rotating device—328 feet long and more than 30 feet high—that roasts cement at close to 1,500°C. "If you look in them," Schuler said, "you're pretty sure you're looking into the gates of hell." But because Solidia Cement requires less limestone and less heat, the kiln in which it cooks can operate at 1,200°C, reducing both production emissions of CO_2 and energy needs.

Because energy amounts to about 30 percent of the cost of cement production, Solidia thus saves about 9 to 10 percent of overall production costs—highly significant for an industry that operates on single-digit profit margins. And with the use of less limestone, even more money is saved. Total cost savings for energy and limestone could be in the 10 to 15 percent range, said Solidia CTO DeCristofaro. "[The industry] should have a huge incentive to make Solidia Cement: lower costs and—if and when carbon becomes an issue—a lower carbon footprint."

Solidia Cement offers other striking advantages when used in concrete. The fact that it is cured into precast concrete with CO_2, not water, makes it much easier to clean the concrete molds and forms after a pour. Whereas conventional concrete starts sticking to the mold or mixer as it cures, Solidia Concrete just sits there until it is exposed to CO_2. "Clean-up time, instead of being hours, is just seconds," Dr. Atakan said. All this reduces concrete waste by 10 percent during block production.

So, it appears that cement makers actually can do well by doing good. According to the company, when the emissions reduction at the cement kiln is coupled with the enhanced CO_2 capture during concrete curing, the carbon footprint of cement used in concrete can be reduced by up to 70 percent.

Solidia 2.0—A Carbon Grabber

When a chemical known as a carboxylate (made from CO_2 and water) reacts with Solidia Cement, a very exciting prospect opens up. With carboxylate, two, four, or even more atoms of carbon will bind to each calcium atom in the curing concrete. (Just how many carbon atoms can be captured depends on which carboxylate compound is used; citric acid is an example of a carboxylate that will bond four carbon atoms to each calcium atom.) Each carbon that bonds to a calcium atom represents one less molecule of CO_2 on the loose.

Understanding the potential impact that Solidia could have on these emissions is a little complicated, so please bear with me. Currently, 4.4 billion metric tons of cement are produced every year. However, according to Nicholas DeCristofaro, "a sizable portion of this cement [about 40 percent] is supplementary cementitious material (fly ash or slag) and limestone," rather than actual cement clinker. If the remaining 2.6 billion metric tons of cement clinker were produced using Solidia 1.0 *and* were turned into blocks, then the industry's emissions would be reduced by 0.78 billion tons annually, as Solidia 1.0 reduces CO_2 by 30 percent. However, currently, only perhaps a third of global concrete production is in the form of the concrete blocks that can be made using Solidia's current process.

By contrast, if the world's 2.6 billion tons of cement clinker per year were instead produced with carbon-hungry carboxylated cement and were designed for on-site concrete pours (the goal for Solidia 2.0), then, said DeCristofaro, "a carboxylate-cured concrete could conceivably offset those CO_2 emissions." This implies that if the common estimate of cement's global carbon footprint is correct, then global use of that kind of versatile carboxylated cement could eliminate on the order of 6 to 8 percent of the world's entire energy-related CO_2 emissions! Having a single process potentially able to avoid such a large fraction of the world's CO_2 would be great news for the climate. And fortunately, carboxylate can be made with renewable electricity, without generating additional CO_2 for its synthesis.

The Next Frontier

Even if Solidia's current process (1.0) were used in every precast concrete plant in the world, however, the cement industry would still be a net carbon source because, as noted, Solidia's current technology can, at best, offset only 30 percent of the emissions from making cement. Remember that two thirds of the CO_2 emissions from cement making come from decomposition of limestone inside the kiln, and a third from the fuel burned to heat the kiln. So, to fully eliminate those latter "thermal emissions," the kilns would have to be heated using renewable energy.

Theoretically, this could be done in a couple of ways. A start-up company named Heliogen says it can commercially produce temperatures of 1,000°C using solar heat from solar thermal power plants. But that is still 200 to 500°C shy of the temperatures needed for cement making. "It could [also] be done with electricity, but it would be more expensive, and cement is a low-value commodity," Schuler told me. "It's not feasible [economically]," Dr. Atakan said flatly. "That's why you don't see a cement plant that is using electricity."

If "green hydrogen" were cheap enough, though, it could cleanly generate enough heat to operate a cement kiln, but without the CO_2 emissions from the fuel used in firing the kiln. Promising new breakthroughs are occurring in electrolysis technology, and so, costs will likely drop. Researchers at RMIT and the University of Melbourne, for example, have recently discovered that the yield of hydrogen from electrolysis can be increased fourteenfold by exposing the reactor to sound waves. This accelerates the separation of water into hydrogen and oxygen while also preventing the formation of troublesome bubbles at electrodes, increasing the stability and conductivity of the electrolyte. Thus, much more hydrogen is produced, and the use of corrosive electrolytes can be avoided. This, in turn, allows engineers to avoid expensive platinum and iridium electrodes and to use cheaper, less-corrosion-resistant electrodes instead.

One straightforward way to incentivize cement makers to use clean fuels, solar heat, or renewable power would be to adopt a national carbon fee to raise the price of "cheap fuels" like coal and natural gas. (Through international negotiations and cross-border price adjustments, it would have to be applied in other nations as well, so as not to render U.S. industry less competitive.)

Eventually, a steep enough carbon fee would motivate cement makers to capture the CO_2 from inside their kilns. If coupled with the use of clean fuels to heat the kilns, this, theoretically, could eliminate the CO_2 emissions from cement making entirely. But the added costs would make Solidia's reduced-CO_2 cement and concrete even more economically desirable. Federal R&D grants (or more investor capital) could also help shorten the time needed to successfully commercialize Solidia 2.0, speeding reductions in GHG emissions.

Growing Cement

Other new, intriguing technologies for reducing cement emissions are on the horizon. Rather than heating limestone in a kiln, bio-

Mason, a start-up based in Raleigh, North Carolina, is using bacteria to grow cement bricks and floor tiles. The company injects microorganisms into ground aggregate, and in about twenty-four hours, the bacteria produce a biocement that turns the mixture into concrete. The synthesis proceeds at room temperature, so no kiln is needed. It all makes sense when you realize that animals like coral already make biocement using CO_2 (from seawater).

Alluringly for bioMason and other tile makers, the market for tiles worldwide is a tantalizing $347 billion a year—almost as large as the $400 billion annual global cement market. Although it's in an early commercialization stage, bioMason has customers and $23 million in investment capital and is building a large factory. It is also developing a type of underwater cement that takes CO_2 from seawater and, thus, grows over time. This could be very useful for erosion-prone shorelines, dams, and bridges.

The general term for what bioMason is doing is creating "engineered living materials." Undoubtedly, this field is opening up a whole new realm of biotechnology: if you continually feed the bacteria with the nutrients and raw materials they need, one bioMason brick can be split and can grow into two, and so on.

Nearly twenty other companies are also actively innovating in this industrial niche. CarbonBuilt, a tiny new Los Angeles start-up based on technology developed at UCLA, recently won a prestigious Carbon XPRIZE for its Reversa process. Reversa takes CO_2 at industrial facilities and—unlike at Solidia—infuses it into wet concrete. CarbonBuilt says its process reduces the carbon footprint of its concrete by 60 percent now—and, one day, will potentially do so by 100 percent.

Sublime Systems, a two-year-old start-up headquartered in Somerville, Massachusetts, is pioneering an electrochemical path to carbon-free cement. The company plans to produce net-zero carbon cement by using renewable power to electrochemically decompose limestone, producing lime without the need for a kiln to cook the limestone and other ingredients. Sublime Systems uses technology developed at MIT and hopes to have a commercial plant in operation by 2026.

Green Steel—and Green Hydrogen

Like concrete, steel is a ubiquitous commodity, vital to construction, manufacturing, and consumer products like the car you drive. It's found everywhere—from the kitchen oven to the factory, from rails to nails to jails, from skyscrapers to drainpipes to scalpels and dentist's drills.

Fueled largely with natural gas and coal, the $2.5 trillion global steel industry is one of the biggest energy consumers in the manufacturing sector and a huge source of CO_2 emissions—at least 1.8 metric tons for each metric ton of steel made annually. Thus, in producing 3.6 billion metric tons of steel annually, the industry releases more emissions than the direct emissions of the cement industry *and* the chemicals industry combined.*

So, if the world is going to be able to hold global heating to 1.5 or 2°C by 2050, as climate scientists say we must, the steel industry's emissions must be slashed by at least two thirds. With some industries, you can slash emissions by making the industries more efficient, so they burn less fuel. But because steel is a mature industry, with blast furnaces operating near their maximal thermodynamic efficiency, the industry's huge CO_2 emissions are not going to be eliminated simply with improvements to the efficiency of conventional technologies. Fundamentally new technology and new investment are needed.

Two very different paths to making truly "green" steel (by decarbonizing its production from iron ore) are being developed. One method substitutes "green" hydrogen, produced with renewable energy, for coal and coke in the steel production process to directly reduce iron, and

---

* In addition, steelmaking plants release particulates that contain heavy metals (such as lead and arsenic), nitrogen and sulfur oxides, and polycyclic aromatic hydrocarbons. (The last occur naturally in coal, crude oil, and gasoline and form or attach to small particles when burned.) Some are carcinogens.

several companies are engaged in this activity. Sweden's SSAB and its partners* are already testing a new plant that uses renewably generated hydrogen to make steel; full commercial operation is slated for 2026. Another company, a start-up called H2 Green Steel, is planning another new hydrogen-fueled steel plant in the north of Sweden. Of note: it will be integrated with its own dedicated green-hydrogen plant. The whole operation is due to open in 2024.

Sweden is attractive to European steelmakers because it has the highest-quality iron ore in Europe. (We'll see in a moment why this is particularly important for direct reduction production lines.) Meanwhile, Europe's largest steelmaker, ArcelorMittal, with four hundred steel plants in seventeen countries, is also planning to open a fossil-free steel plant, by 2025, in Sestao, Spain. The plant's electrical demand (for equipment, such as its rolling mill) will be met with renewable energy, and its hydrogen will be produced in situ by solar-powered electrolysis at several large new solar farms (to be built by a consortium of companies) and then piped to the plant. But, at least at the outset, the company estimates that its green steel will be 60 percent more expensive than conventional steel.

Some observers are therefore skeptical that green hydrogen is the way to go. "The premium is the result of costly electricity to generate hydrogen, store it under pressure, heat the reactor to nine hundred degrees Celsius, [and] run the electric arc furnace," a steel industry expert told me. "Europe has some of the most expensive electricity in the world. So, the production of iron by hydrogen reduction makes no sense in Sweden or Germany. . . . No one wants to pay this for steel. . . . In short, it won't scale and will produce the most expensive iron in the world."

Not everyone agrees. Thyssenkrupp, German's largest steelmaker, also plans a "carbon-neutral" steel plant fueled with hydrogen, if sufficiently large quantities of hydrogen are available.** Outside Europe,

---

* Vattenfall, Sweden's government-owned utility, and LKAB, a mining company, are the partners.
** If not, the company plans to begin operating the plant with natural gas until sufficient hydrogen supplies can be secured.

Fortescue, an Australian iron ore–mining company with thirty billion dollars in annual revenue, plans to convert its entire operations to run on renewable energy and, ultimately, intends to produce vast amounts of renewable electricity. Some of Fortescue's clean power is reportedly to be used to produce hydrogen for export to Europe, to power fossil fuel–free steel mills. Some is to go to the United States to provide clean power for factories, and some to fuel Fortescue's own operations, including its processing plants, haul trucks, drills, and trains.

Through its Fortescue Future Industries subsidiary, the company is already testing a massive hydrogen-fueled haul truck and says it plans to produce steel electrochemically, without hydrogen—the second major pathway to green steel. This steel production technique relies on pure electrochemistry and is called mixed-oxide electrolysis (MOE). It can be powered by solar or other renewable forms of energy.

Despite green steel's promise, the new green steel plants in the works today will initially produce only a small fraction of the steel made in Europe's five hundred steel mills. Prominent among the impediments to the green steel industry's growth is the fact that blast furnaces and other equipment used in steelmaking last for decades and are expensive to replace, and green hydrogen itself is currently six times as expensive as natural gas, though the costs will ultimately come down. Moreover, supplies of green hydrogen today are limited, and the energy required to produce green steel using the direct-reduced iron process is "massive."

Turning Brown Steel Green

Whereas it appears that Europe may have the jump on the United States in producing green steel using hydrogen fuel, a Russian coal, iron ore, and a steel conglomerate known as the EVRAZ Group now owns a steel mill in Pueblo, Colorado, where it plans to produce what could be the greenest and cleanest steel rails, rebar, and continuous steel pipe anywhere in the world.

A coal-fired steel mill was first built on the site where the plant now sits in 1881. The operation merged with a competitor in 1892 to become Colorado Fuel and Iron, a powerful company that owned some sixty mines and was the largest landowner and employer in the state. CF&I helped literally forge the history of the American West, providing much of its railroad track and playing a significant role not only in the West's industrialization but also in the history of the U.S. labor movement.* The mill itself was also the main industry in Pueblo for most of its existence.

In recent years, however, the steelworks had become unprofitable, in part because the cost of the coal it depended on for fuel had become quite volatile in response to global energy price gyrations, especially fluctuating coal demand from China. The mill was therefore scheduled for closure. This would have put about a thousand employees out of work, causing noticeable economic pain in Pueblo, where almost one in four residents lives in poverty. Fortunately, however, EVRAZ, in a partnership with Xcel Energy and solar developer Lightsource BP,** decided to give the plant a new lease on life by finding an inexpensive source of power for it at a predictable price. Their bright idea turned out to be a $285 million solar energy plant, with 750,000 solar panels, just outside Pueblo.

Lightsource BP's Bighorn Solar array now provides cheap, clean power to Xcel. In turn, Xcel provides 90 percent of the EVRAZ plant's energy demand and powers its electric arc furnace, all at just $0.03 per kilowatt-hour, cheaper than the coal power it was using. The plant is, in effect, a steel recycler: scrap iron and steel go into its electric arc furnaces, where they are melted down and refined into finished engineered steel products.

Thanks to the new source of inexpensive power, EVRAZ decided

---

* The company operated the coal mine in Ludlow, Colorado, where, in 1914, the infamous Ludlow Massacre of striking miners and their families was perpetrated by company-backed gunmen and the Colorado National Guard.
** Lightsource is the largest developer of solar power plants in Europe and the third largest in the world outside China. BP, the multinational oil company, is a part owner.

to invest half a billion dollars not only to keep the plant open, saving a thousand jobs, but also to expand it, adding some eight hundred new workers to make new long rails, the length of a football field, for modern heavy-haul and high-speed railways, among other products. The plant expansion is due to be completed in 2023.*

Cheaper, Simpler Steel

Even if the EVRAZ plant expansion goes as planned, steel recyclers and their electric arc furnaces could not possibly decarbonize the entire steel industry alone. There's simply not enough scrap steel to meet the world's steel demand, and the arc furnaces currently release significant amounts of CO_2 when their carbon electrodes oxidize. But mixed-oxide electrolysis, the second pathway to green steel, shows promise for dramatic change. The process isn't dependent on hydrogen, has fewer steps, and is a continuous rather than a batch process. That's important for the efficient utilization of the costly capital equipment needed to make steel. Instead of a blast furnace, MOE relies on electrolyzing iron ore at 1,600°C, in the presence of catalysts, to produce pure liquid iron and oxygen without CO_2.

By contrast, not only does the direct reduction hydrogen process discussed earlier require expensive purification of the iron ore, but the solid iron it produces then needs to be pelletized (another expensive process) before being melted in an electric arc furnace. (That's because all the subsequent production steps in a steel mill, such as alloying, casting, and rolling, require liquid iron.) The many problems with hydrogen reduction are one reason the mixed-oxide

---

* At this writing, its future is uncertain, not because of technological challenges but because of Western sanctions following Russia's 2022 invasion of Ukraine. Russian oligarch Roman Abramovich owns 29 percent of the company stock and, with four other individuals, controls 66 percent of the company. How its financial fortunes and the firm's Pueblo operations may be affected by Western sanctions against Russia and individual oligarchs is currently unknown; following Russia's February 2022 invasion of Ukraine, the company's stock fell 70 percent on the London Stock Exchange. See Roland Head, "Why the Evraz Share Price Fell 70 Percent in February," *Yahoo! Finance*, March 1, 2022.

electrolysis approach to green steel has been generating great excitement. The beauty of this electrochemical approach is its elegance and simplicity—and its total independence from costly hydrogen.

Boston Metal, a small start-up company in Woburn, Massachusetts, is well on its way to commercializing an MOE steelmaking technology that could revolutionize the industry. The company has an interesting origin story: experiments originally designed to extract oxygen from lunar soil for extraterrestrial exploration, followed by a chance encounter, led to its creation and to its game-changing technology.

At company headquarters, I met company founder Donald R. Sadoway, then seventy, and CEO Tadeu Carneiro. Sadoway is a professor in the Department of Materials Science and Engineering at MIT; Chairman and CEO Carneiro is a metallurgical engineer who was CEO of CBMM, a Brazilian mining company that produces 80 percent of the world's niobium. While at CBMM, Carneiro grew its revenue twentyfold, to two billion dollars a year.

Boston Metal is currently developing and commercializing its new, breakthrough steelmaking process, which can eliminate all emissions in steelmaking, except for oxygen, while producing higher-quality steel more cheaply and simply than with current methods. Its new process might be the holy grail of clean-steel production. So, it's not surprising that, when I visited, the company was tripling the size of its facility and had recently raised $25 million in an oversubscribed investment round to fund a demonstration plant.

Boston Metal's breakthrough technologies are based on electro-chemistry research conducted in Sadoway's MIT laboratory. His interest in decarbonizing industry, however, goes back more than thirty years. "In the 1980s," he said, "I had already made the decision that at some point, all industrial chemistry would become industrial electrochemistry. 'Why use carbon to drive a reaction when you can use an electron?' I was already thinking about an electrochemically driven world, which would be a cleaner world."

Upon arriving at MIT, Sadoway started looking for ways to improve the aluminum production process, which uses high-temperature electrochemistry. Currently, the process requires a consumable carbon anode, which means that aluminum smelters make CO_2. But in the course of his aluminum research, Sadoway made the crucial discovery that a chrome alloy would be the perfect electrode both for smelting aluminum and for refining iron.

Why Green Steel Is Competitive Steel

Conventional steel production takes place in large plants, where iron ore is melted in a blast furnace at 1,600°C—think of molten ore in white-hot, fuming, robotically controlled cauldrons producing fiery-hot, toxic fumes full of CO_2. Not only can Boston Metal's new process produce emissions-free clean steel that, when scaled up, should be cheaper than the conventional stuff, but its process is modular—so, instead of requiring a furnace the size of a building, the company will make its green steel, and other metals and alloys, in smaller plants that produce a few hundred thousand tons of steel a year. "In industry, this is the missing link," Carneiro said.* It means that huge plants will not to be required to produce steel anymore.

Boston Steel expects that its steel will be competitive in the market even without a carbon tax, though a price on carbon emissions wouldn't hurt its competitive position. Electrolytic steel takes a great deal less energy to make per ton than blast furnace steel; Boston Metal's electrolytic steel could now be made with only 56 percent as much energy per ton as is required for conventional blast-furnace steel. (The company has conservatively set an initial production-efficiency target of using three quarters as much energy as in the conventional process.) Even at the electricity prices that the aluminum industry pays today, Boston Metal's steel would be 35 percent cheaper than conventional steel.

---

* With current steelmaking technology, the most economical blast furnace is sized to produce three to four million tons of steel a year. The next most economically sized larger plant would cost $4 billion to $6 billion. With Boston Metal's modular process, you can grow capacity continuously.

As noted, blast furnaces are very long-lived capital investments—the last one was built over fifty years ago—and the steel industry itself is very conservative. "Until you can show them that you can pour liquid metal constantly in a steady state [from a mixed-oxide electrolysis facility]," said Carneiro, steel industry people "will bombard you with lots of questions" and express doubts that you're ever going to succeed. "We're not suggesting that the steel industry should take a wrecking ball to all blast furnaces," Sadoway said. "But as those blast furnaces have run through their life cycle and need to be refurbished, then there would be migration [to the new technology]."

Lower Energy Needs

As already discussed, renewably generated hydrogen produced by electrolysis can greatly reduce CO_2 emissions from steelmaking, but because the cost of that clean hydrogen has been so high, the steel industry has had little incentive to adopt it.[*] Direct reduced iron, the hydrogen process, also involves extra steps compared with Boston Metal's MOE approach, because (in addition to the pelletizing process discussed earlier) it means using electricity to first generate hydrogen instead of simply using the electricity directly to do the job of making steel. As Sadoway puts it, "Why not just put the iron oxide into the soup and pass current [through it] and be done with it, and immediately come out with liquid metal?" This technique yields fewer process steps and has lower energy requirements and higher efficiency. Instead of substituting hydrogen for carbon from coke in steel production, just eliminate both![**]

---

[*] Current processes that rely on hydrogen have another issue: they produce a very fine particulate iron laden with hydrogen. That powder then has to be melted in an electric arc furnace, all at a 30 percent cost premium over steel today, Sadoway explained. Boston Metal thus has another likely economic advantage.

[**] Like the carbon in coke, hydrogen is a reducing agent that removes oxygen from iron oxide in the making of steel.

Boston Metal's process is broadly applicable in metallurgy. Using some basic metallurgical science discoveries made in the past decade at MIT by Dr. Sadoway, the company's breakthrough technology has the potential not only to electrify the global steel industry, but also to revolutionize the production of most other metals, including nickel, zirconium, and titanium. Some sixteen different metals either already have been produced or could be produced with the new process. Boston Metal expects to begin commercial operations in 2026 by producing relatively high-value alloys of iron, vanadium, and niobium.

The European Union has a road map for decarbonizing the steel industry that appears to presume a gradual phaseout of blast furnaces by 2050—a godsend for electrolytic steel processors like Boston Metal. "Europe in general," Carneiro said, "don't [*sic*] want to have blast furnaces in the Continent, so they sent the message to the industry: go find a solution." Since my talk with Carneiro, the United States has proposed a climate-related trade deal with the European Union that would create an international market consortium, giving market preference to companies that make lower-emissions steel and aluminum through a system of emissions-based tariffs that would benefit companies like Boston Metal.

Simpler, Cheaper, Better, Smaller

Boston Metal's new process is both more scalable and a lot simpler than what happens in most steel mills today. There, iron ore (after crushing) goes into a blast furnace (which is often five stories high), where it's melted and treated with coke to produce carbon monoxide that rips oxygen off the iron oxide, forming CO_2 and iron. (Both coke, which is made from coal, and coal itself, which is usually used to fuel the blast furnace, produce the CO_2 generated in steelmaking.) Limestone or dolomite is added to the melt to remove other impurities. Next, the molten metal is mixed with recycled steel melted from scrap in an electric arc furnace. The iron ore then needs to be processed in a sintering plant to create pieces of ore of proper size and shape—not a cheap step.

"You need to invest a *lot* in the preparation of the feedstock for the blast furnace," Carneiro explained. By contrast, in mixed-oxide electrolysis, the iron ore could potentially be added to the cell without preparation. The electrical process could also save transportation costs: "If you have electricity, you bring the [electrolytic] cell to the ore, and you don't need to transport the ore, right?"

Another important advantage of Boston Metal's modular electrical technology is that it can be used to expand existing steel plants. "I don't have to go and tell them, 'Build a four-billion-dollar pilot plant of my technology,'" said Adam Rauwerdink, the company's VP of business development. "I can add on [Boston Metal's electrolysis] cells alongside their blast furnace, take their same feedstock . . . and take it directly to my cell, and then combine the liquid metal coming out with the metal they're doing today."

Because, as noted, extra carbon (as coke) is added to a conventional blast furnace to separate the iron oxide from its oxygen, the excess carbon remaining in the resulting pig iron makes the steel brittle, so some of that carbon has to be removed. This means a steelmaker has to transport the pig iron to a basic oxygen furnace, where pressurized oxygen is blown into the melt at supersonic speed. Then, after the roaring furnaces have done their job, the final composition—and hence the properties of the steel—is adjusted by alloying the molten metal with other metals to increase, for example, its strength, hardness, or corrosion resistance.

Boston Metal's method of making steel does away with not only the flaming, CO_2-belching blast furnace but also with the basic oxygen furnace, coke oven, caustic reagents, sintering plant, and tons of process water. That's green chemistry in action. Thanks to mixed-oxide electrolysis, steel can now be produced without CO_2 emissions, with far less hazardous waste, and with less energy. It's not hard to see now why Boston Metal has such a promising future.

Conquering the Global Steel Industry

Once steel can be commercially produced with the new process, Sadoway said, then the producers can "boast that this is green steel." He added: "Can you imagine if there was an automobile producer that made a car with steel made by this process, without CO_2 emissions? People are going to buy that car, and that will put the forcing function on the incumbent [steel producers]," he said, referring to consumer demand that will oblige the industry to produce more green steel.* For all these reasons, a fully electrical steelmaking process could both solve the lion's share of the industry's GHG emissions and end many of its other pollution problems. So, what could be done to accelerate MOE's adoption?

Today, Boston Metal's investors include Breakthrough Energy Investments, a venture capital fund led by Bill Gates in association with some of the world's wealthiest corporations, billionaires, and individual investors, such as Mark Zuckerberg, George Soros, Vinod Khosla, and former New York City mayor Michael Bloomberg. Yet private capital need not and will not play the biggest role in the speedy adoption of clean steel in the United States. Apart from China, the U.S. government and the Pentagon probably have the largest carbon footprint on the planet. If the federal government were to change its rules of procurement to cap the permissible CO_2 emissions per ton of steel it bought, that would be "a really cool policy lever," Sadoway said. Setting a maximum allowable CO_2 limit per ton could ultimately give a new American steel industry a big advantage over traditional steelmakers in China, as it would impel the industry to modernize and become more efficient.

Today, the state of California has taken a step in this direction by requiring that suppliers of steel and certain other products used in state infrastructure projects disclose the CO_2 impact of their steel.

---

* Experts project that demand for low-CO_2 steel will leap tenfold in just ten years. See Artem Baroyan et al., *The Resilience of Steel: Navigating the Crossroads*, McKinsey, April 18, 2023.

"It's part of the bidding process, so it shows up in their bid," Rau-werdink said. Once clean steel is available, he asserted, it also may become virtually a requirement for a builder who wants their buildings to earn a high LEED certification.*

Having had a glimpse of green chemistry in carbon repurposing, the nascent bioeconomy, and heavy industry, we'll turn next to the dramatic progress being made in "front-runner" cities in the United States and abroad that are committed to slashing their GHG emissions as quickly as possible while also growing their economies and improving the quality of life their residents enjoy.

---

* The internationally recognized LEED (Leadership in Energy and Environmental Design) certification, established by the U.S. Green Building Council, rates buildings, homes, and communities according to the extent to which they reduce their energy and environmental impacts.

II

Visionary Mayors, Green Cities

If the world doesn't quickly throttle back its greenhouse gas emissions, many city folk in the not-too-distant future will need waders to get to work and bilge pumps to bail out their ground-floor offices. Two thirds of the world's cities are located on coasts, and tragically, some are already flooding as oceans rise and powerful storms drive torrents of water known as storm surges onto shores. By 2050, global sea levels could rise as much as five feet.

You probably live in a city. I do, and so do four billion other Earthlings—half the world's population. Cities are where the vast majority of our energy is used. So, it's unsurprising that they produce about three quarters of the world's energy-related CO_2 emissions. Moreover, the world is urbanizing so fast that two thirds of all people are likely to be urbanites by 2050. The United States urbanizes a million acres a year. Therefore, if the world is to ever to make a decisive transition to a clean-energy economy, it *must* slash urban GHG emissions.

Can cities mired in fossil fuels transform themselves rapidly into clean-energy havens? I believe they can, if we focus intensely on two truths. First, this *must* happen: if it doesn't, we have failed to ensure the survival of civilization as we know it. Second, it's an opportunity to create trillions of dollars of new wealth, tens of millions of new jobs, and a just transition to a clean-energy economy. The pathway out of the climate emergency runs through cities. They are, as noted, where most energy is used and where most of the world's political power resides.

There's no mystery or rocket science to transforming a city, even a large one. It must be based on scientific principles and, critically, on accountability for the outcomes of the climate policies and choices adopted there. This urban transformation has already begun; a coalition of leading cities around the world is taking thousands of successful actions, calling for a Green New Deal, and adopting declarations recognizing the global climate emergency. More than 11,000 cities and local governments, representing over a billion people globally, have joined the Global Covenant of Mayors for Climate and Energy to help accelerate the clean-energy transition through their voluntary emissions-reduction efforts. Aggregated cumulatively, the actions already pledged by this group are expected to reduce global emissions by 75 billion metric tons of CO_2-equivalent gases by 2050.

Political will, broad-based community support, and excellent management—including a laser focus on driving down emissions—have been key to the successes achieved so far. Yet, even under optimal local conditions, the challenges ahead are so vast that both public and private investments are needed to modernize cities and their infrastructure so they can simultaneously slash emissions and prepare for unavoidable climate impacts.

Whereas the United States may like to think of itself as number one in modernity and innovation, it is *not* a global pacesetter when it comes to making cities climate-safe, climate-smart, or healthy. In part, this is due to the lack of a coherent, and adequately funded, long-term federal climate policy. Yet, despite this profound abdication of federal responsibility, several major American cities—including New York, San Francisco, and Washington, DC—have already peaked their emissions and are making real progress in lowering them toward net zero. Others are lowering emissions from buildings, reducing solid waste, encouraging walking and cycling, and creating new rapid-transit systems. But few are succeeding on every critical front.

A Bumpy Road Toward Net Zero

To find the global front-runners in the race to decarbonize cities, I went to Scandinavia, where several major cities are striving to become models of sustainability and are making notable strides toward that goal. Stockholm, for example, has cut its per capita CO_2 emissions by 57 percent since 1990.

Yet the path to a greener city—even one that is carbon neutral, let alone carbon negative—is a difficult one. Copenhagen, striving to be carbon neutral by 2025, had by 2022 already cut its CO_2 emissions 80 percent compared to 1990 and seemed on track to meet its revolutionary 2025 goal, despite robust economic growth and a 46 percent population increase by 2017. However, disappointingly, the city gave up on its 2025 goal in 2022, having unwisely relied on expensive and poorly performing carbon capture and storage technology as a pillar of its strategy.

Copenhagen is now talking about reaching net-zero carbon between 2026 and 2028, as it is currently projecting a highly significant gap of 430,000 metric tons of CO_2 that likely will still be emitted by 2025. Other ways of closing the emissions gap in the longer term have been suggested, including using waste heat from industrial and commercial sources and even creating a European thermal grid to share energy, as Europe currently wastes enough heat to heat almost all its buildings. Regrettably, rather than reducing emissions more quickly, Stockholm, Oslo, and other front-runner cities have also placed similarly unsound reliance on carbon capture. This lapse in judgment may well also undermine the plans of other jurisdictions.

On the positive side, in visiting some of the world's most advanced, prosperous, modern cities with "radical" emissions-reduction plans, I didn't find anyone complaining about how expensive and onerous fixing the climate was. Instead, I found city leaders who were exuberant about the rapid progress they were making while also improving urban life and invigorating their local economies.

Morten Kabell, for one, was Copenhagen's mayor for technical and environmental affairs when we met in 2015 at City Hall, where he was in charge of the city's climate portfolio. (Copenhagen has six specialized mayors, presided over by a lord mayor.) Like so many Copenhageners who daily cycle to work or school (49 percent), Mayor Kabell looked sturdy and fit. His broad, relaxed face and steady blue eyes radiated confidence and inspired trust. "Making environmental sense makes economic sense," he told me. Thanks to Copenhagen's rapid progress in reducing its carbon emissions, the city had received the European Green Capital Award the year before. Its erstwhile 2025 carbon-neutral target was the most ambitious of any major capital.

City officials had correctly anticipated that the city wouldn't be able to eliminate all its CO_2 emissions by 2025, so, at first, any carbon still produced there at that point was to have been offset with the avoidance of an equivalent amount of emissions elsewhere—for example, by replacing the power from a coal-fired plant in another city with power produced by Copenhagen's wind turbines. The city's own heat and electricity would be produced by wind power, solar energy, geothermal power, and the burning of biomass and urban waste. (The combustion of both waste and biomass would, unfortunately, still release CO_2 into the atmosphere.)

However, at some point, this concept was scotched in favor of having the city's Amager Resource Center, a semi-public municipal utility, capture hundreds of thousands of metric tons of CO_2 from its huge, modern waste-to-energy facility. Then, in 2022, Amager announced that it was ineligible for the required $1.1 billion federal funding for the project from the Danish Energy Agency—and also had misgivings about assuming full economic liability for the plant's possible technical failure, as the grant would have required. This massive default, of course, left the entire net-zero-carbon plan for Copenhagen in deep trouble. The utility is still talking about capturing up to 500,000 metric tons of CO_2 eventually, but its current, more realistic goal is to operate a 4-metric-ton demonstration plant in 2023.

While this reality is disappointing, the fact remains that when it comes to reducing its climate impact, Copenhagen is still far, far ahead of most of the world. This undeniable achievement gap invites a question: What led Copenhagen, in the first place, to establish a world-beating goal that would have enabled it to attain climate neutrality before any other city?

A Shared Vision

Copenhagen's willingness to shoot for the moon in setting its goals for carbon neutrality stems in part from a collective belief among its citizens and the city's government that addressing climate change matters. The inclination to take on this kind of challenge and the plausibility of attaining it are undeniably greater in Denmark, with its widely trusted political system, than in many other places. In the words of another former mayor for the environment, Klaus Bondam, "Denmark is an extremely homogeneous society . . . we think the same politically, we are very consensus-driven and nonhierarchical."

The seeds of this consensus, however, were deliberately planted by individual leaders with a clear vision for change and the power to make a difference. In 1998, political scientist Bo Asmus Kjeldgaard was elected Copenhagen's mayor for technical and environmental affairs (the same post Kabell and Bondam later held), and came into office with the vision of making Copenhagen "the environmental capital of Europe." He began the first of an eventual three terms with a remarkable cleanup of Copenhagen's badly polluted harbor,* which led to the construction of the city's now-famous "Harbour

---

* As an old industrial city, Copenhagen had already begun a revitalization by cleaning up contaminated industrial sites, starting in the 1980s, when the city was in recession and educated professionals and families were moving out. As it began making its buildings more energy efficient, it also redeveloped sections of the city with world-class architecture and plenty of leisure and recreational activities. Gradually, families and professionals began returning to the city, which helped create support for the climate plan from the business community.

Baths" for swimming and sunbathing, and he championed the construction of what was then the world's largest offshore wind farm.

Kjeldgaard also oversaw the reorganization of the city's heat-and-power utility, while preserving its public ownership, and he promoted cycling in general and organic food in public schools. In addition, he organized a club of Danish cities that, like Copenhagen, were committed to experimenting with sustainable city planning. By showcasing sustainability, they hoped to inspire the federal government to create sustainability regulations. Regrettably, although his efforts were popular in Copenhagen, Kjeldgaard, in his first term, was unable to get the city to fund an ambitious clean-energy program. Some colleagues viewed sustainability as merely an opportunity to "greenwash" the city and not as a vital and concrete long-term goal.

Circumstances favorable to Kjeldgaard's approach finally arose in 2006, when Copenhagen was given the honor of hosting the 2009 UN Climate Change Conference. Officials expected the meeting to produce a landmark global climate agreement to succeed the 1997 Kyoto Protocol. By this time, Klaus Bondam was in office as the new mayor for the environment. "We were very happy and proud that Copenhagen was going to host the climate summit," he said. "We like the idea of ourselves as a small nation playing an extremely important role in the world."

The impending summit therefore led to a citywide climate action plan designed to make Copenhagen "the world's most livable and sustainable city." In a historic and prophetic 2007 manifesto, *Eco-Metropolis: Our Vision for Copenhagen 2015*, the city expressed a desire to inspire the world, protect posterity, and help save nature:

> Copenhagen will demonstrate to other capitals how a greener urban environment can enhance the quality of life in practical terms. This will be to the advantage of the citizens both of our city and the world . . . and future generations. . . . Copenhagen will be a major city which lives and breathes because of its concern for the environment, not in spite of it. The City will lead the way and

its citizens will contribute actively to improving the environment through their daily activities. . . . The world will come to Copenhagen to see how to create modern environmental policies in the 21st century. We will show the world and especially other big cities that CO_2 emissions can be reduced effectively without adversely affecting economic growth.

In 2009, the year of the summit, the city adopted another landmark vision statement, *A Metropolis for People*, which proposed that the city redesign itself for walking and cycling, with plenty of greenery and spaces for socializing. The policy not only covered climate goals but also promoted sustainability, noise reduction, air quality improvement, organic food, and cycling.

The city, however, was cruelly disappointed when, on December 18, 2009, the climate conference ended without a binding global climate agreement. It was "a complete disaster," said Bondam. However, in Copenhagen itself, the spirit of eco-enthusiasm remained. The next year, Bo Asmus Kjeldgaard, returned to the post of mayor of the environment, set out to make sure that, come what may, the city would become carbon neutral by 2025.

To avoid the tokenism he had encountered in the 1990s, Kjeldgaard refused to agree to any budget unless he received "a huge amount of money" for implementing the climate plan. He also insisted that every branch of the city government make concrete climate plans covering all aspects of urban life. Through his determination, he secured millions of Danish crowns for the plan—in part because Copenhagen is relatively wealthy, with a well-educated population concerned about climate change. Further, he doggedly insisted that the planning process involve all major stakeholders in Copenhagen: citizens, universities, industries, and other businesses. Ever vigilant against complacency or backsliding, he insisted on annual goals, careful monitoring, and strict accountability for performance in every department.

Whether the city ultimately can even achieve its 2026–2028 target remains to be seen. Danes eat lots of meat, and new cars are expensive in Denmark. Thus, many old gasoline vehicles are still on the roads there; emissions from the nation's cars and trucks are therefore growing. Also, when Danish fuel prices spiked in 2022 because of shortages created by the Russian invasion of Ukraine, the Danish government raised the fuel tax deduction for commuters who drive more than twenty-five kilometers, a populist fossil fuel subsidy that is understandable but won't make meeting climate goals easier. No major city in the world has yet become entirely carbon neutral.

Yet, despite the city's population surge since 1990 and significant growth in GDP, its residents produce less carbon than Danes elsewhere: only 2 metric tons of CO_2 per person now, compared to an average of 4.69 metric tons for the rest of the country and 15 metric tons of CO_2 per person in the United States. Beyond its carbon-neutral goal, the city plans to become totally fossil fuel free by 2050.

While Copenhagen's plan is visionary, experimental, and imaginative, it is also founded on in-depth scientific analyses of each emission problem and its possible solutions. Guided by a climate plan with more than sixty closely monitored initiatives, the Copenhagen Municipality walks the talk: It has green procurement policies and trains its employees in energy-efficient building operation and maintenance. Its new construction has to meet modern, high-efficiency building codes. All major renovations in the city must significantly improve a building's energy efficiency. Builders are encouraged to put in green rooftops and green walls that insulate and beautify.

Whenever new development or redevelopment is planned, the city fosters low energy demand, mixed residential-and-commercial neighborhoods with sufficient density to support public transit, making private cars less necessary. By 2020, almost everyone in the city was expected to be within half a mile of a subway station.

For strategic reasons, however, when Copenhagen promoted its plans to stakeholders and the public, it didn't primarily trumpet its CO_2-reduction goals but, instead, highlighted the economic and quality-of-life benefits of a greener, more prosperous city. Today, the city treats its climate efforts as an organic extension of its goals in virtually every area of urban life, including transport, housing and construction, health, education, social activities, and culture. To build public support for the necessary investments to fund its plans, the city at an early stage repeatedly convened stakeholders (companies large and small, expert consultants, industry associations, the municipal utility, and many others) to confer on plan goals and the business opportunities the plan could generate. Then the city formed partnerships with its utilities, for-profit companies, nonprofits, and research institutions to codevelop the necessary programs.

As a result, Copenhagen is stimulating innovation and green technology investment while radically reducing its CO_2 emissions. It estimates that the climate plan will ultimately catalyze tens of billions of dollars in investment and create tens of thousands of new jobs, proving that cities needn't choose between the economy and the climate.

Putting the Plan in Gear

Plans are relatively easy to put on paper, but how did Copenhagen make them a reality? Consider transportation: The city's pro-cycling policies have been a huge success. About 45 percent of all trips to work or study in Copenhagen are made by bicycle, with three quarters of riders willing to ride even in cold and rainy weather. Using a smart bike-sharing system, people arriving in the city by train can book a bicycle and pedal it to their destination on elevated cycling routes that separate bikes from cars and pedestrians. The city is also part of a twenty-two-city regional network of "cycle superhighways" created to encourage long-distance bicycle travel, reducing air pollution, traffic congestion, and noise.* Cyclists not only avoid high fuel

---

* Cyclists traveling at 12.4 mph on the city's Green Wave route will hit all green lights. The city also has good bike parking options and a modern bike-sharing system for electric bikes; cyclists use tablets to unlock electric bikes at 105 bike stations near busy traffic hubs close to metro, bus, and train stations.

costs but also enjoy extra years of productive life; the avoided health care expenses in Copenhagen due to cycling are estimated at $246 million a year.*

As part of its comprehensive transportation approach, city government has increased the proportion of its fleet of cars running on electricity, biofuels, or hydrogen to 64 percent and is implementing carbon-neutral bus service to replace diesel buses. It intends to have all city vehicles running on electricity, hydrogen, or biofuels by 2025. Copenhagen leads the world in the introduction of hydrogen-powered cars and is investing in integrating its bus, train, and Metro services (both physically and online) to promote easy transfers between transportation modes. It also promotes car sharing and is testing displays of traffic light information so drivers can catch green lights, reducing congestion, fuel waste, and emissions. Instead of mail trucks, 1,800 of the city's mail carriers use electric cargo bikes that can carry close to three hundred pounds of mail over a thirty-mile route. This saves the city nearly $6 million a year in fuel and wages. Moderately priced electric vehicles are already tax-free in Denmark, to encourage sales.

Everywhere energy is used in Copenhagen, the city looks for ways to reduce emissions and make energy use more efficient. It co-generates heat and power from municipal waste and distributes the heat (which otherwise would be wasted by a traditional power plant) throughout the city via an energy-efficient district heating system. Houses and apartments no longer require chimneys; the district heating system produces only half the CO_2 emissions of individual gas furnaces and is cost-competitive with individual systems. In the future, the district system will be able to use heat from wind-powered electric heat pumps, electric heaters, geothermal heat, and heat storage tanks. Then the city will be able to get all its heat and power from renewable energy sources.

---

* According to James Thoern of Copenhagen Design, a pro-bicycling group, cycling daily adds seven years of productive life to the cyclist.

Although biomass fueling can achieve only carbon neutrality, not true climate remediation, Copenhagen's new state-of-the art biomass plant emits a million fewer tons of CO_2 a year than the 600 MW coal plant it replaced. Ultimately, such wood chip–burning plants—which can also cause serious forest damage—will be replaced by cleaner wind turbines, which may, in turn, power electric heat pumps. "We are investing around one billion U.S. dollars right now in wind turbines in and around Copenhagen," Morten Kabell said in 2015, "to make sure that electricity and heat will come from wind turbines in the future."

Copenhagen's sewage treatment plants produce filtered gas that, when mixed with natural gas, makes up almost a third of all the gas used for cooking in the city. The city's utility is also pioneering a novel district cooling system, to replace conventional air-conditioning using cold seawater from Copenhagen's harbor.

Economic Benefits

The payoff for all these initiatives is substantial for ordinary citizens. District heating, for example, is cheaper than heating each structure with its own furnace—no small benefit in a Northern European city. The city also shares the costs of apartment and office building energy retrofits. Not only does this lower occupants' energy bills, but it also adds value for owners, who can earn tax deductions for home energy efficiency and climate adaptation.

Innovative new energy technologies have also brought new products to market, lowered production costs, and made the city's economy more competitive. "Less than one percent of [Copenhagen's] garbage actually goes to landfills," Mayor Kabell noted. The rest goes to modern incinerators with ultra-low emissions, to be turned into heat and power. This virtually eliminates landfill costs. The city removes all valuable materials from its waste so that they can be recycled and reused.

Because of the city's green transformation, Copenhagen as a whole largely escaped the global recession of 2008, Kabell said. Copenhagen's longtime climate director, Jørgen Abildgaard, stated that every time Copenhagen spends the equivalent of a dollar on its climate plan, it generates eighty-five dollars in private investment elsewhere in the city for things like new buildings, building retrofits, mobility services, and new power and heating infrastructure. The city's own economic studies indicate that its climate action plan overall will generate a net return of close to a billion dollars over its lifetime.

There are lessons to be gleaned from Copenhagen's journey: First, the process of eliminating emissions has to start with a scientific assessment of where those emissions are coming from, so priorities can be properly set. Then, from the start, ambitious climate goals and implementation plans must be developed in consultation with all major stakeholders. Goals must span energy and resource efficiency, waste minimization, energy production, green mobility, and smart urban design—and they must not rely too heavily on dubious technologies like carbon capture or outside funding that may not be forthcoming. To win public support, the goals' economic and quality-of-life benefits must be clearly explained rather than presented in terms of GHG emission reductions, which are much harder for laypeople to visualize.

For those who want to learn more about Copenhagen's successes, experts there are eager to teach. For example, Greenovation, a consultancy headed by former environmental mayor Kjeldgaard, has a deep bench of experts eager to advise other cities on advancing toward sustainable solutions. As city leaders foresaw in their prescient 2007 manifesto, the world really does now "come to Copenhagen to see how to create modern environmental policies in the 21st century." The city is showing the world that there need be no trade-offs between ensuring economic growth and a high quality of life and establishing a sustainable future for the earth.

A Red Town Goes Green (Then Reddish . . .)

Some U.S. cities, such as Georgetown, Texas, about thirty miles north of Austin, are also turning green—or at least inclining in that direction. In 2015, Georgetown chose to buy 100 percent renewable electricity to meet all its power needs, right in Big Oil's backyard.

Why would a Republican city council in a small city of 65,000 residents, in a deep-red, oil-friendly state, spurn fossil fuel power for clean power? It wasn't about climate protection or the environment; it was all about economics. The city's need for low, stable energy prices trumped politics, so Georgetown closed out its contract for natural-gas power.

Georgetown's mayor at the time, Dale Ross, describes himself as a "very conservative Republican." Like most of his constituents, he voted for Donald Trump in 2016, but Ross took issue with Trump's belief that fossil fuels could offer Americans a viable energy future. An accountant by trade, Ross wanted stable, long-term energy costs that the city could count on. "How do we get a twenty- to twenty-five-year strategy that creates cost certainty and at the same time mitigates regulatory and governmental risk?" he asked. "The natural-gas providers would only guarantee a seven-year contract. The new contracts [with a solar power provider] guarantee us the same rates for twenty-five years!* It was really easy to sell the people that we were elected to serve on the deal, because it was an economic issue." While getting 100 percent of its power from renewable sources,** the city remains connected to the state's power grid, which functions as a giant battery.

After Georgetown switched to renewable power, the fixed-rate, clean power began to attract major corporations and millions in new

---

* Georgetown signed a long-term power contract in 2015 with Buckthorn LLC, orginally owned by SunEdison, a large solar energy company. (In November 2016, NRG Energy bought Buckthorn from SunEdison.) The Buckthorn contract provides price certainty through 2034.

** The city in 2014 had already signed a twenty-year power contract with EDF for wind power from a wind farm near Amarillo, Texas.

investment to the city. "If you win the economic argument, then you're going to win the environmental argument by default," Ross said. Meanwhile, Ross himself—who in other times might have tooled around town in an SUV—told me when we met that he was looking forward "to be scooting around Georgetown" on a cleanly powered electric motor scooter.

Of course, as with so many claims around climate achievements, Georgetown's newfound identity is not without its complications. To the city's chagrin, although its initial decision to buy renewable energy did deliver predictable energy prices, it ultimately cost the city millions of dollars for several ensuing years, because natural-gas prices fell during those years and lowered the price of grid power while the city still remained obliged instead to continue buying the then relatively costlier renewable power from its suppliers. The city also overestimated its overall power needs and locked itself in to long-term contracts. As a result, local utility bills rose, and citizens understandably griped and tried to scapegoat the renewable-energy decision. However, the problems with the plan were due not to the choice of renewable power but to market conditions and inaccurate forecasts of future needs—a hazard of any long-term contract.

More important, however, the assertion that Georgetown today is still relying on renewable power is dubious when you examine the fine print. The output of the three clean-energy plants with which the city contracts is supplied to the state's power grid, which is operated by the Electric Reliability Council of Texas. ERCOT, however, generated about 80 percent of its power in 2019 from fossil fuels and nuclear power, with only about 20 percent from wind. And today (when power costs from renewables once again cost less than those from fossil fuels), instead of retiring the valuable renewable-energy credits it earns from its renewable generation, Georgetown sells more than 98 percent of them, for about $2.75 million a year, to improve the utility's balance sheet. This means its power mix is only about 20 percent renewable today, as is true for most other ERCOT power users in the state. (The buyer of the city's credits, however, can then use them as an accounting device to defray its own carbon emissions without actually having to reduce them.)

Since my meeting with Dale Ross, the city no longer claims to be reliant on 100 percent renewable energy. As its initial decision to choose renewables was an economic one, it was not inherently driven by environmental values and goals but by the city's quest for energy price stability in a volatile power market. City leaders still abide by that approach. "Our goal was to be financially stable and have competitive rates," Georgetown's current city manager, David Morgan, explained. That type of narrow financial calculus, however, always leaves the door open to future decisions that put profits ahead of the environment, like Georgetown's decision to put its renewable-energy credits on the market.

Morgan today has mixed feelings about the city's choice of renewable power because, for him—as is characteristic for most energy consumers and especially for professional energy managers—economics and convenience trump the environment and the climate. "Renewable energy adds a level of revenue complexity to managing our energy portfolio," he noted. However, it did give the city a green image that helped attract to it a large solar shingle manufacturing plant, which today employs three hundred workers.

As to what comes next for Georgetown on the path to clean energy, only time will tell. Today, the city's power costs have stabilized, Morgan said, due in part to improved management of its energy portfolio. Ultimately, it appears that Georgetown will decisively shift to 100 percent renewable power only when ERCOT, and thus the entire state of Texas, is able to do so. In the real world, progress is sometimes complicated and may involve two steps forward and one step back. All too often, decisions are made for relatively minor short-term economic advantage instead of to prevent far greater and costlier economic harm.

The Greening of San Francisco

As the case of Georgetown, Texas, illustrates, city leaders understand that most Americans are willing to go along with climate protection efforts as long as those efforts are neither inconvenient nor an economic sacrifice. Green is fashionable. Hybrid cars are cool—the environmental benefits come with lower fuel bills, as is true for electric cars (also cool). Solar panels and home insulation reduce utility bills. But woe to the American politician who tries to impose higher taxes to fund all-out climate-protection efforts.

Thus, even cities that have progressive climate programs, like San Francisco, generally are not proposing any carbon taxes. Nonetheless, San Francisco has recently adopted a very aggressive, data-driven climate action plan based on broad consultation and collaboration across city departments, residents, community organizations, and businesses. The plan addresses not only emissions but also racial and social equity, public health, economic recovery, and community resilience. The driving force behind the plan, the city said, was a need to tackle "the interwoven and widening climate, equity, and racial justice challenges we face." As part of San Francisco's effort to foster a more just economy, the plan calls for the construction of at least five thousand new units of housing per year, with no less than 30 percent of them affordable and with a focus on the rehabilitation of existing housing.

With respect to climate change, the plan sets forth an ambitious goal of eliminating fossil fuel use by 2040 and, by 2025, deriving all electricity in the city from renewable sources. New construction in San Francisco already must be zero-emissions. By 2035, all existing large commercial buildings must be zero-emissions, and all buildings of any kind must be zero-emissions by 2040.

San Francisco's 324-page *2021 Climate Action Plan* outlines in detail how the city intends to cut greenhouse emissions by 60 percent by 2030 and zero them out entirely by 2040. The plan goals include powering homes, vehicles, and businesses with 100 percent renewable energy; cutting landfill waste to zero; making 80 percent of all San Franciscans' trips car-free by 2030; and ensuring that 25 percent of all vehicles are electric by 2030 and 100 percent of them are so by 2040. The plan is unusual in that it also proposes to remove carbon from the atmosphere. In this and many other ways, San Francisco is in the vanguard of U.S. cities committed to "deep decarbonization."

As of 2019—six years ahead of schedule—the city had already reduced its emissions by 41 percent compared with 1990, while at the same time, its population grew 22 percent and its GDP almost doubled. Also in 2019, 83 percent of the electricity supplied to San Franciscans came from GHG-free resources. The city is reportedly "well on its way to achieving 100% renewable electricity by 2025."

Though the city government itself produces less than 3 percent of the city's emissions, it, too, pays close attention to minimizing them. It buys 100 percent renewable power for government buildings, and the new ones it builds are energy efficient. To reduce travel emissions, the city government encourages its employees to use public transit, van pools, and bikes to get to work. BART, the electrified independent rail system, which links cities around San Francisco Bay, gets 97 percent of its power from zero- or low-carbon sources and is supposed to be carbon-free by 2035.

The city plans to have a zero-emissions transit system soon—its buses and municipal public transit are all supposed to be electric by 2035 and are already "almost all electric," according to one city official. All products the municipality buys are screened according to green purchasing guidelines. (However, like most other cities, when San Francisco tallies up all its emissions in products consumed, it neglects to account for "embodied emissions." Thus, for example, the emissions produced in China from manufacturing and shipping steel to San Francisco for its buildings are ignored, although they affect the climate just as much as if they had been released in San Francisco.)

To reach 100 percent renewable power by 2030, the city runs a

Community Choice Aggregation program: for less than three dollars on a monthly bill of seventy dollars, city residents, in under five minutes (online or by phone), can choose to buy 100 percent renewable power. To reduce San Francisco's waste to zero by 2030, the city conducts a "world-renowned" waste recycling, recovery, and composting program to prevent organic material from decomposing in landfills and releasing methane, a potent greenhouse gas.* Under its Roots Program, it encourages street tree planting, living roofs, permeable landscapes, and stormwater management. New buildings must have rooftop solar panels or vegetation; the green roofs provide valuable open spaces and help purify the air while absorbing runoff.

Keeping Cities on the Up-and-Up

As the stories in this chapter demonstrate, cities have a lot of power to lower greenhouse gas emissions, but they have to fully exercise that power. They can levy taxes, regulate planning and zoning, raise capital by issuing bonds, provide public transit, and educate the public. They can also mobilize public opinion to build consensus on critical issues. But some cities are doubtless still on the sidelines or slow-walking, unwilling to take political risks on behalf of the climate by asking residents to do anything inconvenient or unpopular.

Even when citizens support climate remediation, meaningful accountability for the decisions that are made, and whether they are acted upon, is key. Energy consultant Sam Brooks, the former director of Washington, DC,'s energy division, has been critical of cities that claim that they are "100 percent renewable" or "100 percent green-powered" when they are still churning out carbon emissions and simply paying to pollute by buying renewable energy credits. Some cities make scant use of solar power or fail to significantly improve building energy efficiency. Others have successfully

---

* Over a twenty-year period, methane is more than eighty times as potent a GHG as CO_2.

improved efficiency but, due to urban population growth, are still increasing their emissions or not accounting for energy embodied in imported goods and services or emissions occasioned by travel into and out of the city. Brooks believes that all cities should annually report their energy use (not just their power consumption) and their energy sources. The public could then track progress, or lack of it. It would also be helpful, he suggested, if all commercial and public buildings over a certain size were required to report their energy and water consumption publicly, as Minneapolis now requires.

Although the economic benefits of a clean-energy municipal economy speak for themselves, governments should not be afraid to emphasize the necessity of a livable climate for its own sake (and ours). Cities must not only eliminate all CO_2 emissions as rapidly as possible—as San Francisco seems to be doing—but should also promote public education in schools and the town square about the gravity of the climate crisis and the urgent need for climate solutions. As far as practicable, cities should also adopt green roofs to reduce building air-conditioning loads and atmospheric GHGs. In addition, they should protect and enhance open spaces within their borders, using trees where appropriate and along streets to reduce summer temperatures and ambient CO_2. To the extent that cities have jurisdiction over plastics—95 percent of which go unrecycled—they should use their authority to regulate them, while lobbying state and federal authorities to implement more sweeping restraints. Strong and effective recycling programs for household and industrial refuse should be implemented, too, and organic materials should be composted for return to the soil, to prevent methane-producing decomposition in landfills. In general, U.S. cities have become remarkably active on climate issues, but they should continue broadening their engagement with them while raising their climate ambitions across the board.

To truly protect the climate, however, we clearly will need all levels of government everywhere to be fully engaged, not just municipalities. And in addition to eliminating fossil fuel emissions, efforts must also be made to pull excess carbon out of the atmosphere. In the next chapters, we will look at some ongoing efforts to do so on the farms and in the forests of the United States.

PART III

Beyond the City: Natural Climate Solutions

12.

Reinventing Agriculture

What if there were a way to safely pull billions of tons of carbon out of the atmosphere, to substantially reduce or even eliminate global warming? What if this approach cost relatively little and could be used around the world? What if it also put billions of dollars in cash into the hands of countless working Americans and people world-wide? What if it even slashed fossil fuel consumption and made the world more resilient to climate stress?

Well, it turns out there *is* a system that can do all that. It's called carbon farming, and it just might be key to restabilizing the climate. In the process, it can revitalize rural economies while also producing healthier, more nutritious crops. And, amazingly, it is low-cost, low-tech, and low-risk.

The carbon farmer works with simple inputs: land, seed, compost, moisture, and sometimes specially selected microorganisms that speed a depleted soil's return to health. Carbon farming also doesn't pull land out of production; nor does it disturb natural ecosystems. It's a "down-to-earth" solution to global warming that employs nature's omnipresent carbon cycle, which constantly shuttles mole-cules of carbon into and out of the atmosphere, soil, freshwater, and ocean. Yet carbon farming is still neither widely known nor widely practiced today.

A School of Hard Knocks

In well-worn jeans and a plaid shirt, Gabe Brown looks like the North Dakota farmer-rancher he is. But if you therefore assume that Brown, fifty-seven, practices normal U.S. production agriculture, you would be totally wrong. Brown has an iron will, a deep religious faith, a tremendous capacity for hard work, and what he describes as a calling: to bring hope to struggling farmers and ranchers while providing healthful food to consumers. Unlike most farmers, though, he's not as concerned with yields per acre or dollars per pound as he is with soil health. How soil became "top of mind" for Gabe Brown—and how Brown himself became a rock star of regenerative agriculture—is a tale of good tidings for the climate, the planet, and American agriculture.

Early in his farming career, Brown endured modern-day trials of Job. In 1995, he wasn't much different from many farmers he knew: a young man with a new family, a struggling farm, and a large operating loan to service. That year, a hailstorm wiped out 1,200 acres of his spring wheat the day before he was to start harvesting it. Because hail had been uncommon and mild during the previous thirty-five years, Brown had no hail insurance and was financially devastated. The bank stuck with him, though, and loaned him more money—but, once again, the following year, hail destroyed his entire crop. At that point, the bank refused to provide a similar new loan.

"My wife and I had a decision to make," Brown said. "Do we give up farming and ranching, or do we find another path?" The Browns chose to stick it out. "When I had these hardships," Brown said, "my neighbors—and I wish no ill on any of [them]—were just waiting for me to fail so they could buy the land. None of them really stepped up to offer to help. That just made me more stubborn. . . . There's no way I'm going to give someone the satisfaction of seeing me fail and lose my ranch."

Brown now somehow had to figure out how to ranch and farm

without all the expensive chemical fertilizers, herbicides, pesticides, and genetically modified (GMO) seeds on which neighboring farmers and ranchers depended and which he now had no money to purchase. In those days, no one bailed the grass in the roadside ditches into hay for cattle, because of the garbage and rocks found there. "It was a pain to do," Brown recalls. But his ranch was relatively small, and he could no longer afford to buy forage for his cattle. So, he went from neighbor to neighbor and asked if he could put up the hay in their ditches.

"They just laughed and said, 'Sure.'" I would mow it and rake it and bale it. Then I'd carry those small square bales out of that ditch [and] onto the road. At night, my wife would drive along with the kids in the car seats with a flatbed trailer behind, and I'd throw those bales onto that trailer one at a time. They probably averaged about seventy or seventy-five pounds, and I remember years we did seven thousand of them. . . . That's a lot of steps up and down a road ditch."

The work also had to be done at night, because both Brown and his wife had had to take minimum-wage, off-farm jobs. They would get home from work at 5:30 p.m., and then Brown would grab a bite to eat and do his farmwork until 1 or 2 a.m. "During that time frame," he said, "I had a choice: I could either [keep] my farm or sleep. [When] there's work to do, I just do it."

The next year was extremely dry. Brown and his wife were just able to scrape enough feed together to keep the cattle, but once again, he had no crop income. "So, you just keep digging a bigger hole, because we had land payments to make," he explained.

The next June, another hailstorm cost Brown 80 percent of his crop.

Those four years, Brown said, "were hell to go through. I wouldn't wish it on anybody, but in the end, it was the best thing that could have happened, because it forced me to change my mind-set. . . . I realized, 'I have to look at my whole operation . . . from the eyes of nature and how nature functions.'"

During the years of hail and drought, Brown had often wondered how the two thousand acres of unplowed native prairie on his land could grow so much forage naturally every year, without synthetic inputs. It always had live roots in it, was always protected by vegetation that sealed in moisture, and was extraordinarily rich in species. To figure this out, Brown went to his local public library. There he read the journals Thomas Jefferson had kept about agricultural practices on his plantation at Monticello, Virginia, where Jefferson planted turnips and vetches to improve degraded soil. Brown also read the journals of Lewis and Clark, who had wintered at native Mandan villages in North Dakota—just north of Brown's own ranch—in the early nineteenth century. The Mandans were planting "the three sisters"—corn, beans, and squash—along with tobacco. They were really focusing on the synergies of nature, said Brown. They got a legume, a grass, and the squash plant "all working in harmony to benefit each other." He took note.

He also noticed that when the third hailstorm pounded his crops onto the ground, it armored his thirsty soil, sealing in its moisture against drought. This was important, because his ranch has no irrigation and gets only ten to twelve inches of rainfall a year, plus another five inches of moisture from melted snow. (It snows there during every month except July.)

Informed by his new knowledge of Mandan agriculture, Brown decided to try planting legumes and grass, cover crops that would thrive synergistically through the residue of the hail-killed crops. His intent was both to raise feed for his livestock and to add organic matter to the soil. Then, not even having money to buy the twine to bail hay, Brown simply let his livestock graze off the cover crops. The livestock got a free meal, and their manure enriched the soil. "That started the act of livestock integration on cropland."

Through his efforts to survive and keep his farm, Gabe Brown gained crucial insights into how ecosystems function and the importance of livestock to a healthy soil ecosystem. Surmounting the challenges he faced forced him to create a new, "carbon-friendly" agriculture that was as economical as it was creative and unconventional. At a time when many family farms were succumbing to competition from industrial agriculture, Brown was able to avert bankruptcy by throwing out the prevailing business model. Instead of the soil-depleting, additive-heavy, financially draining agricultural practices he had learned in vocational school, his current farming techniques mimic nature, heeding soil biology and integrating profitable enterprises in an agricultural ecosystem in which little is wasted; the by-products of one operation are cleverly used as the inputs or feedstock of another. As a result, the 137 different products Brown sells today include organic, grass-fed beef and lamb; pastured pork and pigs, poultry, honey, fruit, and heirloom vegetables in season, as well as border collies. "Don't tell me there's not money in production agriculture!" he says. "There's a myriad of opportunities."

With multiple enterprises—they include leasing or selling hunting and grazing rights and selling grain, feed, and seed for cover crop mixes—Brown always has something profitable, even when the price of a particular crop plummets. The family is part owner of a meat-processing plant that handles their animals once a month and has helped set up a food co-op in town—not coincidentally, organic food grown on healthy, rich soil is tastier and more healthful than food farmed on conventional, industrialized factory farms. Amazingly, Brown runs this financially successful, complex operation with a tiny crew: himself, his wife, his son, and his daughter-in-law, plus two summer interns. The family does all its own direct marketing. Recently, the Browns for the first time hired a full-time employee.

Today, instead of baling grass in ditches at night, Gabe Brown is on the road almost eight months a year to consult and to lead regenerative-agriculture workshops through the nonprofit Soil Health Academy, in which he is a partner. "I really believe that my purpose is to give people hope. . . . By that I mean farmers and ranchers and now, more so, consumers . . . . We're trying to regenerate everything, including climate." One reason for Brown's success is an ability to observe the world keenly from multiple perspectives. When it comes to soil and land management, he thinks like an ecologist. When it comes to ranching, he makes hardheaded business decisions. Fortunately, these, too, turn out to be climate friendly and good for the planet. In contrast to a typical agribusiness, Brown teaches practices that use no fertilizers, no irrigation, no pesticides, no herbicides, and no fungicides. Instead, the focus is on creating and maintaining soil health. "If you have healthy soil," he says, "you're going to have healthy water, healthy air, healthy plants, healthy animals, and healthy people."

When he's not out on the road, Brown still likes to rise early and visit the vast native prairie on his ranch as the sun comes up, to enjoy the birds, the other wildlife, and even the insects. Whereas, in his days as a conventional farmer, he used to wake up every morning deciding which weed or pest he was going to kill that day, today, he told me, "We do everything we can to promote life."

What Makes Soil Ecosystems "Tick"

To understand what Gabe Brown is up to, one has to understand how soil ecosystems operate: they run on carbon, the same way fuel powers an engine. Carbon-rich organic matter not only gives rich, fertile soils their dark color and clumpy texture, but also nourishes soil organisms and plants. Carbon-poor soil is less able to support life, producing lower crop yields, less forage, and less biodiversity. So, as it turns out, soil health is like a magic elixir for climate health.

Brown's new approach to farming was not originally aimed at mitigating climate change. He simply noticed that the cover crops he grew when his fields otherwise would have been fallow significantly raised the soil's water-holding ability and put more live roots into it year-round, as on the native prairie; when those cover crops died, their roots decomposed and increased the soil's organic matter content, nourishing other plants and soil organisms. So, the organic matter Brown added to nourish his crops and livestock also had the unsought benefit of boosting the soil's carbon concentration. (Organic matter is more than 50 percent carbon.) Even in harsh, dry North Dakota—where it's sometimes -40 degrees Fahrenheit in winter—Brown's agricultural techniques have captured vast amounts of valuable carbon. And that carbon, removed from the air and packed away in the soil, provides climate benefits.

Brown's ranch is the subject of a meticulous soil survey operation designed by Professor John Norman, an environmental biophysicist. Norman analyzed the carbon and nitrogen in the top four feet of soil on the ranch. Early results indicate that those horizons contain ninety-two (U.S.) tons of carbon per acre. "That's a really, really big number," Norman said. "The amount of carbon that he's sequestered in this soil is staggering." Even digging four feet below the surface wasn't deep enough for Norman to record all the extra carbon. Moreover, he said, "The deeper you bury the carbon, the longer it's going to be in there." That's important for climate stability, because if the carbon moves back into the air right away, it hasn't been purged from the atmosphere for the long term. "[Gabe]'s built a remarkable soil in a couple of decades. . . . A wise farmer," Norman concluded, "can grow soil a lot faster than Mother Nature."

To further increase his soil's organic matter, Brown nowadays inoculates his seeds with mycorrhizal fungi, and he plants a diverse mix of cover crops to keep the soil from overheating in the summer as the plants capture carbon from the air. Mycorrhizae form a relationship with the roots of vascular plants and are critically important to the development of soil structure, fertility, and water-holding capacity; they also aid plants in using soil nutrients and in resisting disease. By promoting plant growth and health, they help increase soil organic matter. "Nature is more collaborative than competitive," Brown believes.

Ultimately, Brown's cover crops are incorporated into the soil after frost-kill and decomposition, or when Brown "mob-grazes" a pasture. That's when cattle trample much of the forage into the soil, where it protects the soil against wind and water erosion and helps insulate the ground from temperature extremes, thereby also improving warm-weather water retention. "The hotter it gets, the less water is available for plant growth," Brown says. "At seventy degrees, one hundred percent of the water is available for plant growth. At one hundred degrees, only fifteen percent is used for growth, and eighty-five percent is used for evaporation. At one hundred thirty degrees, one hundred percent of water evaporates; at one hundred forty degrees, soil bacteria die."

"[Soil] structure is built by a living system of microorganisms—little animals and the roots of the plants," John Norman says. "They basically make a house for themselves and maintain that structure under a condition that's high yield for the whole system."

Conventional farmers are addicted to fertilizers, pesticides, and herbicides. They see nature as more competitive than cooperative, so they try to remove or poison anything they see as competing with their crops—thereby killing beneficial insects and soil life, including the helpful fungi. In addition, conventional farmers often leave the ground bare in the spring, allowing the soil to erode under rushing snowmelt water and pounding rains that can seal its surface, increasing runoff and decreasing water storage. In Brown's regenerative agricultural system, by contrast, plant residues are left on the ground to decompose, and tiny organisms come up to the soil surface. "They increase the infiltration rate by a huge factor," Norman said. This is important not only for allowing adequate moisture to soak in, to carry plants through dry spells, but also for farmers and ranchers as they try to adapt to climate change. As the climate gets warmer and the frequency and severity of flooding increase, permeable soil is more important than ever to absorb the heavier rainfall. "Gabe Brown's soil can take a foot of water in an hour with no runoff," Norman reports. "That's unheard of in a conventionally tilled agricultural soil."

According to Norman, the structure of the soils that Brown has created in North Dakota is even more remarkable than that of the most productive former prairie soils in the world. The techniques Brown uses make it possible to take a light-colored, depleted, eroding soil that has lost much of its carbon content and in a relatively short period of time turn it into dark, chocolate-colored, carbon-rich soil that looks to Brown "like black cottage cheese."

Gabe Brown's success through carbon farming is not an anomaly. The ecological principles he has used to improve his soils and make his ranch profitable, while also protecting the climate, are being validated in many other places; they clearly could be scaled up further nationally and globally. That is a hopeful sign, because the abuse, degradation, and poisoning of land is widespread on millions of acres in the United States and throughout the world.

The Marin Carbon Project

Although John Wick is not a farmer like Gabe Brown, in the same spirit of love for the land that Brown is manifesting in North Dakota, Wick—along with other ranchers, land managers, consultants, organizers, and scientists—is currently working in Northern California to refine carbon-farming techniques and scale up the technology throughout the state and beyond.

The fulcrum of all this intense activity is the Marin Carbon Project. Based in Point Reyes, California, this ambitious consortium of agricultural institutions, ranchers, farmers, researchers, government agencies, and nonprofits has, over the past decade, initiated more than $15 million worth of field and laboratory research and demonstration projects that have conclusively linked restorative land management practices with increases in durable soil carbon. These findings have proven instrumental in influencing state agricultural policies in favor of carbon farming.

John Wick, a cofounder of the project, provided a significant amount of its start-up funding. A retired building contractor, he's a kindly, smiling fellow with short brown hair and a Mister Rogers–style sense of wonder. Although he now co-owns a ranch, he is an "accidental rancher." The transformation came about when he gave up contracting to help with his wife's children's book business. Needing a larger studio for her and a pleasant environment in which to work, the two acquired a former cattle ranch in Nicasio, California. Neither had an interest in raising cattle; instead, they wanted to be good land stewards by creating a habitat for ground-nesting native songbirds and enhancing local biodiversity.

Their game plan was simple—or so they thought: evict the cows, leave the land alone, and let it return to nature. So, they kicked out the cows, but to their surprise, their laissez-faire land management approach gradually produced disastrous results. A veritable plague of weeds arose on the land, including woolly distaff thistle, a spiny, invasive pest species. They "sprayed and prayed," but all their efforts to control the weeds failed.

Wick eventually consulted Dr. Jeffrey Creque, a thoughtful local rangeland ecologist. Creque's advice to Wick was to encourage species he wanted instead of trying to eradicate ones he didn't want. Creque then advised him to introduce some "disturbance" in the form of light, intermittent grazing—the idea being to give hardy, native perennial plants a competitive advantage over the introduced annual grasses and weeds that were taking over the land. Cowless now, Wick had to borrow a herd of 250 local beef cattle. In just five weeks, he moved them to sixty-seven different locations on his ranch to mimic the relatively gentle impact of a large, unfettered herd of wild grazers. "It was the hardest I'd ever worked in my life," Wick said.

The results, however, were spectacular. Without planting a seed, said Wick, "we started to actually see whole systems of native [perennial] plants appearing on their own. This became really exciting to us." Purple needlegrass, the state's official grass, came out in force, joining California oat grass, blue wild rye, and many others. (Native California perennials are well adapted to the local Mediterranean climate and sink their thick roots deep into the soil, tapping hidden moisture to remain green year-round, unlike the introduced annual grasses that turn brown every summer.) Eventually, the natives would make up fully a quarter of all the grassland plants on his range.

Dr. Creque then had the idea that the resurgence of these deep-rooted perennials might also be increasing the soil's durable carbon content. Grasses—which, like other plants, combine carbon dioxide from the air with water to produce the carbohydrates in their leaves and roots—have been likened to straws through which the soil sips and swallows carbon. Because the world's billions of acres of range-lands are, in effect, a giant carbon reservoir, their perennial grasses, when they are healthy and all sipping together, can serve as a giant carbon pump to replenish soil carbon. Like a battery that can be charged and discharged, the soil carbon provides plenty of energy to sustain life—vigorous plants growing in a fertile, nutrient-rich soil ecosystem. The soil carbon that the plants don't absorb and the decaying soil organisms they don't consume can then remain locked in the soil for a long time in stable chemical compounds, comprising a valuable carbon sink. If agricultural lands like this could be returned to health on a grand scale, Dr. Creque knew this could potentially be significant from a climate change perspective and could also generate salable carbon credits for landowners.

Scientific Investigation and Validation

Even before he met John Wick, Dr. Creque was already a believer in the then-novel idea that carbon-removal processes operating in natural and managed ecosystems could be enhanced, often very cost-effectively, through proper stewardship. As an agronomist for the McEvoy olive ranch in Petaluma, California, Creque had raised the soil's organic matter content from 1 percent to over 12 percent.

The prospect of managing for carbon struck a chord in Wick, who had become horrified about climate change after reading David Spratt and Philip Sutton's *Climate Code Red: The Case for Emergency Action*. "That [book] terrified me," Wick admitted. "I couldn't sleep at night, and that motivated me to try and do something." Wick had also recently heard an Australian researcher named Darren Doherty assert—quoting work by soil scientist Rattan Lal—that raising soil's organic matter just 1.5 percent in all row-cropped fields in the world would remove all the carbon added to the atmosphere since the Industrial Revolution, thereby returning its atmospheric concentration to less than 290 parts per million—a healthy, normal condition.* Given that there is far more rangeland in the world than cropped land, Wick believed that—if Lal's estimate was accurate—raising the carbon concentration on just a portion of rangeland, even by 0.75 percent, was all that was needed to bring the worldwide concentration of atmospheric carbon dioxide back to safe, pre-industrial levels. Moreover, both rangeland and cropland could simultaneously draw down atmospheric CO_2, adding extra oomph to a global CO_2-removal effort.

Creque knew he could raise soil organic matter by far more than 0.75, or even 1.5, percent—he had already done eight times better in his work on the olive plantation. He and Wick now posed three important questions: Were Wick's livestock management practices causing a significant, measurable increase in soil carbon? Could other management practices, such as composting, further increase the soil's ability to capture and hold carbon? If so, would that gain in carbon endure?

The two soon discovered that by applying a thin dusting—initially a half-inch layer per acre—of organic compost, a natural fertilizer containing carbon and nitrogen compounds, they could increase the soil's capacity to capture and hold even more carbon. Moreover, the nitrogen in the compost was in a stabler form than in synthetic fer-

---

* Each part per million of CO_2 in the atmosphere is equivalent to 2.13 gigatons (1 billion metric tons) of carbon. Each molecule of CO_2 is 3.67 times the weight of the carbon atom it contains; therefore, if 1 GT of carbon is in the form of CO_2, the CO_2 must weigh 3.67 GT and, accordingly, 2.13 GT of carbon in the form of CO_2 will weigh 7.82 GT. The atmosphere now contains 422 ppm CO_2, according to CO_2.earth (using Mauna Loa Observatory data). To reach 290 ppm thus requires the net removal of 132 ppm of CO_2, or the net removal of about a thousand GT—about a trillion metric tons—of CO_2 from the atmosphere. However, because of complexities in the carbon cycle and the behavior of carbon sinks in particular, reaching that target will actually require removal of more than the desired net removal of atmospheric CO_2, because, as the concentration of atmospheric CO_2 falls, so does the rate at which it is absorbed by the ocean. Suffice it to say that the eventual return to atmospheric normalcy would be an enormous undertaking. Still, the possibility is there.

tilizer, so it was less prone to be released to the atmosphere as nitrous oxide, a powerful greenhouse gas. Although the initial results from the compost application seemed almost magical, they were rooted in well-established principles of soil science and biochemistry. "It was like putting medicine on the poor soil. It quickly became healthy and, on its own, started supporting more plant growth, which started to sequester more carbon," Wick explained.

Wick and Creque, assisted by UC Berkeley soil geochemist Dr. Whendee Silver, obtained measurements of the results of the compost application over a five-year period. Their field data showed a remarkable result with potentially global implications: The initial compost application jump-started the soil's microbial community. The healthier soil energized the plants, which grew larger and pulled more carbon out of the air through photosynthesis—so the soil's carbon content kept increasing, even without further composting. The scientists whom Wick and Creque next consulted concluded through computer modeling that a one-time, one-metric-ton-per-hectare (about 2.5 acres) application of compost to the soil would result in a one-metric-ton increase in stable soil carbon (not counting the mass of the compost), *year after year*, for between thirty and one hundred years. It was a virtuous circle.

If this kind of composting could be done on a large enough scale, Wick and Creque knew, it could ultimately take billions of metric tons of CO_2 out of the world's atmosphere. Moreover, any organic material—food waste, livestock manure, vegetation, even human waste—can easily be turned into healthy compost to jump-start soil ecosystems. As Wick is fond of saying, "This beautiful solution, which is photosynthesis . . . is everywhere around us."

Scaling Up Soil Carbon Capture

The idea of striving to protect the climate by enhancing the soil carbon bank, Wick and Creque understood, was therefore not rocket science. It's a technique even an illiterate subsistence farmer in a developing country can be taught—and potentially can be helped to afford. But science definitely has a large role to play in optimizing the process so that it can be cost-effectively scaled up.

As yet, Wick doesn't profess to know whether the greatest agricultural opportunities for storing atmospheric carbon lie in grazed rangeland or, perhaps, in row crop production or even in urban soil systems. "We have four million acres of lawns in golf courses in America," he noted. "Think about all the ball fields and the parks and all the managed turf. . . . Which system should we look at to harvest enough atmospheric carbon in time to cool the planet?" he asked. "That's an exercise we need to complete."

Spreading compost on rangeland is also not the only scientifically sound technique for restoring soil carbon. Other techniques include planting cover crops, reducing tillage, leaving crop residues on the land, and managing grazing—all practices that Gabe Brown has employed in North Dakota to triple the carbon content in his soil. The global potential for drawing down carbon has barely been scratched for any one of these techniques, much less for several of them in tandem. California alone has 40 million acres of grazed land. Global rangelands cover 15 to 18 billion acres, yet only a relatively small number of farms and ranches currently practices scientific carbon management of any kind to increase their soil carbon.

Given the huge amount of rangeland and cropped land, large-scale carbon farming and ranching clearly could make a vast contribution to solving the global climate crisis. Scientists, for example, project that compost applied to just 5 percent of California's rangelands could capture 28 million metric tons of carbon dioxide a year, an amount equivalent to taking six million cars off the road. That would be a good start.

Spreading the Gospel

John Wick and Jeff Creque's carbon ranching produced many local benefits. Plants on Wick's ranch became more robust; forage yields increased by 50 percent. The richer, darker soil retained more water and organic matter. Below the surface, worms, fungi, and other beneficial soil microbes flourished. Aboveground, the healthier, more biodiverse sward of plants drew more diverse insect and bird life. Soil erosion decreased, and with the reduced sediment load, water quality improved.

In 2008, Wick and Creque founded the Marin Carbon Project with the intent to spread carbon capture and GHG emissions–reduction practices to rangeland, agricultural, and forest ecosystems across the state. This meant enrolling and training other ranchers in carbon farming and developing formal carbon-offset protocols, so ranchers could earn money for each ton of carbon sequestered.

Eventually, the Marin Carbon Project found an invaluable ally in Calla Rose Ostrander, an environmental planner with the air of an urban cowgirl. Ostrander brought the project high-level organizational skills and a passion to take carbon ranching to the next level. After six years as climate change projects manager for the City and County of San Francisco, she understood how local government worked. Her mission was to figure out how to apply that knowledge to rapidly scale up carbon farming from Wick's Nicasio ranch to the entire state.

Ostrander soon realized that the Marin Carbon Project should link up with existing agricultural services, so she and Wick built alliances with the U.S. Department of Agriculture's Natural Resources Conservation Service (NRCS) and with California's Department of Food and Agriculture (CDFA). With help from both the nonprofit Carbon Cycle Institute and NRCS offices, the Marin team created carbon ranching planning projects in thirty-three NRCS districts from Northern to Southern California.

"In three years," Ostrander said, "we took a working model and we scaled up to the state." She and Wick were instrumental also in getting a five-pronged statewide Healthy Soils Initiative adopted in 2015, under the authority of the CDFA. The initiative promotes the development of healthy, carbon-rich, sustainable soils on California farms and ranchlands, with an emphasis on increasing their carbon content. It helps farmers and ranchers with funding and promotes composting statewide. Wick and Ostrander, meanwhile, are promoting carbon farming as widely and rapidly as possible, on every scale from the macro to the hyperlocal. "This action starts at home," Ostrander declares during presentations. "It starts with you, it starts with the people you interact with, it starts with your community."

Some California ranchers are already on board with the Marin Carbon Project and are exploring its potential, each in their own way. Loren Poncia, forty-three, is a stocky fourth-generation cattle rancher who owns the thousand-acre Stemple Creek Ranch in Petaluma, California. His parents were conventional ranchers and "price takers," selling their cattle for what little they could get at auction. After Poncia finished his degree in agricultural business and dairy science, he found himself looking for a day job—and ended up working "for the dark side" at Monsanto.

But after more than a decade, he was able to save enough money to buy his parents' cattle, to lease their land, and, beginning in 2014, to ranch full time. He had the ranch certified organic and created his own brand of organic beef and lamb, so he could set his own prices, instead of depending on cattle auctions. His "grass-finished" meats now find their way to the tables of the best restaurants in the San Francisco Bay Area.

Poncia was one of the first ranchers to implement the Marin Carbon Project's carbon farming plan. Stemple Creek Ranch is a carbon farm demonstration site and has been forever protected from development by a local land trust. When I visited in early fall, the ranch's rolling coastal prairie hills were a dry brown, dotted with black cattle and rocky gray outcrops. Deer, wild turkeys, coyotes, badgers, and smaller mammals share the land with the livestock, a poultry-raising operation, and five hundred beehives. Occasionally, a river otter splashes in the creek, where Poncia has planted willow to bring back riparian vegetation.

By focusing on the soil as John Wick and Gabe Brown are doing, Poncia has been able to grow more grass and forbs (any herb other than grass), supporting more cattle. To rejuvenate the soil, he establishes perennial plants that store more carbon by keeping live roots in the soil year-round. Those long roots access moisture inaccessible to annual grasses, keeping his land green year-round—which is important to Poncia, as his ranch typically experiences six months of drought a year. The long, powerful roots also bring valuable micronutrients to the surface and help decompact the soil so air can penetrate and nurture a rich community of other plants and soil microbes. Natural decompaction is particularly important for Poncia: although plowing also decompacts the soil, Poncia doesn't plow; doing so releases carbon into the atmosphere. Instead, he raises and releases worms that obligingly tunnel into the soil and aerate it, providing additional natural decompaction while leaving rich, nutritious castings behind for plants and other soil organisms.

Today, Poncia—who now calls himself "a soil farmer"—has realized his dream of being a rancher, but in a way that benefits the environment, the livestock, the soil, the wildlife, and, ultimately, the consumer, with healthy, nutritious food.

Soil Geotherapy

Whereas Wick, Creque, Poncia, and others with the Marin Carbon Project work to increase the carbon stored in rangeland, a researcher at New Mexico State University, David C. Johnson, is working to bring about a revolutionary increase in soil carbon capture by manipulating soil microbes in cropland. Dr. Johnson is a trim, friendly sixty-seven-year-old agricultural scientist with a ready smile, a pleasant voice, and neatly cut gray hair. He's the kind of guy who's as comfortable using the latest metagenomics technology in a microbiology lab as he is shoveling cow manure into a composter.

Recently, Dr. Johnson had conducted a series of precise soil biology experiments that promise to revolutionize the use of agricultural systems to protect the climate. Paradoxically, his cutting-edge experiments employ an ancient art that has been practiced for thousands of years, from Amazonia to Asia: composting. But whereas Gabe Brown's regenerative agriculture relies on livestock to help restore microbes to the soil, Johnson, who has a PhD in molecular biology, inoculates the soil with microbes in compost, thus making microbial soil therapy available to farmers who don't raise animals. "I am trying to imitate what was happening in the Great Plains for fifty-three million years, with the grazers inoculating the soils as they moved."

Johnson believes that if we managed global agriculture properly, using this unique approach, we could store the entire world's carbon emissions. Experimenting at New Mexico State University's Leyendecker Plant Science Research Center, near Las Cruces, he reports coaxing each acre of specially treated cropland to remove up to sixteen metric tons of CO_2 from the air each year there—roughly ten times the increase other scientists have reported in many different soils and climates, all without chemical fertilizers, pesticides, or herbicides. This is an astonishing amount—and, as we will see, a controversial result. At the removal rate reported, about a quarter of the world's arable cropland, theoretically, could sequester the entire world's annual human-caused carbon emissions, Johnson said.

Johnson doesn't claim to know how you get the farmers who control that farmland to use his approach, but according to his calculations, this acreage could be treated profitably. Each treated acre, he said, would yield net benefits on the order of five hundred to six hundred dollars, when credits were provided for carbon capture and when other benefits (including reduced minimum tillage and irrigation needs and increased soil fertility) were accounted for. There are roughly 4.2 billion acres of cropland worldwide, so if Johnson is right—or at least in the ballpark—this strategy could generate a globally significant reduction in atmospheric CO_2 and would be of worldwide interest.

Johnson ascribes the high carbon capture rates, along with large increases in crop yields, to improved soil health stemming from the application of the microbes from his compost and an ensuing increase in the soil's fungal-to-bacterial ratio. Other scientists have expressed a range of views, from cautious interest to skepticism. Professor Rattan Lal of Ohio State University, widely regarded as a leading authority on soil carbon sequestration, says he was "intrigued" by Johnson's outcome. "I want to understand why he's getting such exceptional results." Lal thinks that further, larger-scale trials are needed to validate Johnson's work. Keith Paustian, a professor of soil and crop sciences at Colorado State University, says he has seen some "quite high rates of carbon accrual" in degraded croplands that were converted into productive perennial grass systems. But he has not seen strong evidence that the same outcome could be produced by adding microbes.

Harold van Es, professor of soil and water management at Cornell University's School of Integrative Plant Science, is one of Johnson's severest critics. Of particular concern to van Es is Johnson's past record of presenting his results in seminars, reports, and other "gray literature," rather than waiting until they were published in peer-reviewed journals. (Johnson has few peer-reviewed studies.) "In science, we strongly believe that research should be subjected to peer evaluation," van Es said. As for Johnson's methods themselves, the fungal-to-bacterial ratio of soil is indeed important, van Es agreed. "But there are many ways to increase that ratio," not just Johnson's approach. "Reducing tillage has similar effects, and this has been much more widely documented."

To arrive at his sensational global carbon capture numbers, Johnson projected results from cropland plots of three to seventy-five acres of various soil types in five states. As large an undertaking as that is, it is still a fairly limited sample. Henry Janzen, a research scientist at Lethbridge Research and Development Centre in Alberta and a professor at the University of Manitoba, cautions that such a projection is risky. "Every ecosystem is unique," he says. "A practice that elicits soil carbon gain at one site may not be effective at another. And always, the rate of carbon gain will depend on a host of interactive factors, including soil properties, previous management practices, climatic conditions and the vagaries of human whims."

Thus, three years into a cotton farm research project in Turkey, near Izmir, it was no surprise that Johnson found that the soil seemed to be capturing about six metric tons of CO_2 per acre per year using a method he calls Biologically Enhanced Agricultural Management (BEAM). But nor is this result discouraging—although less than half the rate of carbon capture reported in his previous experiments, this is still a very significant increase that was also associated with an 85 percent reduction in nitrogen fertilizer use, the elimination of herbicides, and large reductions in the use of pesticides, diesel fuel, and irrigation. Johnson also reported an increase in the soil's photosynthetic capacity, as measured by carbon captured by plants, and improved soil health, as indicated by a reduction in the amount of CO_2 "exhaled" from the soil. Even more important for the farmer was a $360 to $445 additional profit per acre, thanks to a reduction in input costs.

Key to these as-yet-unpublished results was the injection of liquid compost into the soil with each seed and the use of multispecies cover crops in the interval between the commodity crop, along with reduced tillage. "Most farming systems only have one crop per year and fallow their soils for the winter," Johnson said. Notably, these positive experimental results are nonetheless consistent both with Dr. Janzen's critique that different environments, soils, and climates produce different results, and with Dr. van Es's words of caution: because the BEAM approach uses both microbial compost *and* reduced tillage, it's hard to know exactly which is having the greater effect on the soil's fungal-to-bacterial ratio.

Following the Microbes

Through his research, Dr. Johnson is in essence trying to mimic how nature built up the deep, carbon-rich soils of the American prairie. "We're talking six feet or more in the profiles of some of these soils," he said, indicating that the extra carbon had worked its way from the atmosphere where the plant absorbed it down into the deeper layers of the soil. In one experiment, Johnson inoculated the soil at his Leyendecker Farm research site with four hundred pounds per acre of microbe-laden moist vermiculture (worm) compost, created in a unique, low-tech homemade reactor made from wire mesh, landscape cloth, a wooden pallet, and a perforated plastic drain. "This reactor design," Dr. Johnson said, "allows the material to be composted aerobically, allowing complete biological breakdown of compost materials and resulting in a microbially diverse, fungal-dominant compost product." After nine to twelve months in the reactor, the end product is dark, with a claylike consistency, and can be applied as a liquid extract or directly to a field or furrow for seeds or as a seed coating to increase germination. The fungi, bacteria, and protozoa in the compost feed a soil food web of nematodes, micro-arthropods, and other beneficial organisms. Overall, the process increases plant health, growth rates, and crop yields.

In the earlier field experiments to see how much he could increase soil carbon and plant productivity—the trials in which he reportedly hit jackpot levels of stored carbon—Johnson sowed a series of diverse cover crops and after they had grown for a while, he "disked" (harrowed) each of them back into the soil two to four times a year before sowing another crop. He also conducted a series of greenhouse experiments to learn how the increased soil carbon he had been able to get into the soil affected the composition of the soil microbial community. Microbes have a lot of influence over what happens to the carbon—and where it goes after plants remove it from the atmosphere. Some of the carbon fuels the soil's microbial community, which respires like other organisms, releasing CO_2 back into the air. But Johnson has concluded that a lot less microbial respiration needs to occur than previously thought, provided that the fungal-to-bacterial ratio of the soil is tipped more toward fungi. To the extent that the respiration can be reduced, it's a good thing, as "exhaled" carbon doesn't enrich the soil. Soil managers' main goal is to get more carbon stored in a durable form in the soil and incorporated into plants, which invariably improves their health, nutrient status, and yield. As the concentration of soil carbon rises, the availability of soil nitrogen and trace elements also tends to increase. "The [nutrients] exist in the soil parent material," Johnson said, and he has confirmed that the right microbial community composition is required to extract those nutrients from it and make them available to plants. He even knows (through his metagenomic analyses) exactly which microbes are doing that work.

"In a poor soil that's bacterial-dominant and low-carbon, we're only getting about three percent of that carbon flow going to the plant, and you get ninety-seven percent of that flow going toward the soil to improve that soil microbial community," he said. By contrast, in a healthy soil ecosystem, he explained, up to 56 percent of the carbon flow that the plants get from photosynthesis actually goes into the plant. That's more than *five times* the amount that's incorporated in plants in a typical agricultural system reliant on chemical fertilizers, herbicides, and pesticides. For all these reasons, Johnson says, the higher the ratio of soil fungi to bacteria, the better. "There's a lot of interactions between plants and fungi," he explained. "The management technique that we've adopted in agriculture over the past ninety-some-odd years destroys the fungal community."

Fungi are an integral part of the soil and, thus, of the entire agricultural system. They handle the logistics of supplying goods and services to the plant; in addition to handling the transportation of nutrients and water, they also take care of communication between the plant and other microbes. "They transit materials both ways," Johnson said. "You actually see microbes moving up and down the fungal hyphae.... The fungal-to-bacterial ratio is your thirty-thousand-foot view. It's important. I won't say it's an end-all, be-all ... but if you don't have that fungal community, then you don't have this interrelationship between the other microbes and the soil and the plant.... There are some other symbionts that work directly with the root," he said, "but a lot of this goes through the fungal community, especially your mycorrhizal fungi." The mycorrhizae produce filaments known as hyphae that envelop the plant root, sometimes penetrating it, and thereby furnish the plant with nutrients and water. In return, the plants provide the fungi with carbohydrates from photosynthesis. "I've seen a quintupling in production of plant material," he said, "just by improving the soil—improving the amount of carbon and changing it from bacterial dominant to fungal dominant."

Other researchers are documenting similar processes on rangeland, Johnson asserted. "It needs to be stressed that this is a *system* we're restoring," he noted. "We can't just throw in those certain fungi; they depend on a lot of other players as well. It's just like us in this society. Everybody has a different talent or forte, and we depend on everybody in this society to work in the best way, and it's the same thing in the soil."

Johnson is proud of another important research finding he's made, pertaining to how efficiently carbon is used in the soil. It, too, is a challenge to conventional wisdom: "Every scientist is going to tell you that they don't care how much carbon you can put into the soil.... The more carbon you put in, the more microbes you have in there, the more respiration you're going to have," the thinking goes. If that were true, he said, it would mean that instead of being stored in the soil, the increased carbon would wind up being quickly recycled back into the atmosphere as carbon dioxide. What Johnson has found instead is that as soil health increases, with increased carbon content and a shift toward fungi over bacteria, the amount of carbon respired into the atmosphere drops by 80 percent. "The healthy [microbial community] ... puts more of the carbon into new microbes or new biomass instead of respiring it," he said. That's a big win for the climate.

The Practicalities

Questions remain about how Dr. Johnson's research could be scaled up, how it compares to other forms of "geotherapy," and to what extent it can be replicated elsewhere. The work of the Marin Carbon Project still has similar issues to explore. How does the amount of carbon stored through composting and grazing compare with the carbon and methane released by producing and transporting the compost and by the methane emissions of the livestock doing the managed grazing? Could enough compost be made readily and widely available internationally, over millions of acres? What would that program cost? How long would it take? How much energy would be required? Can the results obtained in California and New Mexico be replicated elsewhere? Can the newly stored soil carbon be protected in the long term, so it doesn't later return to the atmosphere and nullify the benefits gained? Perhaps most significantly, who would plan, manage, and implement the kind of international program that would be needed, and how would farmers and ranchers the world over be compensated for their climate-saving work?

Where there's a will, there's a way, and these questions can be answered, but clearly a great deal of work remains to be done before a vast national or international rangeland composting program could be successfully mounted. It does seem clear right now, however, even without a lot of head-scratching, that a global system of carbon credits could supercharge an international network of carbon farmers and ranchers. If they knew that by following a verifiable carbon-storage protocol, they could promptly get a generous check to pay for groceries or new farm equipment, the program would likely become oversubscribed as fast as news of it spread.

Assuming sufficient quantities of compost, the costs of a massive national or global composting effort would not be trivial. Nor would the logistics be inconsiderable: much of the world's rangeland is remote and inaccessible, and many countries are struggling with poverty and conflicts. If each country were expected to fund its own program, many would require financial aid. Regrettably, the global community of developed nations has so far been remarkably tight-fisted in supporting small-scale farms (which represent 84 percent of all the world's farmers) to practice carbon farming. Yet because the farmers who work these farms are often extremely poor, it probably wouldn't take a lot of money to incentivize them to adopt carbon farming techniques. It could prove to be a very cost-effective way to help stabilize the climate.

At some point, the developed world will realize it has little choice but to provide much more adequate financial support to other nations to protect the climate by every available, cost-effective means. Over the long term, the cost of catastrophic climate change is likely to be far greater than the cost of mitigating it in advance. A recent estimate by the Intergovernmental Panel on Climate Change put the cost of climate disruption at $69 trillion, and that could well turn out to be an underestimate; the estimated cost has been climbing steeply over the years. Moreover, the costs of palliative climate adaptation (where even achievable) to fend off extreme weather, ecosystem carnage, or sea level rise, are often far greater than timely climate change mitigation. And, on the spectrum of climate change–mitigation measures, carbon ranching and carbon farming are likely to be among the most cost-effective known to humans—and nature.

As noted earlier, the world has far more rangeland than cropland, so if carbon ranching were employed on a significant fraction of rangeland, even less cropland would need to be managed for carbon storage. The good news is that useful and profitable crops and livestock can still be successfully raised on agricultural lands and rangelands while they are simultaneously managed for carbon capture.

Could the world afford to implement a system of carbon capture like Dr. Johnson's on cropland? Johnson's calculations suggest that at a carbon price as low as $20 per metric ton (far less than the social cost of carbon), the cost of incentivizing the capture of 37 billion tons—current global industrial carbon dioxide emissions—would be $740 billion, spread over years to decades.* For comparison, world GDP in 2021 was $84 trillion, so Johnson's figure is less than 1 percent of a single year's global GDP—an incredible bargain, considering what's at stake.

Rewarding Farmers in the Developing World— and Beyond

Could farmers and herders, even in remote areas, be expected to participate in recarbonizing the soil? When I asked Dr. Johnson, he replied that he had spoken with a lot of people in the southeastern United States who had already adopted regenerative herding systems on rangeland (such as his "multi-paddock grazing management system"). "They say they will never go back to the way they did it before. This makes so much more sense to them: It's more profitable. They don't need any equipment. They let the cows—or whatever they're grazing—do the work for them." This system will work at any level, from developing world to developed world, for both grazing and farming, according to Johnson. "You get into Third World countries—they can't afford the fertilizers. But they can afford to do something along this line."

Even at a very modest carbon offset price, which could be provided by a carbon offset payment, farmers who copied Johnson's methods could do so very profitably. All it would take to create demand for the carbon credits would be a national carbon fee.** Not only would

---

* *Total* global GHG emissions from all human activities reached an all-time high of 58 billion metric tons in 2022 and are projected to reach 62 billion metric tons by 2030, according to Homi Kharas et al., "Tracking Emissions by Country and Sector," Brookings Institution, November 29, 2022, https://www.brookings.edu/blog/future-development/2022/11/29/tracking-emissions-by-country-and-sector/.

** Dr. Johnson lobbied for farmers to receive payments or tax credits for capturing carbon in their fields under the Obama administration's Clean Power Plan, but was unsuccessful. (Ultimately, the Obama plan was invalidated in 2016 by the U.S. Supreme Court.) Ironically, a perverse recent change in the tax code known as the 45Q tax credit gives oil companies and electricity producers fifty dollars per metric ton in direct tax write-offs for any carbon dioxide they place in long-term geological storage, and thirty-five dollars per metric ton pumped into the ground to enhance the rate of oil recovery, even though using carbon dioxide to produce more fossil fuel just makes global warming worse. See *Credit for Carbon Oxide Sequestration*, U.S. Code Title 26 (2018), §45Q.

farmers using his system be removing atmospheric carbon, but they would also be packing their soil with valuable nitrogen, phosphorus, and potassium, valued at hundreds of dollars an acre. Farmers and agribusiness would also benefit through savings on reduced tillage, reduced irrigation, and avoided fertilizers and other chemical inputs. Monetizing all the benefits of carbon farming to the soil, the public, and the farmer could do wonders for the United States' rural economy. If farmers were properly rewarded, they could do well by doing good, and the increased cash in their pockets would help rebuild struggling small towns throughout America's farmland.

While we're at it, why not create a national market for *nitrogen* captured, similar to the carbon offset market? Nitrous oxide is a powerful greenhouse gas that has more than three hundred times the global-warming impact of carbon dioxide over a one-hundred-year period and also destroys stratospheric ozone. Both carbon and nitrogen emissions could be controlled in a sensible cap-and-trade system or under a fee-based GHG-management system. If, along with today's nascent carbon credit market, we also created a nitrous oxide market, with penalties for the gas's discharge and rewards for its capture, then current market externalities—damages and benefits to the soil and climate—could be brought into the market and controlled more economically and efficiently. The same could be done for methane, the third major climate-disturbing gas—whose global-warming impact is eighty-two times greater than that for carbon dioxide over a ten-year period.

✣ ✣ ✣

Clearly, when it comes to representing the full cost of GHG emissions in our economy, we need to do better—and later in the book, I will provide further ideas about what we can do legislatively to encourage carbon trading, using fees and credits to encourage cuts in emissions while rewarding those who take out of the atmosphere carbon that has already been emitted. The next chapter, however, focuses on people who are laboring to make sure that carbon that has already been removed and stored, for centuries or more, in forests isn't wantonly or brazenly disgorged into the atmosphere.

13.

Forest Savers and Fixers

Apart from carbon farming and ranching, there's another clever, inexpensive way of removing atmospheric carbon, all without new technology. It relies on a durable, complex, self-replicating device that lasts for hundreds of years and requires zero maintenance. If we had invented it yesterday, it would be a big "Wow!" on the front page of every major newspaper. Actually, though, it's not exactly hot news. In fact, it's been around for 360 million years and, unless you live in the polar regions or on the tundra or steppes, you've seen it all your life.

It's called—you guessed it—a tree. And it's been patiently bioengineered by nature through evolution to flourish and do its job under myriad environmental conditions. What precisely is that job? Well, a tree provides many useful ecosystem services, but one of its primary functions is to perform photosynthesis, combining water and carbon dioxide from the air to make oxygen and carbohydrates. This it does remarkably well.

In a forest community, trees—pillars of carbon—can hold the equivalent of up to seven hundred metric tons of carbon dioxide per acre in their trunks, roots, and branches, with even more carbon stored in the soil beneath them. Moreover, they can store it for hundreds or even thousands of years. Protecting these carbon warehouses (and growing more) is among the least costly and safest ways of cleansing the atmosphere.

Danna Smith, a feisty, outspoken forest protection activist, believes that for only a few dollars per ton of carbon stored, forests can be a major force for climate protection and carbon removal. They represent, she says, our best means for removing and storing carbon. "We know they work—and at a global scale. [They also] give us many gifts: beauty, clean water, fresh air, natural flood control, and air-conditioning." But in the southeastern United States alone, we annually lose 5 million acres of forest to logging, according to Smith. That total rises to 11 million acres nationwide—and the nation suffers serious consequences. "Many people don't readily understand that the loss of forest cover to logging has huge climate impacts," Smith said. In addition to forests damaged by logging, 1.5 million acres of forest are converted to other uses each year, and their climate-protection value is therefore likewise compromised or lost.

Fighting for Forests

During her career, Smith has responded to forest carnage in the United States and abroad by leading multiple campaigns to obtain precedent-setting commitments from some of the world's largest paper producers and corporate consumers to buy sustainable wood fiber and paper. She and the organization she founded also got these players to commit to increasing the recycled-fiber content of their paper and to bringing more forests into compliance with high, sustainable forest management standards established by the Forest Stewardship Council.

When Smith began her forest protection efforts in 1994, hundreds of thousands of acres of natural southern hardwood and pine forests were being converted into industrial pine plantations every year; logging rates were four times that of the heavily logged South American rain forests. All that logging in the United States and abroad released millions of tons of stored forest carbon and obliterated diverse native wildlife.* In the United States, chip mills, which prepare trees for pulping, were popping up across the South as part of a paper industry expansion.

While still a young woman in the 1980s, Smith already knew that her calling was to protect the wild forests of the Deep South, where she lived. But to do that, she needed to find a way to stand up to a timber industry that was providing jobs and tax revenue to local communities in or near those wildlands. That was a colossal challenge.

The roots of Danna Smith's activism lie deep in the marshy soil of lowland South Carolina. She grew up "a free-range wild child" on an islet near sleepy, scenic Daufuskie Island, off the South Carolina coast near Hilton Head. Accessible only by boat and festooned with ancient oaks draped in Spanish moss, her home was a balmy childhood paradise nestled into the floodplain of the Savannah River, close to a wildlife refuge. "We lived by the rhythm of the tides there," Smith said. "I roamed the pine and palmetto forests." She remembers the feel of pine needles under her bare feet, the pungent smell of marsh, and the wings of a butterfly lit by the setting sun as she picked blackberries. Those gentle surroundings forged her steely commitment to protect southeastern forests.

Smith's parents helped foster her love of nature. Her mother was an environmental reporter and self-taught naturalist; her father, an engineer and avid outdoorsman. He often took Danna hunting and fishing and gave her a double-barreled shotgun and a baby raccoon as a pet on her tenth birthday. His day job, however, was to oversee pipe fitting on paper mills and power plants. (By Smith's account,

---

* Cutting down a long-established natural forest and replacing it with rows of evenly spaced, rapidly growing pine trees releases carbon into the atmosphere in several ways. The physical disturbance of the forest soil that occurs during logging brings about an exhalation of stored carbon; so does the use of herbicides and the destruction of the forest understory to prepare the ground for rows of seedlings. Merchantable trees that once took large amounts of carbon from the air are turned into lumber for construction, or even into wood pellets to be burned in stoves or power plant furnaces. The new seedlings that are planted, though they grow rapidly, take decades or longer before they are capable of storing as much carbon as the large trees that were logged. Lumber used in construction also will ultimately relinquish its carbon to the air when the building is torn down; the average building has a life span measured in decades, rather than the centuries that old trees stand in a forest.

in his later years he recognized the environmental impact of these industries and supported her work.)

Although Smith is a lawyer and once had a law practice specializing in construction contracts, she found the career unfulfilling and, so, went to work as a campaigner for Greenpeace in Atlanta, Georgia. When the Greenpeace office closed, she participated in a Rainforest Action Network demonstration. It was March 4, 1996, and RAN was protesting the arrival in Savannah's harbor of a ship carrying a cargo of illegally logged mahogany from Brazil. Smith boarded the ship and tied herself to a railing. Then, with another person, she rappelled over the side, unfurling a banner that read, "Ban Mahogany Imports, Save the Amazon." For six hours, she hung suspended there while news helicopters filmed the event for broadcast and Coast Guard vessels zipped back and forth, trying to figure out how to get the demonstrators down. All the while, Smith's view from the ship was of the forests across the Savannah River, where she grew up.

RAN's demand to meet with the Georgia Port Authority was ultimately met. As Smith walked off the ship, she felt a close connection between the forests of Brazil and the South Carolina forests of her youth. At that moment, she made a commitment to spend the rest of her life protecting the forests of the southeastern United States. Today, she directs the nonprofit Dogwood Alliance, which she cofounded in Asheville, North Carolina, shortly after the RAN demonstration. Its goal from the outset has been to save the forests and communities of the Southeast from industrial logging.

This has proven to be a difficult task: In one of Dogwood's earliest efforts to block the construction of a new wood chip mill, Smith went alone to the small town of Laurens, South Carolina, where she gave a presentation at the local library on the impacts of wood-chipping facilities on forests and communities. Before going, she had done her homework and had written a book-length report called *Chipping Forests and Jobs*. This didn't impress the community, however. Local landowners, foresters, and loggers in the audience yelled at her during her talk, until she broke down in tears. She was still crying when a policeman pulled her over for speeding as she drove home.

"What's wrong?" he asked.

"I'm just trying to save some forests, and they're yelling at me," she told the officer.

"Slow down," he said. "Everything's going to be all right."

In retrospect, the trauma of the Laurens presentation proved important to Smith's development as an activist. She realized that she should have tried to engage the community in a dialogue, rather than coming in as an outside expert. The ordeal, she said many years later, "only made me a stronger advocate and campaigner."

A New Threat to U.S. Forests Arises

Smith's goal of stopping forests from being pulped has evolved over time into a broader effort to put forest protection and restoration high on the list of potential solutions to climate change. Yet, even as the world's climate crisis has deepened, new forest threats have blossomed, requiring new tactics. Over the past several years, for example, a new market has emerged that threatens southern U.S. forests: European utility companies are shifting from burning coal to burning wood in their power plants—a move many environmentalists initially supported, because wood is a renewable energy source. Just in the past five years, the U.S. South has become the world's largest producer of wood pellets to feed this rapidly expanding new market.

Dogwood Alliance responded by incorporating the latest scientific findings and expanding its work beyond the United States, with Smith creating an international coalition of organizations aimed at halting the burning of wood pellets in power plants. That coalition has helped persuade the United Kingdom and other European countries to stop subsidizing imported wood as fuel. Although this may, at first, seem counterintuitive, burning wood releases *50 percent more* carbon than coal per unit of electricity.

Smith and her colleagues have also assembled a coalition of twenty-five organizations committed to protecting wetland forests, and some years ago, she played a crucial role in exposing wood pellet companies that claimed to be using wood waste for pellet fuel while actually logging mature wetland forests. More recently, Dogwood Alliance has been instrumental in mobilizing two hundred organizations, scientists, and elected officials in support of a "Stand4Forests" declaration. "Climate science shows that we cannot stop a climate catastrophe without scaling up [forest] protection," the declaration states.

Forests' Carbon Capture Potential

A pivotal step in the campaign against wood pellet burning has been outreach to the European Union's government. Danna Smith and climate scientist Dr. William Moomaw held a joint briefing for members of the EU Parliament on the forest and climate impacts of the wood pellet industry. Moomaw is a physical chemist and professor emeritus of environmental policy at Tufts University. He specializes in climate change mitigation and has been a lead author of the IPCC's prestigious global climate assessment reports.

Together, Moomaw and Smith wrote a report titled *The Great American Stand: U.S. Forests and the Climate Emergency*. The report concludes that we could reduce annual carbon dioxide emissions by 75 percent in the next half century using natural forests. "A number of years ago," Moomaw told me, "I began to realize that even if we stop putting carbon dioxide into the atmosphere, we'll still be at a much higher concentration than we would want to be." At that point, he began to explore possibilities for removing the excess atmospheric carbon dioxide and was impressed by what he learned living systems could do.

Until fairly recently, most mainstream scientists had generally assumed that there was no practical (read: cost-effective) way to use forests for this purpose on a large enough scale. Moomaw now disagrees. "If we allow previously cut forests to grow to their full size," he noted, "we could probably remove about two billion tons of carbon each year from the atmosphere. . . . If we were to stop land use change worldwide—meaning the conversion of forest lands into agriculture and urban areas—that's another 1.3 billion metric tons." In addition, deforestation results in the release of another 3.1 billion metric tons of carbon annually.

Currently, we are emitting about 9.4 billion metric tons of carbon a year globally from fossil fuels and industry. "Yet," said Moomaw, "concentrations [of carbon] in the atmosphere increase by just 4.7 billion metric tons a year—because forests, wetlands and soils, and the oceans remove and sequester the difference." So, if we just reduced our annual fossil fuel emissions by half, to 4.7 billion metric tons, that alone would eliminate any net annual increase in atmospheric carbon, according to Moomaw. In short, on an annual basis, we would be officially carbon neutral planetwide!

Almost the same result could be had if we entirely halted deforestation and land use change, which together put about 4.4 billion metric tons of carbon into the air each year. "Halting all deforestation [and land use change] is equivalent to reducing our fossil fuel emissions by half," Moomaw noted. If we did both—namely, slashed fossil fuel emissions in half and *also* halted deforestation and land use change—then "that would lead to annual reductions in atmospheric concentrations of carbon dioxide of about 4.6 billion metric tons per year," he said.

Moomaw thinks that, realistically, we can remove at least three but possibly as much as four billion metric tons of carbon a year from the atmosphere through forests alone. Nowadays, however, U.S. forests hold only 10 to 50 percent of the carbon they're capable of storing. Moomaw's conclusion is that protecting and restoring forests, along with other ecosystems, is the only way we can begin getting excess carbon out of the atmosphere and cooling the planet "on any reasonable timescale, like the next fifty to a hundred years."

Battling the Chip Mills

Initially, back in the late 1990s, Danna Smith and her colleagues called for a moratorium on all new chip mill permits in the Southeast until state and federal agencies had completed comprehensive studies of their ecological and economic impacts. They believed, perhaps naïvely, that if they got the threats to southeastern forests properly documented and widely publicized, then the powers that be would adopt forest protection policies and perhaps even needed legislation.

Dogwood Alliance began doing some basic research of its own to get an accurate picture of logging regionwide. The alliance then managed to get the federal government to do the first comprehensive multi-agency study of forest sustainability for the southern United States and also persuaded the governor of North Carolina to commission a similar study, by North Carolina State University, on the economic and ecological impacts of the chip mill industry. The alliance provided the state researchers with assistance in scoping the study.

Public hearings were also held around the state to solicit input on the state's study. With high hopes, the Dogwood Alliance "put lots of time and energy into educating people, [and] getting people to show at the hearings." To the activists' dismay, however, when the government studies came out, they asserted that the biggest problem facing North Carolina's forests was urban sprawl. "It was extremely frustrating to us," Smith said, "because there was plenty to talk about from the findings in those reports about the impacts of the logging, but it felt like the government politics were really, like, sweeping the impacts under the rug and pointing the finger elsewhere, to development."

Because influencing government now seemed like an almost insurmountable challenge, she began looking for new tactics. "That's when we started to think about the market-based approaches to campaigns, and that's when I started connecting with folks like the Rainforest Action Network again"—as it was running campaigns to put public pressure on the timber industry and consumers.

Out of the discussion Dogwood held with the Rainforest Action Network came a market-based campaign approach focused on the paper industry. Dogwood developed a "trickle-down" strategy to pressure large U.S. producers and consumers of paper with consumer-facing brands to adopt higher environmental standards for their paper production and purchasing. "If we can get large customers to agree to set high standards for their paper, and then hold their suppliers accountable to them," Smith reasoned, "then their suppliers [will] have to change the way they're sourcing wood on the ground, and that means that private landowners will have market incentives to improve their forestry practices."

To make sure that this requirement was met, Dogwood asked paper suppliers to buy only Forest Stewardship Council–certified wood. (The council is widely recognized as the most reputable certifier of sustainable forestry practices.) Dogwood soon targeted Staples—which, by 1996, had become a three-billion-dollar office supply company—as the focus of a grassroots campaign.

Confrontation or Collaboration?

It was a classic David-versus-Goliath struggle: Dogwood was then a small grassroots group of seven or eight people; it had a budget of only half a million dollars. Staples was a Fortune 500 company. Smith knew, however, that Staples was very concerned about protecting its brand's positive image, and she sensed that the company would also be eager to show that it was taking meaningful steps to help protect the environment. "It really is a brand issue more than it is a dollar-for-dollar profit issue," Smith surmised. So, Dogwood went to work, and over the next two years, it and its partner ForestEthics (now Stand.earth) organized more than 620 demonstrations, picketing in front of Staples stores across the South. "We were in the national media, holding Staples accountable for essentially destroying forests in the southeastern U.S. for its paper."

Staples eventually came to the table, in 2002, and made a commitment not to buy paper from endangered forests and to buy only Forest Stewardship Council (FSC)–certified paper or recycled paper. "To our credit as a company," said Mark Buckley, Staples's environmental vice president, "we met with Danna and others and started to develop a better understanding of the impacts of our supply chain."

But that wasn't the end of the struggle. According to Smith, most big paper companies—like Weyerhaeuser, International Paper, and Georgia-Pacific, were involved in large-scale clear-cutting and conversion of forests to plantations. "The FSC-certified fiber sourcing was a bottleneck in the South," Smith said, "and it was really challenging to figure out how to get landowners to become FSC-certified." Little FSC-certified paper was available in the southeastern United States, because few landowners were then able to meet the FSC's rigorous environmental standards. Most of their holdings were tiny. Unaided, these landowners couldn't afford the costs of certification—nor, more important, the forgone income occasioned by leaving more trees standing during logging operations. So, where was Staples going to get its paper?

Dogwood now recognized that Staples and other big paper companies faced a real problem. The grassroots group had a decision to make: "Are we the arm that's just about pushing companies, or are we going to be really trying to work cooperatively here? ... If we are willing to [take them to task] and call a company out," Smith decided, "at the end of the day, we've got to be willing to roll up our sleeves and try to be a partner in solving some of these problems." At this point, she and Mark Buckley both sensed an opportunity to tie forest conservation to climate protection. Smith's goal was to show how improved forest management could reduce carbon emissions. She and Staples embarked on what became known as the Carbon Canopy Project.

Dogwood thus had to figure out a way to get money to forest landowners to enable them to get their forests FSC-certified and keep them that way, year after year. If landowners were adequately remunerated for the carbon the trees they left standing would remove from the atmosphere, that would be an exciting, triple crown victory for the landowners, the paper companies, and the climate. At that point, Dogwood took a major step toward solving the problem by recruiting as a partner the Forestland Group (TFG), a timber management organization and one of the largest industrial forest landowners in the United States. Together, they structured a pilot project to manage a 9,000-acre tract of TFG land in southwestern Virginia for the next one hundred years to increase its carbon storage.*

The Carbon Canopy Project is no longer ongoing—the price of carbon credits was insufficient to make participation attractive to most owners of small-scale southern forestland. Dogwood Alliance, however, is continuing to work with big supplier companies to improve their procurement policies. As a result of the progress made with Staples, Dogwood was able to create a domino effect among large corporate paper consumers, obtaining similar procurement policies from Office Depot, and then Office Max, and then FedEx, along with some big consumer product companies like GlaxoSmithKline and Johnson and Johnson, which use lots of paper for packaging. Looking back on the campaign, Staples's Mark Buckley said, "Danna has been a really wonderful partner. . . . It was a really great collaborative effort that helped us to build the first paper procurement policy in the industry."

"We came away [from the Carbon Canopy Project]," Smith said, with the realization "that while carbon markets could be a real solution, the current regulatory and policy environment and market conditions did not exist for it to be able to scale." She concluded, therefore, that a voluntary approach would never work to satisfac-

---

* As a direct result of this pilot project, TFG went on to enroll an additional 240,000 acres of its forestland for sustainable management, bringing the total to nearly a quarter million acres. According to Dr. Moomaw's calculations, protecting close to 250,000 forest acres will keep roughly an additional 30 million extra tons of carbon dioxide from the atmosphere (the amount that would have been released if the forests had been logged, the logs turned into wood pellets, and the pellets burned for heat or power). While 250,000 acres is a large amount of land, it is still but a small fraction of the 200 million acres of forestland in the U.S. South under threat from industrial-scale logging.

torily protect forests and the climate. Nonetheless, her work with Dogwood Alliance directly impacted paper producers serving a third of the U.S. paper market and has, thus, kept millions of tons of carbon dioxide out of the air.

Going Global to Stop the Wood Pellet Scam

Unfortunately, the beginning of widespread change in the American paper industry was soon followed by a rise in international demand for wood pellet fuels from U.S. forests—a climate problem disguised, ironically, as a climate solution. In 2013, when Dogwood decided to take on the wood pellet industry, its first impulse was to do what it knew how to do and target the biggest players in the marketplace. Those turned out to be Enviva, the biggest wood pellet exporter from the southern United States, and its large UK utility customer Drax. However, soon after meeting with both firms, Smith understood that it was actually government policy that was driving the utility company to convert from coal to wood: "They were getting huge subsidies to do it," she said, "and they were trying to meet regulations for reducing carbon emissions."

EU government policy was increasingly calling for the burning of trees for "carbon-neutral" electricity and was providing vast amounts of subsidies to companies that were required to reduce their carbon emissions by switching to pellet fuels. "So, we are seeing this really crazy juxtaposition," Smith said, in which the forest solution to climate change is letting trees grow, while "the forest-destroying wood pellet market is being propped up by lots of government regulations and subsidies and is, therefore, spreading like wildfire across the Southeast. Forests' carbon simply should not be used for offsetting fossil fuel emissions and credited with reducing them."

Smith and her colleagues understood, therefore, that a market-based strategy alone would not suffice against the wood pellet industry. Difficult as it was to battle national governments, Dogwood and its allies realized that they would now have to somehow influence national policy at an international level. This insight prompted Dogwood to seek European partners in their struggle to get government to recognize both the emissions caused by industrial-scale logging and the climate benefits of healthy forests.

Smith also was looking for what she called "the smoking gun." She wanted to show that the wood pellet industry was not using only wood waste, as companies contended, but was, in fact, logging old, mature wetland forests along rivers. Dogwood launched its own investigation, following logging trucks from Enviva's wood pellet facilities to harvesting sites. The day in 2013 when it began its campaign, it launched with a front-page story in the *Wall Street Journal* exposing how Europe was now burning wetland forests in North Carolina as a "renewable-fuel climate solution." "This was huge," Smith said. The story received major international media coverage in Europe and, as a result, the United Kingdom commissioned an investigation. More recently, the British government announced that it would no longer subsidize any power plant conversion to biomass burning if the plant used imported wood. Belgium has issued a similar announcement, and the Netherlands has adopted some restrictions.

Meanwhile, in the United States, Dogwood is again working with local communities to oppose pellet mills and has been successful at stopping or delaying several. But Dogwood has thus far not succeeded in getting the European Union to adopt a policy similar to the United Kingdom's restraints on subsidizing wood burning; it currently still treats the burning of wood for electricity as if it were truly carbon neutral. Moreover, despite opposition from Dogwood and its partners, in a recent decision, the European Union opted to extend that policy from its current 2020 expiration to 2030. "The EU vote hit hard," Smith said, "and I think we all felt it."

The international coalition against wood pellet burning that Smith instigated is still intact, and it has convinced many organizations that using wood as a substitute for coal is not a good idea. "We really

fundamentally shifted the dynamic within the environment movement as a whole [on this issue]," she said. Dogwood Alliance has
also gathered strong scientific support. As a result of the coalition's
work, "over eight hundred scientists wrote a letter to the EU government last year urging them to change their policy," she reported.
Furthermore, the alliance now has a growing worldwide coalition of
organizations resisting the burning of wood for electricity and has
made progress at the EU member state level in a couple of countries
that had been major consumers of imported wood from the American South.

Dogwood Alliance is also expanding its network into Japan and
elsewhere in Asia as nations there begin to convert coal plants into
wood ones, creating new forest threats. Meanwhile, Smith has come
to realize that to expand forest protection to the scale of conservation needed in the twenty-first century, activists must now focus on
linking forest protection to the cause of reducing emissions from
forest logging or loss. This may create opportunities to include
carbon credits as a new market for protecting forests and the carbon
they store. "We will not solve the climate crisis without a massive
scale-up in the protection of forests," Smith contends. The conversation about climate change in the United States, she argues, has
been almost exclusively focused on getting off fossil fuels. "We need
to put forest protection [on a par with] moving to solar and wind
power as a national climate imperative."

Currently, Dogwood Alliance is working to defend forests not
only with the international coalition against wood pellets but with
two dozen other groups in a relatively new and proactive Wetland
Forest Initiative. The aim is to protect 35 million acres of wetland
forests rich in stored carbon through lobbying, fund-raising, and
education.

The Private Forest Protection Bonanza

The Pacific Forest Trust's Laurie Wayburn also is wrangling with some of the regulatory and policy issues that concern Danna Smith.* "All forests work," Wayburn told me, "but only some forests get paid." Wayburn and cofounder Connie Best set up the Pacific Forest Trust in 1993 with the intention of helping to protect forests by seeing that more of them "got paid." And thanks to PFT, more of them now do. Although it is a lean organization with an annual budget of just under $4 million, PFT nonetheless is an environmental powerhouse that has protected more than three hundred thousand acres of forestland valued at between $750 million and $900 million.**

As long as forests were undervalued for the public benefits they provided (in particular, their climate benefits and carbon storage), Wayburn and Best knew that private landowners—who control four out of five U.S. forest acres—would continue having a strong incentive to manage the land to maximize their profits. Therefore, they would tend to value it mostly for its timber and development rights and would often clear-cut their property and then manage it as a tree farm, destroying the forest and releasing its carbon. From a climate-protection standpoint, this is a huge problem for the United States and other nations.

Wayburn knew that forests were more than just a collection of "sticks in the ground." She understood that simplifying a forest by turning it into a single-species, single-age tree plantation not only released carbon dioxide but also degraded the forest, robbing it of its ability to perform its "watershed functions," such as storing water, reducing the intensity of floods, mitigating droughts, and sup-

---

* Through the organization she cofounded, the Pacific Forest Trust, Wayburn was a critical partner in the nine-thousand-acre pilot carbon storage project in Virginia that Dogwood Alliance developed with the Forestland Group. She and her team developed a rigorous certification system to ensure that any additional carbon storage ascribed to the Dogwood-TFG project would be real and verifiable; the resulting project qualified for FSC certification and carbon credits.

** Such valuations fluctuate according to market conditions.

porting wildlife. As Wayburn points out, "Well-managed forests are the world's most effective carbon sinks." Conversely, deforestation and land use changes currently account for 20 to 25 percent of all human GHG emissions, according to Wayburn—second only to the emissions from fossil fuels.

Laurie Wayburn is the daughter of the late and storied medical doctor and conservationist Dr. Edgar Wayburn and his wife, Peggy, also an avid conservationist and an outdoorswoman. Elected president of the Sierra Club five times in the 1960s and also its longtime director, Dr. Wayburn, who died in 2010 at 103, was instrumental in the establishment of Redwood National Park and the Golden Gate National Recreation Area and the expansion of Point Reyes National Seashore and Mount Tamalpais State Park, all in California. Most notably perhaps, he was, in Laurie's words, the "tip of the spear" for the effort that protected one hundred million acres of Alaska through the passage, in 1980, of the Alaska National Interest Lands Conservation Act, and he was awarded the Presidential Medal of Freedom in 1999 for his conservation work.

In the 1960s, while Dr. Wayburn was at the helm of a tense campaign to establish Redwood National Park, Laurie, then a child, would sometimes answer the family's home phone to angry calls from people bitterly opposed to the park and making violent threats against the family. The raw emotion the calls evoked left a mark on her and later led her to realize that additional ways to protect forests were needed.

As she grew older, she became interested in finding a way to enable private landowners to hold on to and conserve their forests profitably, without having to set the land aside as parks. The way to do this was not obvious to her at first, however, and so, following her graduation from college, she went overseas for a decade of international environmental work on behalf of the United Nations Environment Programme (UNEP) and the United Nations Educational, Scientific and Cultural Organization (UNESCO). During those years, she saw how private land management and conservation could coexist—and, in fact, be synergistic—in places like Kenya and Uruguay, and she surmised that this approach might work also in the United States.

An ah-ha moment on applying the approach in the United States arrived when she first recognized that if the Pacific Forest Trust got verifiable forest carbon offsets included in California's climate legislation, these would have value in the marketplace and could be used as a tool to make forest protection and sustainable management economically profitable for private forest owners. To accomplish this, Wayburn and PFT proposed an independently verifiable process through which forest owners received payment for the additional carbon they stored on their land through conservation management that led to older, more natural forests. They reasoned that this would reduce clear-cutting and the conversion of forest land to other uses, thereby helping protect the climate while preserving the many public benefits healthy forests provided.

In 2002—after years of hard work that started in 1993 and involved expert consultants—Wayburn and her team at last secured passage of California Senate Bill 812, which, for the first time, included forests under California climate law. SB 812 allowed private forest landowners in California to join a unique and powerful emission-reduction certification system that enabled those who qualified to earn and receive marketable carbon credits for reducing carbon emissions.

Then, working with the California Climate Action Registry (now the Climate Action Reserve), PFT, with the advice of many scientists and forestry experts, led a broad coalition of stakeholders in the development of what came to be known as the California Forest Protocol. The protocol represented the first state standards for forest-generated carbon dioxide reductions—mandating the use of standardized, transparent accounting practices so third parties can verify stated gains in net forest carbon against forest baseline conditions and prevailing industrial forest management practices—and required that participating landowners enhance and maintain native forests to increase their carbon storage. Subsequently updated and improved over time, the protocol became part of the regulatory framework under which the California Air Resources Board, or CARB, today manages forest carbon offsets.

Thus, any forest owner who today enrolls in the program can earn and receive certified and marketable forest carbon offsets. And anyone who wishes to reduce their carbon footprint—or is under a regulatory requirement to do so—can now purchase a forest carbon offset. Each credit is good for the emission of one metric ton of carbon dioxide. Moreover, the stored carbon offsetting the emission must remain stored for a minimum of one hundred years.

At PFT's urging, forest carbon offsets were also included under the state's Global Warming Solutions Act of 2006 (AB 32), which required the state to reduce its GHG emissions to 1990 levels by 2020.* Holders of forest carbon credits can now use those credits under the act to help fulfill their regulatory burdens to reduce GHG emissions. (Would that the protection of forest carbon did not ironically require sanctioning carbon emissions elsewhere, but no other practical means seemed available for financing forest conservation wholesale.)

Together, the SB 812 program and AB 32 laid the groundwork for an ongoing multibillion-dollar U.S. forest carbon market. Today, more than five million acres of forest in twenty-nine states are enrolled in programs collaborating with the California system and are thereby protected by the availability of forest carbon credits. Forest owners receive modest-to-significant economic incentives in return for managing their forests sustainably to increase their stored carbon over a one-hundred-year period. Carbon offset credits sold against the growing stock of carbon pay the owner for the time it takes for the forests to grow old and become carbon rich again. The credits thus create strong financial incentives for forest owners to manage their lands according to sound conservation principles that result in ever more absorbed carbon.

---

* The state did so in 2016, four years early.

Now, because of the California Forest Protocol, Laurie Wayburn says, "We can manage our forests so that we have more of the characteristics that are found in primary, undisturbed forests and get more timber out of them as well as the carbon benefits, the water benefits, the biodiversity benefits, and so on."

During their careers, Wayburn and Connie Best have also played key roles in developing the concept of working-forest conservation easements. These easements not only prevent development of the forest for residential and other uses but also promote forest restoration and carbon storage and guide forest management. Landowners who voluntarily agree to have perpetual conservation easements attached to their title deeds are then able to monetize the easements by selling them to qualified conservation organizations, land trusts, and government agencies.

Now attached to millions of acres of forest nationwide, forest conservation easements are generating significant additional revenue for conservation-minded forest owners. Also, when charitably donated, the easements generate significant tax deductions and, in some cases, tax credits. This framework successfully aligns forest owners' financial interests with the public's interest in protecting forests for their climate and other environmental benefits. PFT, which holds the conservation easements for some participating forests, was thrilled in 2007 to send the first six-figure check from the first carbon project to a private landowner—a very significant amount of money for the owner of a small property.

In conceptualizing and helping develop a robust national market for previously unpaid-for stored carbon and climate benefits, Wayburn and Best have played a critical role in creating a readily scalable market mechanism that can serve as a global model for forest protection. "The existence of both public and private markets for carbon credits and for conservation easements has made it possible for forest owners to cut fewer trees and grow older, more natural forests while receiving payments in return," Wayburn said. Her current ambition is to see that forest owners are financially rewarded not only for the additional carbon their forests store but also for the many other important, public benefits that forests provide that have long been taken for granted.

While much of Wayburn's career has been devoted to ensuring that the important role of forests in storing carbon is understood— and that private forest owners are compensated for that role—she has in recent years been educating lawmakers and the public about these much broader forest climate benefits. "Offsets did not embody the full value of forests for climate," she explains. "When you look at the impact of climate change, what are you seeing?" she asks. "Drought alternating with flood. What mitigates both drought and flood? Forests!" In an intact state, forests reduce flood intensity by 25 to 40 percent. "They also milk water from the air—not just rainfall, but water in passing clouds, fog, or mist." This adds another 25 percent to the water balance, she said.

The Van Eck Forest Win-Win Model

The 2,200-acre Van Eck Forest, in Northern California's Humboldt County, is a prime example of Wayburn's approach to sustainable forestry. The forest is a magical, peaceful, mossy place of stately, towering trees; sunlight-dappled ridgetops with lovely views; and, in summer, the sugary scent of fir. The mostly redwood forest with lush, green stream corridors is home to mountain lions, black bear, the rare Pacific fisher (a type of weasel), threatened spotted owl, salmon, and many other species. Because the trees are large and long established, they are mighty bastions of stored carbon. PFT harvests timber there annually, removing almost one million board feet of timber a year. But, at the same time, timber volume on the forest has more than tripled, so, the forest, the public, the local community, and the economy have all benefited.*

---

* Because working forests with conservation easements are all managed sustainably, even though some timber is cut periodically, more timber grows back between each harvest cycle than is removed.

When Pacific Forest Trust first began working with the van Eck family, their forest properties, like nearby forests, were at risk of being overexploited in short-rotation logging operations, or even being sold off for ranchettes. Concerned forest owner Fred van Eck, however, contacted Wayburn to discuss having PFT manage the property and develop a working-forest conservation easement. Several years after his initial contact, the Van Eck Forest in California became the first in the nation to be protected under California's rigorous Forest Protocol and, thus, became eligible for carbon offset credits. The Van Eck Forest Project now serves as a model for how sustainable forest management can produce significant and marketable climate benefits.

PFT manages another Van Eck Forest—a 7,200-acre plot in Oregon, near Corvallis. When it began managing it in 2002, the land was a plantation with trees that averaged twenty years of age. During the twenty years it has been under PFT's management, the trust has quadrupled the amount of carbon in the forest and, despite taking out a couple million board feet of timber every year, has quadrupled its total timber, and forest carbon, inventory. Whereas commercial forest operators who focus on trees only as a crop would plant a property like this in a crowded monoculture plantation of Douglas fir, which they would clear-cut every thirty-five to forty years, PFT does not clear-cut forests and is restoring the native fir, hemlock, spruce, and cedar, reintroducing species diversity to forest openings via the selective cuts it creates and growing and retaining large old trees on the property.

PFT is demonstrating that managing to sustain a natural forest is not only much more productive for water, wildlife, and the climate but also far less risky for the forest owner. The forest openings—i.e., the introduction of more space between trees—and the associated species diversity this fosters through its management reduce fire risk. By contrast, overcrowded, small young trees on a plantation are highly susceptible to disease and fire. "If one of them gets sick," Wayburn said, "they all get sick." More space between trees and greater species diversity make disease transmission more difficult and slow the spread of insect pests like the spruce bark beetle, which has devastated millions of acres of forest around the West. Wider spacing also allows for more light and water per tree, resulting in a healthier, more resilient forest. The big downed, dead logs that PFT leaves on the ground "absorb a ton of water," Wayburn said, and so, also help mitigate wildfire risk.

PFT is especially enthusiastic about retaining and reintroducing hardwoods like alder, oak, and maple. "Those hardwoods are incredibly important for biological diversity because they're all among the biggest food sources in the forest. And they're also more fire-resilient than young fir," said Wayburn. The group is also starting to experiment with forest canopy restoration in redwood forests, to learn how to do it in ways that are cost-effective and replicable. The forest canopy is a critically important part of a natural forest ecosystem. "Robust canopies hold perhaps twenty percent more water in large fern mats than industrially managed forests," Wayburn told me. While they don't have immediate financial value, they do make the forest moister and cooler and "are a key to forest health and productivity."

In early 2022, the van Eck family's premium, certified carbon credits sold for eighteen dollars for each metric ton of stored carbon they represented. Since the project began in 2008, the landowner has earned well over a million dollars from these credits, which add about two thousand dollars per acre to the net present value of the forest.

Forests as Infrastructure

Far from resting on their laurels, Wayburn and PFT have set their sights on seeing the native forests along the Mississippi and Missouri Rivers restored. PFT is also working to preserve the unprotected, privately held half of the 12-million-acre Klamath-Siskiyou-Cascade Ecoregion, which are the most biodiverse coniferous forests in the world. PFT is currently partnering with many other forest owners across the Pacific Northwest to protect key watersheds for the threatened coho salmon. In addition, PFT is working with states and federally across the United States to ensure that high, consistent, durable, and meaningful standards are established for forest carbon credits.

PFT is also making headway in helping forests receive credit—and payment—for their benefits beyond carbon sequestration. Among PFT's legislative accomplishments, it was a lead sponsor in 2018 for California Assembly Bill 2480, which recognizes forest watersheds as key infrastructure of the state's water system; it therefore makes forests eligible for the same funding and financing as California's conventional brick-and-mortar water infrastructure. In 2020, PFT also played a key role in the passage of AB 2551, which has forest- and watershed-restoration provisions and recognizes specific forested watershed regions for their value as sources of California's water. The state's 2022 budget now includes hundreds of millions of dollars earmarked for comprehensive conservation in these key watersheds.

With Wayburn at the helm, PFT also led in the development of California Assembly Bill 1757. Known as the state's Natural and Working Lands Bill and signed in 2022, it requires California to set a carbon-removal target for those lands. And as if all this were not enough, in 2021 PFT led a coalition that won over two billion dollars from the state for forest fuel and fire management.

Wayburn believes that public policies maximizing and protecting the broader public benefits forests provide are essential to a low-carbon future for the planet. The key, she believes, is to provide public subsidies, such as cost sharing, for whole-forest restoration and management. "Managing for the whole forest," she said, "gets you a wide array of forest products—large-to-small timber, fiber, pulp, and chips—plus the water, wildlife, cooler temperatures, health and recreation benefits, plus more stable employment, as well as increased timber yields." In short, it's a grand slam home run for the climate of the kind we need more desperately every day.

PART IV

Political and Legal Reform and Other Challenges

14.

New Laws and Policies

Success in saving the earth and reestablishing a safe climate requires effective and democratic governance. Although at first it might seem tangential, the nature and quality of governance strongly influence what can be achieved in climate protection; neither a government ruled according to the whims of an unchecked autocrat nor an ailing democracy subject to the designs of special interest groups can be relied on to protect the environment, the climate, or human rights. In addition, the specific design of climate laws and policies affects not just climate change but also people's daily lives. Thus, insofar as economic, social, and environmental injustice are all exacerbated by climate change, they are all proper concerns of the climate activist. Marginalized and disadvantaged groups are important allies in the struggle to save the climate—partners whose interests should be protected not just for the sake of fairness and humanity but also because their interests, if ignored, might understandably lead them to oppose climate-protection policies.

Hurdles in the Race to Climate Protection

With these larger realities in mind, let's examine some of the current barriers to the kind of fair and comprehensive climate policies that would speed us toward a safe, stable earth and a more just society. Of these obstacles, the anonymous "dark money" spent to influence elections or legislation, given by donors whose identities remain secret, is one of the most serious. The U.S. Supreme Court, in its 2010 decision in *Citizens United v. Federal Elections Commission*, ruled that corporations, like individuals, were entitled to free speech and that political donations were a form of speech; therefore, corporate political donations could not be banned or capped. Since *Citizens United*, billions of dollars in dark money—more than two billion dollars in 2020 alone—have flowed into election campaigns at the federal, state, and local levels.

In addition, *Citizens United* led to a proliferation of "super PACs," a type of political action committee partially funded by dark money and able to accept unlimited contributions from individuals and corporations and from groups donating dark money. Although barred from contributing directly to candidates or coordinating their activities with candidates (rules not hard to subvert), super PACs are allowed to buy political ads supporting or attacking candidates, thereby directly influencing elections without voters even knowing where the money for the ads is coming from. Apart from super PACs, current election laws allow certain types of nonprofits to spend unlimited sums purchasing ads on behalf of elected officials without disclosing who the donors are or the size of their donations.

Much of the money spent to influence elections, official policy, legislation, or judicial appointments comes from a relatively small number of extremely wealthy individuals. In the largest known such donation, a single anonymous billionaire was discovered to have legally donated $1.6 billion to a tax-exempt Republican political organization that spent tens of millions lobbying for the appointment of three conservative U.S. Supreme Court justices. We now have a supermajority of justices on the Supreme Court whose views on many important issues are not representative of mainstream voters and who have chosen to reject the long-standing nonpartisan judicial tradition of stare decisis, which holds that previous judicial decisions should be adhered to unless they contravene the principles of ordinary justice.

These concerns about dark money are highly relevant both to the unfettered operation of democracy and to climate protection. The public, when polled, supports the restoration of a safe climate. Currently, however, special interests—corporate lobbyists, super-wealthy fossil fuel–friendly individuals, and the industry itself—all can and do politically sabotage clean-energy and climate-protection policies behind a veil of secrecy. For governments to be able to mount exemplary, sustained climate-protection programs, the undue influence of highly concentrated wealth has to be checked.

Removing corporate and dark money from politics would likewise reduce the influence of wealthy individuals and companies intent on avoiding higher taxes and, hence, on blocking effective government action to create a more just society and an egalitarian economy. Thus, though social and economic justice may seem independent from climate protection, these ends are closely interrelated, given that curbs on dark money and corporate political donations would benefit both causes, as well as for other reasons. For example, if housing in a city is unaffordable for people who work in its service industries, those people will have to live far from work and may well have to commute long distances every day, generating tailpipe emissions. If special interests are able to keep marginal tax rates low on high-income earners and corporations, permitting both to dodge much of or all their tax liability, government will be deprived of revenues it needs to build low-income housing or to subsidize lower-income workers with financial incentives or direct assistance so they can make their cars, homes, and appliances clean and energy efficient. Government also will be less able to make necessary investments in modern, efficient public transit, or in helping utilities modernize their electric grids, or in encouraging clean-tech industries, or in funding clean-energy research and development.

By contrast, if government can fund this work generously, perform it successfully, and do so in highly visible ways, it can regain public trust and elicit cooperation in the energy transition and broad-based public support—versus facing a divided, polarized citizenry. National unity is especially critical for addressing climate change, because the epic national climate-protection effort now needed ultimately will require a shared sense of both national purpose and common destiny. Instead, to no small degree because of special-interest political manipulation, many ordinary Americans now see a government and an economy operating in the interests of the well-to-do at their expense.

The current distribution of income and wealth in the United States tilts steeply toward a small fraction of the population, providing substantiation for this perception. In 2018, for example, the richest 1 percent of U.S. income earners held $25 trillion—as much wealth as the bottom 80 percent of the population. Those who hold great concentrations of wealth can block social action and climate action alike, to serve their own economic interests. Ordinary Americans pay the price: Hundreds of thousands are homeless. Twenty-nine million live in poverty; three times as many have no medical insurance or are underinsured. Life expectancy is declining.

Even if it currently cannot count on tax reform and more generous tax remittances from the ultra-rich, the U.S. government will need somehow to find the means to address widely unmet social needs for adequate nutrition and for good, affordable housing, medical care, jobs, education, and transport. These economic justice and equity concerns must be prioritized by people organizing to protect the climate and need to be tightly woven into the fabric of a new climate action plan. Fusing those issues with climate protection will make it possible to secure broad public support for the plan and will help restore broader respect, and support, for democratic government itself. This is absolutely essential if the United States is to take concerted action—unhampered by special interests—to swiftly slash emissions and lead a global decarbonization effort.

National Mobilization Amid Crisis: The New Deal

Happily, there is historical precedent for a similar comprehensive mobilization of government and citizenry in dire circumstances. In 1933, when President Franklin Delano Roosevelt took office during the Great Depression, things could hardly have looked bleaker. Seven thousand banks had failed. Millions of people were hungry, desperate, and hopeless. A quarter of the workforce was unemployed; vast numbers queued up at soup kitchens. Initially, many people were skeptical that Roosevelt could accomplish a massive economic turnaround, as people today are skeptical that we can combat climate change. But the president understood that the profound problems the nation faced required pervasive reforms. He therefore resolved to address an accumulation of fundamental, systemic problems in American society, problems affecting agriculture, banking, industry, natural resources, social welfare, and international trade. With the strong support of Congress and the American people, he launched massive administrative, labor, governmental, and social welfare reforms that put the nation back on its feet and created 12 million federal jobs. New Deal programs restored the nation's morale— which proved to be a necessity for the rapid transition to a wartime footing demanded by Japan's 1941 attack on Pearl Harbor.

With the arrival of war, Americans who had only just begun to leave the Depression behind nonetheless accepted rationing. They observed thirty-five-mile-per-hour speed limits to save fuel and rubber. Mines and factories operated in double shifts. Millions of women rushed into the workforce. With astonishing speed for a demilitarized nation, the United States began turning out hundreds of thousands of aircraft, tanks, jeeps, and ships. This is just one example of the scale and intensity of the response needed today to reverse climate change: a concerted nationwide industrial effort to produce millions of solar panels, wind turbines, batteries, carbon capture machines, and zero-emission vehicles (ZEVs). The nation and the world are now in a situation even more perilous than the Great Depression or World War II. Rising seas pose serious threats to major U.S. cities, including Boston, Charleston, Miami, and New York. Millions of acres in the Midwest have been flooded in climate-related extreme weather events in recent years. Had a foreign enemy inflicted the kind of damage on the United States caused by these abnormal floods, or the firestorms that swept across California and the Pacific Northwest in 2020, or the hurricanes and droughts the nation has begun experiencing with increased frequency—the United States would have immediately mobilized for war with that enemy. We must do the same for the climate crisis.

A National Climate Plan

As Antoine de Saint-Exupéry wrote, "A goal without a plan is just a wish." One of the first things that needs to be done in any crisis is to create a plan for solving it, and the climate crisis is no different. A national climate action plan is therefore needed.

During the Trump administration, which vociferously embraced climate science denial, progress on controlling U.S. emissions devolved to states, cities, and individual businesses and institutions. Over the long term, however, climate policy is too important to be left to a hodgepodge of uncoordinated laws and policies haphazardly and inconsistently applied across the fifty states and thousands of cities. The United States needs a much more coherent and properly coordinated effort, one guided by scientific and technical analysis. The plan must be based on an ambitious but attainable set of GHG-reduction quotas linked to target dates and enforcement mechanisms. This structure would provide measurable emission-reduction goals, subject to frequent reviews so that midcourse corrections could be instituted as necessary.

The point of the plan would not be to supersede the climate agendas of any city, state, or group, nor the aspirations of H.Res. 109, the Green New Deal. It would simply be to provide a reliable inventory of the measures and policies that will, without fail, drive U.S. emissions into the negative territory. The plan would also include specific targets, milestone dates, and analysis of costs and benefits of its main measures.

Whereas the climate plans of the 2020 presidential candidates and the Green New Deal provided thought-provoking input and appropriately raised the issues of environmental justice and equity requirements, none is a substitute for a scientifically validated national plan prepared by teams of top-notch government scientists and engineers with proper public input and stakeholder consultation. U.S. government scientists and independent researchers have for decades produced reports on the impacts of climate change. Despite this, the United States has no comprehensive climate action plan with clear, enforceable targets, time lines, and road maps for climate protection and restabilization. To fill the void, the nation needs a scenario study that might be titled *America's Energy Transition: Achieving a Clean Energy Future*. Such a study, building on previous authoritative studies, would analyze renewable-energy-generating and distribution technologies in terms of their costs, commercial readiness, potential output, resource constraints, and efficiencies.

Much of this research has already been done in numerous studies by scientists and engineers using various assumptions. These "technology assessments" are available as inputs to the even larger scenario study that would be needed. But rather than simply assessing technologies in isolation, researchers developing a national plan should employ a systems perspective, formulating and modeling competing scenarios with clusters of complementary technologies.* Each scenario would require different policies for its implementation.

---

* Each cluster would be described by its components and their individual performance characteristics and constraints, costs, manufacturing requirements, and deployment lead times; each scenario would thus be based on a unique combination of generation, distribution, storage, and efficiency technologies, with unique implementation pathways dependent upon varying renewable resource requirements, regulatory issues, and implementation time requirements.

The official scenario study should also include evaluations of carbon-removal technology clusters to optimize GHG drawdown by engineered and natural systems.

We would learn from this exercise which scenario would deliver the greatest GHG reductions most cost-effectively and quickly, with the fewest unwanted impacts, while meeting our energy needs. We would also learn what would be required to bring each cluster to scale in terms of natural resources and raw materials, costs and financing, and manufacturing capacity. (Some of this work likewise has been done by university groups and nonprofits, but none represents official policy.)

After examining the science, technology, and public policy aspects of each scenario and outlining the policy options required to achieve each one, the government should then invite extensive public input and stakeholder consultation, to build trust and engagement in the final plan. Depending upon the outcome of this process, the policy road map would be designated as the "optimal energy path scenario" for the nation and provided to Congress, so it could use these findings as a basis for funding and implementing new legislation.

Because each region of the country has different resources and presents different opportunities and challenges, regional advisory councils comprised of scientists, engineers, businesspeople, and major stakeholder representatives should offer recommendations on how best to adapt the plan to local conditions in five or more U.S. geographic regions.* The climate plan that President Biden outlined during his campaign for the presidency in 2020, and which has evolved during his administration, recognizes that climate challenges vary by region.

---

* These regions might include the Northeast, the Southeast, the Midwest, the Southwest, and the West, and perhaps subregions such as the Mid-Atlantic, New England, the Pacific Coast, and the Gulf Coast.

Rethinking Government (at the Top)

Vital to the execution of Franklin Roosevelt's ambitious program of reform was the governmental reorganization he spearheaded. Similarly, today we must add some new federal agencies and coordinating mechanisms to win the global campaign against climate disruption. During two of FDR's terms in office, his administration created sixty-nine new federal offices, including many agencies and programs we now take for granted, such as the Federal Communications Commission, the Federal Deposit Insurance Corporation, the Securities and Exchange Commission, and the Social Security Administration.

A necessary early step in mobilizing the nation is the creation of a powerful, cabinet-level Climate Protection Department or White House Office of Climate Mobilization to help coordinate all federal action. This office would supervise the reduction of U.S. greenhouse gas emissions and implement the national climate action plan.

Some who adamantly favor smaller government may rail against the creation of new agencies and offices, even to address a problem as intractable as climate change. They might argue that we could just assign new tasks under the national climate plan to existing federal agencies, such as the EPA or the Federal Emergency Management Agency. However, with most federal agencies already heavily tasked with multiple missions, they probably could not devote the necessary laser-like focus to an urgent, new critical mission, given all their other conflicting mandates. Domestic U.S. climate policy is currently coordinated through the White House Office of Domestic Climate Policy (aka Climate Policy Office) within the Executive Office of the President, assisted by a multi-agency National Climate Task Force, all created by executive order. Such arrangements lack permanence and could be scuttled by a new administration. New, permanent offices and agencies with clean slates and relatively simple, specific mandates should be created, some within a new national Climate Protection Department. These could include a:

* Climate Emergency Agency, to coordinate government and private-sector efforts to harden vulnerable cities, regions, and sectors of the economy against extreme climate impacts;

* Climate Risk Information Office, within the just-mentioned Climate Emergency Agency, to provide risk assessments to support the setting of federal standards for flood and wildfire protection and other matters;

* Clean Transition Redevelopment Agency, to coordinate economic redevelopment efforts in fossil fuel–dependent communities and facilities and the retraining and reemployment of workers with newly obsolete jobs;

* Climate Education Agency, to educate the public about the nature of the climate threat and its remedies;

* Clean Infrastructure Agency, to collaborate with the private sector in the creation of a smart electrical grid capable of delivering clean energy from sunny, windy regions to urban centers, where it is needed most;

* Clean Industry Agency, to collaborate with the private sector in decarbonizing industrial production processes and to coordinate efforts across industries;

* Clean Transportation Agency, to expedite the electrification of transit;

* Clean Energy Bank, to fund the development and deployment of clean-energy technologies;*

---

* This concept is discussed in more detail later in this chapter and, previously, in chapter 2.

* Building Modernization Agency, to coordinate the retrofitting of the nation's millions of buildings with energy-efficient windows, lights, and heating and air-conditioning systems, using clean sources of electricity;

* Carbon Capture and Storage Agency, to provide support to industrial efforts to capture and store carbon in geological formations;

* Natural Resource Restoration Bureau, which would be responsible for restoring forests, wetlands, rangelands, and agricultural soil ecosystems to remove and store atmospheric greenhouse gases; and

* Civilian Climate Corps (modeled on the New Deal's Civilian Conservation Corps, which employed millions of unemployed Americans), a robust and fully funded agency responsible for hiring millions of underemployed and unemployed Americans to restore natural resources, in furtherance of the Natural Resource Restoration Bureau's work. President Biden created a Civilian Climate Corps by executive order on January 27, 2021, but major funding for it failed to win congressional approval.*

The United States should spare no effort to reorganize and coordinate its federal agencies to prioritize climate stabilization. If this government reorganization resulted in accelerating the

---

* The program currently exists within the Bureau of Land Management, where it is mainly a low-key continuation of a partnership program for young people that the bureau has had for thirty years with nonprofit youth and other conservation corps around the country. A Civilian Climate Corps for Jobs and Justice Bill was introduced in the U.S. Senate in April 2021 by Sen. Edward Markey, but it did not advance. A thirty-billion-dollar Civilian Climate Corps program was also included in the Biden administration's Build Back Better Bill, which was defeated in Congress, and the program was excluded from the administration's successful Inflation Reduction Act of 2022. (See Maxine Joselow, "The Civilian Climate Corps Was Dropped from the Climate Bill. Now What?" *Washington Post*, September 8, 2022.)

U.S. transition to a clean-energy economy, it would be money well spent.

Powerful Federal Actions

In the absence of a formal national clean-energy transition plan, new federal climate legislation is nonetheless needed. For decades, attorney, engineer, and "green cowboy" S. David Freeman was one of the most forceful advocates of strong, immediate federal action to oblige utilities, automakers, builders, and building owners to reduce emissions quickly and phase out fossil fuels. A top adviser to three American presidents and to governors of California, New York, and other states, Freeman managed five major electric and water utilities during his career; he also chaired the powerful Tennessee Valley Authority, established as part of the New Deal.* In 2019, Freeman and I collaborated on an opinion piece that summarized our views on the need for aggressive federal action to curb greenhouse gas emissions. It recommended a series of new federal mandates to smooth the transition to a clean-energy economy:

First, the U.S. government should require *all* new power plants to use zero-emission renewable-energy technologies. In addition, it should require utilities to reduce all carbon pollution from existing fossil fuel plants by 8.5 percent of 2020 emissions per year, so as to reach zero in twelve years. As fossil fueled generators are phased out, needed electricity would be provided to the grid from solar, wind, hydro, efficiency investments, and geothermal energy, with backup energy-storage systems and regional grid interties (i.e., interconnec-

---

* As the nation's first national energy adviser, under President Lyndon Johnson, Freeman coordinated energy policy across the entire federal government, laying the groundwork for restructuring U.S. government energy agencies. He also played an important role in the creation of the U.S. EPA and left his mark on the nation's energy efficiency policy. His seminal 1974 Ford Foundation energy policy study, *A Time to Choose*, pointed the way for legislation improving auto efficiency by instituting what are known today as CAFE (Corporate Average Fuel Economy) standards; it also laid the groundwork for President Jimmy Carter's support for energy efficiency and renewables.

tions). Meanwhile, as those recommendations were phased in, many consumers and commercial establishments would have their power needs taken care of by millions of rooftop residential and commercial solar power systems. As older power plants were phased out—in keeping with the tenets of our first recommendation—demand for power would be greatly reduced by extensive adoption of energy efficiency technologies (and practices), which invariably cost less than new power generation.

Our second recommendation was that all new buildings must be required to obtain their power entirely from renewable-energy sources within two years. In 2020, California already required that all new buildings of up to three stories be equipped with solar panels; if the building is also energy efficient, the solar panels could make it a net-zero-power building (depending on its location and the roof's orientation). All new California homes also currently must be equipped with solar energy. And, starting in 2030, the sale of new natural-gas water heaters and home furnaces will be prohibited in the state. Nationally, a net-zero-energy home saves homeowners more than $25,000 in cumulative energy and maintenance costs.

Just making a typical home extremely energy efficient might add ten thousand dollars to its price and perhaps forty dollars to the monthly mortgage, but it would provide eighty dollars in monthly savings. Adding solar panels might add another 6 to 12 percent to the home's cost, but the panels can be leased and financed with utility bill savings. So, Freeman and I stipulated, as part of our second major recommendation, that all existing buildings be required to reduce their GHG emissions by at least 5 percent of 2020 emissions per year, thus cleaning up building sector emissions entirely within twenty years.

A third federal law we recommended would require—just as in Norway and the Netherlands—that all new cars, trucks, or buses sold domestically be ZEVs within ten years. At least 10 percent of each auto manufacturer's vehicles would have had to be zero-emission in the first year, and the mandate would increase by an additional 10 percent annually, to 100 percent in a decade.

California governor Gavin Newsom issued a similar but weaker executive order in 2020, requiring that all new passenger cars and light trucks be ZEVs or plug-in hybrids by 2035, with a 2040 deadline for medium and heavy trucks. The 2035 rule was formally adopted in 2022 by the CARB. Because about 40 percent of California's emissions are from transportation—50 percent if all refinery-related emissions are included—the rule will reduce the whole state's emissions by an impressive 35 percent.

How the law will impact light vehicles in California provides an idea of what would happen if it were implemented nationally: During the run-up to 2035, government tax credits and preferential targeted financing for ZEVs will stimulate consumer demand for all the new required California EVs. Because the life cycle costs of EVs and plug-in hybrids are generally much lower than those of gasoline and diesel vehicles, California's zero-interest loans could be repaid by the borrower profitably over a vehicle's lifetime, or sooner. Owners who scrapped gasoline and diesel vehicles could receive a cash rebate to further cushion EV costs.

Because some seventeen other states follow California's lead on air quality, the new rule is likely to reduce all U.S. gasoline and diesel light-duty vehicle sales by 40 percent, transforming the U.S. transportation sector. Major auto companies are also planning on phasing out the production of internal combustion engine cars. A federal ZEV mandate could bring about these benefits sooner and to every state in the nation.

A fourth law Freeman and I proposed would establish a Federal Railroad Electrification Authority to finance and build a cleanly powered, all-electric high-speed rail system connecting major U.S. cities. This would provide an efficient, economical alternative to regional air travel.

A fifth proposed law would require, by January 1, 2025, that airlines reduce their emissions annually by 5 percent of 2020 emissions through greater efficiency, renewable fuels, and hybrid or electric propulsion (if available), until those emissions reach zero.

Our sixth legislative proposal was that Congress should establish a national green bank to provide loans to implement all the various mandates outlined here, as well as low- to zero-interest loans to cement, steel, and other industries to decarbonize their production. These investments would not just cut GHG emissions but would also more than pay for themselves over time by making these industries more efficient and competitive.

Our seventh proposal is necessitated by that fact that plateauing annual carbon emissions will not stop the heating of the planet already under way due to the excess carbon in the atmosphere, and its long atmospheric lifetime. Therefore, a national carbon-credit plan is needed; farmers, ranchers, foresters, other resource managers, and private investors should be made eligible for tax credits or payments of up to one hundred dollars for each properly certified ton of carbon removed from the atmosphere. (Payments could be raised over time to keep up with inflation and market conditions.) Carbon credits under carbon-trading regimes would have market value to investors. (As described elsewhere, a similar credit scheme could be developed for nitrous oxide.)

The Carbon-Tax Argument

Apart from carbon credits, carbon taxes are the other logical way to put a price on carbon. Many environmentalists and economists support the idea, and a carbon-tax bill was introduced in the U.S. Senate by senators Bernie Sanders and Barbara Boxer in 2013. Currently, 15 percent of global emissions are covered by (generally very inexpensive) carbon-pricing systems. Many public officials, economists, and business leaders recognize that fossil fuels today are artificially cheap because their environmental, climate, and public-health damage costs are not reflected in their price. A national carbon tax could fix that problem, they argue, and greatly accelerate the transition to clean energy by imposing a simple fee at the mine mouth, wellhead, or port of entry into the United States that would be gradually raised, squeezing fossil fuels out of the market.

Over the past fifteen years, a retired Northern California physician named Peter Joseph has become virtually a full-time carbon-tax advocate and climate activist. When Dr. Joseph saw former vice president Al Gore's climate change documentary, *An Inconvenient Truth*, in 2006, he was shaken by it. "I thought to myself, I'm an emergency physician. This is the mother of all emergencies. Am I going to sit this out, or am I going to get involved?" Dr. Joseph went to Nashville in early 2007 to take Gore's Climate Reality Project training program and went on to give "over two hundred, maybe two hundred fifty" slideshow presentations on climate change—but he was troubled. "Invariably, at the end of the talk," he said, "basically describing the end of civilization as we know it, the end of the world, the first question would be, 'Well, what can I do?' And all I had for them was all these really virtuous personal things that people could do, like eat less meat. . . . People would experience a cognitive dissonance between putting solar on your rooftop versus the end of the world. The scale of the problem versus the solution was a total mismatch."

Giving talks wasn't going to mobilize people fast enough, Joseph concluded. Congress had to be mobilized. After a lobbying trip to DC, he was introduced to Citizens' Climate Lobby, a grassroots non-profit advocacy organization dedicated to educating people about climate change and mobilizing them to lobby Congress for a carbon tax. There were three people in Joseph's local CCL chapter when he began leading it in 2011. Today, the chapter has nine hundred members, and CCL has two hundred thousand members nationwide, with representation in every U.S. congressional district.

A federal carbon-tax bill, H.R. 763, currently has the support of CCL and at least 82 congressional Democrats (but only a single House Republican). The proposed tax would start at \$15 per metric ton of CO_2 and be raised by an inflation-adjusted \$10 a year. After about twenty years, it could exceed \$200 a ton, but the tax would be eliminated once emissions fell 90 percent.

A coalition of 106 environmental organizations coordinated by the US Climate Action Network, however, has raised concerns that if carbon prices are not set high enough, they will not motivate polluters to achieve zero emissions. Even a $200-per-metric-ton tax on carbon dioxide would raise gasoline taxes by only $1.76 per gallon, and consumers might grumble about it but would ultimately take it in stride. According to the Intergovernmental Panel on Climate Change, the price must be $245 per metric ton. But raising carbon prices fast enough to reach net-zero carbon by 2050 will be "an almost insuperable political challenge," according to Vox energy and climate reporter David Roberts.

Many environmentalists argue that a $15-per-ton tax would have little initial impact on U.S. emissions and that political opposition would prevent it from ever getting high enough to turn our $20 trillion economy away from fossil fuels. David Freeman, for one, was dubious that carbon taxes would ever be high enough to really bite into fossil fuel consumption. "The problem with a partial solution is if you passed it and people thought that was going to solve the problem, you're doing more damage than good," he said. "When we had lead in toys, we didn't think about putting a tax on toys with lead. We just outlawed it. We have not yet accepted the fact that fossil fuels are poisoning our environment and need to be outlawed. . . . I don't think the Democratic Party, much less the Republican Party, has fully accepted that."

Dr. Joseph and CCL, however, want a low initial carbon fee to avoid disrupting the economy, as long as fees rise predictably and steeply thereafter. Even a tax on carbon of $50 per metric ton, however, would reduce 2030 emissions in the transportation sector by only 2 percent, according to a report by the Rhodium Group. The US Climate Action Network doubts that a carbon tax would reduce U.S. emissions sufficiently without "a suite of [other] regulations and investment policies. These should include a ban on new fossil fuel extraction, exploration, and processing facilities plus a phaseout of existing facilities." Dr. Joseph concurs: "There are certain sectors of the economy where a price signal [alone] does not penetrate," he said.

Another controversial aspect of H.R. 763 is a puzzling lack of urgency: From 2035 to 2050, the bill's carbon-reduction target is a mere 2.5 percent a year. Yet costs of renewables and energy efficiency adaptations are likely to be lower by 2035 than today, which would offset some of the tax's economic impacts. Also, by 2035, the economy will have had more time to adjust to decarbonization, so there's even less justification for soft-pedaling and slow-walking. Finally, even by 2050, the bill never aims at net-zero carbon nationally, much less at an eventual drawing down of atmospheric greenhouse gases to safe levels.*

What to do with the revenue from a carbon tax is also a controversial issue. Some proposals would return all revenues directly to the American people, a concept known as "revenue neutrality." (Citizens' Climate Lobby and others call this a "fee-and-dividend" program, though the payments are not literally dividends.) H.R. 763 embraces this approach and would return those revenues in equal shares to all people, regardless of how much they paid in carbon taxes. Rebating the tax in full will be politically popular, CCL maintains, and the collections and rebates could be handled by existing government agencies (such as the IRS) without expanding government, which appeals to fiscal conservatives.

Others critical of the dividends believe that the public prefers using carbon revenue to reduce carbon emissions rather than to distribute rebates. The issue is broader than any particular carbon-tax plan: If a sizable chunk of the trillions of dollars a carbon tax would raise were *not* returned to the public, it would be a lot more difficult to muster political support for the tax's passage. Repurposing *some* of the revenue to combat climate change, while reserving some for rebates for low- and middle-income taxpayers, would accelerate the clean-energy transition without sacrificing most of the rebate's widespread appeal.

---

* Lest they be misconstrued, these comments about H.R. 763 are not intended as criticisms of carbon taxes in general but in support of a substantially strengthened version of the bill. Needless to say, carbon taxes are not a magic bullet, but they would certainly be enormously helpful in speeding up the energy transition as part of a comprehensive approach to the climate crisis.

However they may feel about the way tax rebates are spent, U.S. manufacturers are likely to be comfortable with H.R. 763's trade provisions, because the bill has an important cross-border adjustment section that applies the domestic carbon tax (like a tariff) to imported carbon fuels and carbon-intensive products. Thus, it protects U.S. carbon-intensive products from competition with imported products not subject to carbon taxes in their country of origin. And under a tax-forgiveness provision, U.S. exporters of carbon-intensive goods would also be reimbursed in whole or part for their carbon taxes if they exported goods to a country without a carbon tax or with one lower than that in the United States; their exports would thus not be at a competitive disadvantage. The downside is that not all domestically generated carbon would be taxed under this bill. To raise the price of fossil fuels internationally and curb global emissions, countries with a carbon fee could band together and charge trade tariffs to other nations that had none; this, in effect, would be an international "cross-border adjustment."

Appealing to Big Oil and Big Business

The industry-supported Climate Leadership Council (CLC) and other business and government leaders strongly support a revenue-neutral carbon tax, although one even weaker than that proposed in H.R. 763. They also support an international cross-border adjustment and assert that it will "put America in the driver's seat of global climate policy and encourage other large emitters—such as China and India—to follow America's lead and adopt carbon pricing of their own." This would be an important benefit. However, the council acknowledges that even a well-designed cross-border carbon adjustment would cover only about 80 percent of all goods traded.

The CLC is an important climate policy lobbying group comprised of more than 3,500 prominent business leaders and economists*—including past Federal Reserve chairs and many Nobel laureates. Revenue neutrality suits its large, mainstream corporate members, including oil companies like Exxon, who dislike higher taxes and oppose government regulations. The CLC contends that a full, direct rebate of all tax proceeds would benefit the American people and would "for the first time . . . align . . . the economic interests of working Americans with climate progress." A family of four, the CLC estimates, would receive $2,000 in "cash dividends" in the first year of its plan. That is its allure to the public. The drawback is that if the entire $288 billion were given back to the public in year one, none would be available for climate mitigation or climate adaptation.

The CLC proposes "a series of grand bargains"** that would trade acquiescence to a gently rising tax on carbon in exchange for sweeping away the EPA's current and future power to regulate carbon emissions by utilities and industry. It would also "suspend, repeal or preempt" future federal low-carbon fuel standards and standards for certain non-road vehicles, including farm equipment. In what could be described as wishful thinking, the CLC's concept was to craft a plan that would trade "the most ambitious carbon price enacted by any leading emitter nation for regulatory relief, thereby appealing to environmentalists, businesses and conservatives at the same time."

Vox reporter David Roberts calls the proposal "first and foremost a bid by oil and gas and nuclear to secure the gentlest and most predictable possible energy transition." The US Climate Action Network disagrees emphatically with such a bargain: "A carbon price must not be coupled with deregulation, liability limits, or other

---

* The CLC's founding members are corporations with a combined market cap of over $4 trillion and more than three million employees, including five oil and gas "supermajors"; Ford and GM; AT&T, IBM, and Microsoft; and Goldman Sachs and JPMorgan Chase. Two of the largest household products companies in the world, Procter and Gamble and Unilever, plus the world's largest health care company, Johnson and Johnson, and the largest insurance company in the United States, are also among the major company founders.

** The CLC proposal is called the "Baker-Shultz Carbon Dividend Plan: A Bipartisan Climate Roadmap." See "Bipartisan Climate Roadmap," Climate Leadership Council, accessed February 14, 2023, https://clcouncil.org/report/bipartisan-climate-roadmap/.

policies that make it more difficult to mitigate the climate crisis or other environmental harms," the group says. USCAN, however, sees carbon pricing as a complement to other climate and environmental policies, not a replacement.*

The proposed regulatory preemption sought by oil and gas goes a long way toward explaining the support from mega-corporations for an industry-flavored carbon tax. (Agribusiness and the military would also be insulated from the carbon tax, as the Treasury would fully reimburse them for any tax paid; so, too, with certain facilities that safely, permanently, and lawfully store carbon emissions.) Under its approach, the U.S. government, with its regulatory hands tied, would be left without ready recourse if the carbon tax proved less effective than projected. Moreover, if carbon taxes are imposed at the wellhead and coal mine, then fossil fuel companies can pass these costs on to businesses and consumers through higher prices; carbon taxes would also allow the oil and gas companies to increase their competitive advantage over coal (given that coal emits almost twice as much carbon per unit of heat as natural gas). Finally, no fuels would be phased out under the CLC's plan, and current subsidies to the oil and gas industry could continue.

This business-as-usual road map responds to the climate emergency through market mechanisms. But although it seems unlikely, with its five-million-dollar annual industry-supported budget and its massive political influence, the CLC might conceivably gain traction with this plan. And if it were adopted by Congress, it would be a roadblock to future federal climate legislation. That would be a huge setback for the climate, because some portions of the economy can never self-regulate into a zero-carbon world and will need federal help or pressure to restructure. So, carbon taxes and regulations are both clearly necessary.

---

* Three large mainstream environmental organizations—the World Resources Institute, the World Wildlife Fund, and Conservation International—are among the CLC's founders, attracted by the political feasibility of its carbon-tax gambit due to its mighty corporate support.

Other Possibilities

The far-ranging discussion of an all-out national mobilization to combat climate change in this chapter reveals only a glimpse of the many legislative and policy complexities of the coming clean-energy transition.* If an American president were blocked from taking meaningful legislative action on climate change and barred from new environmental regulations by a rightward-leaning Supreme Court, an administration determined to advance climate policy on its own would still have various options, through presidential executive actions, the Federal Reserve Bank, and the Department of the Treasury.

The secretary of the treasury chairs the federal Financial Stability Oversight Council, an intergovernmental body charged with protecting the nation's financial security. In addition to the Treasury, its members include representatives of the Federal Reserve System, the Securities and Exchange Commission, and the Commodity Futures Trading Commission. The council can regulate banks' reserve margins and force those with large exposure to fossil fuel investments to hold more reserve cash and to meet higher financial risk–disclosure standards than is currently the case for fossil fuel investments. Such stringent regulation could make it more difficult for fossil fuel facilities to secure financing. In addition, the council can set restrictions on financial disclosures and investments for fossil fuel industries themselves by designating them as "systemically important financial institutions," which the council is empowered to regulate. This could enable the council to set capital reserve requirements for the oil and gas industry and even limit their further investment in fossil fuels if it deemed those investments a risk to the nation's financial security.

Even without special congressional authorization, the United

---

* Other vital topics not explored here for lack of space include curtailment of federal fossil fuel leases and subsidies; fossil fuel exports; the financing of federal "buy-clean" requirements; carbon removal; industrial decarbonization; and clean-energy R&D support.

States, through its membership in the World Bank, the International Monetary Fund (IMF), the Export-Import Bank, the Group of Seven (G7), and the Group of Twenty (G20), can have very significant impacts on the availability of global funding for fossil fuel projects. Climate change is a global issue that no single nation can solve in isolation. Therefore, as the next chapter shows, the pursuit of international collaboration is essential to augment national climate policy efforts. The United States must maintain a strong international leadership role on climate change and assertively enroll other nations in robust climate-protection efforts to ensure that climate action becomes globally ubiquitous.

15.

Climate Diplomacy: Getting to Yes

Imagine a strong, legally binding, science-based global climate agreement. Such an agreement would set the world on a smooth, speedy course toward eliminating dangerous emissions. Emissions would first plateau. Then they would decline, year by year, to net zero. Finally, activists' nirvana would be attained: the world would become carbon negative. In this scenario, every nation would do its fair share. Poor nations would be encouraged to participate with generous funding by wealthier ones; all nations would be ambitious about reducing their emissions. They would accurately and transparently report their progress, and any default on their treaty obligations would bring swift and harsh economic consequences.

In recent years, those who have yearned for this idyllic scenario have had their hopes repeatedly dashed at a long series of global climate summits. Multilateral agreements reached have been weak, and progress toward them has been frustratingly slow. Even once agreements are signed, implementation is sometimes halfhearted or worse.

What's more, these disappointments are not entirely accidental. According to two astute observers with deep knowledge of climate negotiations, "The chaos and inefficiencies in the UN [climate negotiating] process serve the interests of the major emerging economies, petroleum-producing nations, and others. These nations consider delay acceptable or even desirable." In 1990, shortly before global climate meetings became an institution, annual global emissions were 22.5 billion metric tons of CO_2. Today, after three decades of global climate negotiations and worsening climate disruption, they are 60 percent *higher*.* In this chapter, we will explore why these gatherings have failed to deliver the desired results and what alternatives the United States and the world could pursue to make faster progress on decarbonizing.

Sounding the Alarm

"Climate diplomacy" usually means high-level statecraft to get nations to reduce their greenhouse gas emissions. It can also refer to negotiations over assistance to other nations with climate adaptation, climate-related economic development, or emissions reductions. The United States has been an indispensable pioneer in climate diplomacy since the field originated in the late 1980s, in no small part in response to testimony from climate scientist Dr. James Hansen, then director of the NASA Goddard Institute for Space Studies.

Back in those days, most climate scientists exercised great restraint in discussing climate change; they feared that speaking publicly about a contentious political issue would tarnish their scientific reputation. But in 1987 and 1988, Hansen courageously testified before the Senate Committee on Energy and Natural Resources, where he made it clear that human-made climate change had begun and would cause heat waves and other major global impacts. His unequivocal testimony made headlines and brought climate change to the world's attention. From an obscure scientific concern about

---

* The concentration of CO_2 in the atmosphere is now 422 ppm, an extremely dangerous level that is increasing more quickly than ever. The concentration of *all* greenhouse gases (CO_2 and non-CO_2) is even higher. See CO2.earth, https://www.co2.earth/daily-co2; and "The NOAA Annual Greenhouse Gas Index (AGGI)," NOAA Global Monitoring Laboratory, updated Spring 2022, https://gml.noaa.gov/aggi/aggi.html.

what *might* happen to the planet in the far-off future because of excess gases nobody could see, Hansen's testimony boldly elevated the issue of climate change to a topic for kitchen table conversations. Climate diplomacy suddenly became a global imperative.

Ever since then—with the exception of the four years of the Trump administration—the United States has been deeply engaged in climate diplomacy. U.S. experts, diplomats, and policymakers have had key roles in drafting the major climate agreements that have served as milestones in global climate diplomacy to date.* Noteworthy as they are, however, these U.S.-led multilateral efforts have failed in their main mission: to prevent dangerous climate change by arresting the rise of greenhouse gas emissions. Decade by decade, not only have atmospheric GHG concentrations been growing, but their growth has been accelerating.

Not unrelated, from 2016 through 2019, international banks invested $2.7 trillion in fossil fuel companies, adding fuel to the fire, so to speak. Among the multinational banks supporting fossil fuel expansion, four U.S. banks—JPMorgan Chase, Citi, Bank of America, and Wells Fargo—are dominant. Large subsidies have also continued to support fossil fuel sales and development around the world,** and multilateral funding bodies, such as the World Bank Group, are still financing fossil fuel development—though this support has slowed, and the Biden administration is committed to opposing these activities by the development banks. In terms of climate change, however, we're still getting what we're paying for rather

---

* The United States drafted the UN Framework Convention on Climate Change under the George H. W. Bush administration and the 1997 Kyoto Protocol under the Clinton administration. Under the Obama administration, the United States participated in drafting agreements for adoption by the 2009 United Nations Climate Change Conference, in Copenhagen, but no meaningful new global treaty was adopted. Nonetheless, the Copenhagen meeting did ultimately lead to the Paris Agreement, adopted during the Obama administration.

** Direct and indirect subsidies for coal, oil, and gas in the United States exceeded Pentagon spending in 2015, according to the International Monetary Fund. See James Ellsmoor, "United States Spend Ten Times More On Fossil Fuel Subsidies Than Education," *Forbes*, June 15, 2019, https://www.forbes.com/sites/jamesellsmoor/2019/06/15/united-states-spend-ten-times-more-on-fossil-fuel-subsidies-than-education/?sh=58595a3c4473; and "U.S. Military Spending/Defense Budget 1960–2023," *Macrotrends*, 2023, https://www.macrotrends.net/countries/USA/united-states/military-spending-defense-budget.

than what most climate diplomats want. For the United States to be more successful in future efforts to slow global warming, something new has to be tried.

One way forward is to fully integrate climate diplomacy into U.S. foreign and defense policy, especially into international trade and foreign aid policies. President Biden's 2021 executive order on "Tackling the Climate Crisis at Home and Abroad" boldly outlined his plans to do so by putting climate action at the center of U.S. foreign, domestic, and national security policy. The president, the secretary of state, and the president's special envoy for climate change, former secretary of state John Kerry, have committed themselves to galvanizing action within all branches of government and federal agencies. Kerry and the State Department are simultaneously responsible for coordinating international climate policy within the State Department and internationally across all U.S. embassies and consulates. Kerry further defined his mission in 2021 as one of persuading the nations of the world to raise their independent national emissions-reduction commitments. As we'll see, this strategy is necessary but still not sufficient to produce a climate-safe outcome.

Thirty Years of Climate Diplomacy

Other than the secretary himself, perhaps no one has a better feel for what Kerry will face than Jonathan Pershing, a tall, bearded geologist and geophysicist who is now a top Kerry climate deputy. Pershing himself was special envoy for climate change in the Obama administration, with a leading role in negotiating climate deals with China, India, the European Union, Canada, and Mexico. He and I spoke at length in the light, airy headquarters of the William and Flora Hewlett Foundation in Menlo Park, California, where he headed the foundation's Environment Program before joining the Biden administration.

Dr. Pershing is an earnest, attentive guy who, like many tall people, walks with an almost imperceptible stoop. With his serious yet pleasant demeanor and well-pressed, pinstriped shirt, it's easy to imagine him in a dark suit, seated at some highly polished horse-shoe-shaped table in a foreign ministry or at the United Nations, gravely discussing climate policy. While an undergraduate, he minored in political science, philosophy, and English literature. After receiving his PhD, he first went to work for the State Department as an American Association for the Advancement of Science fellow. As they say, the rest is history. His arrival there coincided with the dawn of U.S. climate diplomacy, and as early as 1990, Pershing was participating in preparations for the United Nations Framework Convention on Climate Change (UNFCCC), which, in 1992, established the process for all subsequent major global climate negotiations.*

Despite his intimate engagement with climate diplomacy, Pershing assesses its achievements quite soberly. "There are people like myself," he told me, "who've been involved with this for thirty years or so, and we haven't succeeded. . . . We had some substantial success collectively, but you'd have to argue that we also had some significant failure, in the most objective terms that you could use." The most significant yardstick to use, he said, "is what happened to greenhouse gas concentrations," emissions, and policies to reduce and regulate them. "As a global matter," he said, "[the policies] have been insufficient." The question for U.S. climate policymakers and diplomats now is how to arrest rising emissions and get them to plunge steeply downward. As climate activists and others often say nowadays, "We are out of time."

---

* The first meeting of the UN Framework Convention on Climate Change took place in 1991, and negotiations led to the signing of an international treaty in 1992, by which the signers agreed to a framework under which they would cooperate to stabilize global emissions of greenhouse gases to limit climate change, to avoid "dangerous" human interference with the climate system, and to cope with the by-then-inevitable consequences of climate change. The treaty went into force in 1994.

How and Why Negotiations Fail

There's no doubt that getting the 195 nations of the world to meaningfully reduce their greenhouse gas emissions is extremely difficult. World leaders and high-ranking diplomats may make grandiose statements about the gravity of the climate crisis but still be unwilling to authorize game-changing action; as a result, frontline negotiators are therefore left to squabble over semantics and diplomatic processes: form bereft of content. Even though the leadership of a nation may be fundamentally uninterested in committing financial or political capital to the task of decarbonizing, they may gladly accede to all kinds of high-profile, public diplomatic events that create the perception that really important progress is occurring, as opposed to those events being merely another round of climate pageantry with captivating show-and-tell sessions.

These challenges create perverse incentives for diplomats to spend vast amounts of time working on matters that are at least manageable but that don't in themselves result in actual on-the-ground or in-the-air emission reductions. For example, negotiators have gotten very good at research and analyses of climate problems and creative information exchanges. Global climate diplomacy generates an ocean of meetings, conferences, council sessions, "high-level side events," dialogues, joint consultations, working groups, debates, salons, panel discussions, briefings, exhibitions, white papers, think tank collaborations, and online "knowledge platforms." It's understandable: all this is much easier than actually getting countries to reduce their emissions.

Creating voluntary agreements that nations are willing to support, let alone enforceable treaties their leaders are willing to sign and implement, requires a granular description of exactly what's going to happen when and how. This is unquestionably very hard work—even more so in today's fraught political environment. It is common for seemingly endless haggling to occur over hundreds of word choices and phrases in a single proposed text describing a subsidiary matter, so that resolution of some difficult issues can take years. And actual climate negotiating texts are acronym-studded treatises written in the dense technical language of international law, replete with tangles of references to prior international conventions, annexes, regulations, provisions, modalities, and procedures—as with this text from one climate policy paper: "The conditions for access to GSP+ include ratification of 27 international conventions listed in Annex VIII of Regulation (EU) 978/2012, the implementation of which is regularly verified by the GSP monitoring body." You get the idea. It's a miracle not only that *anything* gets done but that the participants stay sane.

Whereas all these activities would have a role to play on the way to a legitimate and enforceable climate treaty, nothing said here is meant to detract from the honorable intentions of the many dedicated, hardworking, patient diplomats and politicians engaged in the negotiating process. Even failed negotiations may create significant benefits, as will be described shortly. But the end result remains that climate negotiations have moved toward global consensus at a glacial pace, while the melting of actual glaciers has been accelerating. In short, despite thirty years of diplomatic fanfare, actual emissions reductions relative to an all-out, business-as-usual scenario, have been far too little and too late.

The immense danger with all these well-intentioned diplomatic exercises is that they become a self-perpetuating theatrical production divorced from the necessity of taking real actions commensurate with the scale of the emergency. As George Marshall puts it in his book *Don't Even Think About It: Why Our Brains Are Wired to Ignore Climate Change*, "The climate negotiations are always beginning, or in their favorite cliché, 'setting the stage' for the drama to come. The U.N. declared that the Vienna climate talks in 2007 'set the stage for the Climate Change Conference in Bali.' The U.S. Council on Foreign Relations explained that the 2009 Copenhagen conference 'set the stage for ambitious action' . . . [and] the Durban negotiations reached an agreement to 'set the stage for the big deal in 2015.'"

The Trouble with Gay Paree

I attended that "big deal," the 2015 Paris climate conference, at which the stage was again set—this time for the agreement reached there by the parties to go into force in 2020. However, the Paris Agreement depended entirely on voluntary pledges by participating nations to reduce their emissions through "intended nationally determined contributions." These INDCs were designed to keep the world from warming by more than 2°C, and preferably by no more than 1.5°C, by 2100, even though it was clear to many climate scientists that holding the increase to 1.5°C was already very unlikely and would not be especially safe, either.* It is challenging even to keep the global average surface temperature from increasing by 2°C by the early 2060s, without drastic action, given that the atmosphere currently contains more than 520 ppm of carbon dioxide–equivalent greenhouse gases.**

Another big difficulty with the Paris Agreement was that the INDCs fell short—way short. When climate experts initially did the math and ran their climate models, they could readily see that, in the aggregate, the initial INDC emissions-cut pledges were so weak that, *even if they were all fulfilled*, the world would *still* heat up by 3.5°C by 2100—and more thereafter. (Some of these pledges have since been strengthened, but scientists project that even if all the new pledges are achieved, the world will still experience an increase on the order of 2.5 to 2.7°C by 2100.) Moreover, as we have seen in ensuing years, many of even these already inadequate pledges remain unfulfilled.

---

* This is due to the nature of current infrastructure and to inertia in current energy use practice. The United Nations Environment Programme, in its Emissions Gap Report 2022, states that there is currently "no credible pathway to 1.5°C in place" and that "only an urgent systemwide transformation can avoid climate disaster." Emissions Gap Report 2022: The Closing Window, UNEP, October 27, 2022, https://www.unep.org/resources/emissions-gap-report-2022.

** Carbon dioxide–equivalent gas is a measure of the global-warming potential of all gases released, converted to their equivalence in terms of the quantity of CO_2 that would produce an equal amount of heating.

Nonetheless, upon the summit's conclusion, the Paris Agreement was hailed with celebratory pomp and circumstance as a great success, and there was talk soon after of raising the ambition of the INDCs in the years 2015 to 2020. So, once again, "the stage was set" for action at the next global climate meeting, the Glasgow climate talks in November 2021. But in 2021, after five years had elapsed and a scheduled two-year "Global Stocktake" of emissions pledges began, neither at the 26th UN Climate Change talks that year in Glasgow, Scotland, nor the next year at the 27th UN Climate Change talks in Sharm el-Sheik, Egypt, did participants produce collective emission-reduction commitments sufficient to ensure that global heating remain below the 1.5°C Paris target. So when the 28th UN Climate Change conference opens in November 2023 in Dubai, United Arab Emirates, the struggle to get the participants to strengthen their emission-reduction commitments will continue, eight years after the Paris Agreement.

Taking stock of the Paris deal today, one can see that some of the participants subsequently did strengthen their INDCs, but the collective pledges still remain far short of what's needed to meet the goals repeatedly set forth from 1992 to the present. The United States' exit from the agreement under President Trump was a further blow. However, on his first day in office, in early 2021, President Biden rejoined the Paris Agreement on behalf of the United States, and the Biden administration subsequently reinvigorated global climate diplomacy by bringing an inspiringly ambitious U.S., INDC pledge that, by 2030, it would reduce its emissions 50 to 52 percent relative to 2005. The Glasgow talks also produced an agreement by one hundred countries responsible for 70 percent of the world's methane emissions to reduce those emissions 30 percent from 2020 levels by 2030, complementing a U.S. Methane Emissions Reduction Action Plan.

Equally important were significant but still inadequate increases in climate funding from developed nations to developing ones so the latter could meet their climate needs and reduce coal dependence and deforestation. (The one hundred billion dollars in annual funding pledged in 2009 to the developing nations by 2020 still hasn't fully materialized.) In addition, a highly significant twenty-five-member "buyers' club" was inaugurated at Glasgow, including some of the world's largest companies, to create early market demand for innovations across eight "need-to-abate" sectors—steel, trucking, shipping, aviation, aluminum, concrete, chemicals, and direct air capture. (Today, these sectors represent more than a third of global CO_2 emissions, a fraction likely to grow.) Multinational lenders were also put on notice to "green" their acts. But, in another sense, there was no beef on the bun. Thanks to the roughly 250 billion additional tons of greenhouse gases of all kinds emitted from 2016 to 2021, the earth's average surface temperature continued rising and is now 1.2°C higher than in pre-industrial times, with 2022 not surprisingly being one of the ten hottest years in the modern instrumental record.*

Paris, Je T'Aime

Yet, despite all its shortcomings, the Paris Agreement is what we have, and it would be inadvisable to throw the baby out with the bathwater and jettison the Paris process. It is part of the solution, and the world currently has nothing better that engages 195 nations. The UN climate negotiations provide a forum for the global sharing of knowledge, insights, and expertise, and for discussion of climate risks and conflict avoidance. They're also an opportunity for tens of thousands of civil-society participants to build partnerships and coalitions and to showcase inspiring solutions. So, while UN negotiations have failed to bring down global emissions and concentrations of greenhouse gases, they do provide incredibly valuable services to the world community and the global effort to slow climate change.

Thus, although in some respects the Paris Agreement and process

---

* The nine warmest years on record are 2014–2022.

set back the cause of climate protection by perpetuating a decades-long illusion that such protection was ensured and that a meaningful cut in emissions was imminent, the process has prompted some uncommitted nations to commit to reducing their emissions and has motivated others to increase their level of ambition. All told, it has furthered global cooperation on climate protection and made the world a better place than it would be had these meetings never happened. Its accomplishments also include:

* Advancing authoritative, peer-reviewed climate science and normalizing reliance on it;

* Developing protocols for gathering national emissions-reduction commitments from virtually all nations of the world;

* Insisting in its deliberations on proper monitoring and reporting of results obtained in emissions reductions by its signatories;

* Setting up a process for strengthening national emissions-reduction goals every five years; and

* Helping to mobilize financial and technical support for developing nations.

In addition to raising global awareness of climate problems, the UNFCCC negotiations shed light on unresolved bottleneck issues and foster continuous engagement by major emitting nations. The talks also bring global expertise to bear on difficult technical issues, such as those pertaining to carbon accounting, carbon risk disclosure, and standardizing the reporting of mitigation actions, along with how global climate financial assistance is to be allocated. Furthermore, the Paris process brings countries together each year and encourages them to take stock of their progress toward their INDCs, urging them to ramp up their ambition and, meanwhile, build domestic climate resilience. It also shines a spotlight on warnings about climate change consequences issued by the IPCC.

So, despite their weakness and lack of enforcement mechanisms, the UNFCCC and its Paris-style climate agreements nonetheless are the embodiment and symbol of the world's commitment to combating climate change. In addition, as professors Charles F. Sabel and David G. Victor aptly put it, "Paris enjoys unique legitimacy among international organizations as the authoritative voice of public opinion in global climate affairs."

An Awesome Agenda, an Overarching Concern

In the absence of a truly satisfactory global climate negotiation process, how should the United States independently conduct its international climate policy? The Biden administration's climate plans are designed to produce both short-term global emissions reductions and net-zero global emissions no later than mid-century. The policy explicitly recognizes that international climate action "is more necessary and urgent than ever," now that new scientific evidence and current emissions trends demonstrate that the speed and scale of action needed are "greater than previously believed." Thus, the president's 2021 executive orders direct all U.S. agencies with significant international programs to integrate climate concerns into their work, their strategies, and even the management of their facilities; the orders also commit the United States to working bilaterally and multilaterally "to put the world on a sustainable climate pathway." By rejoining the Paris Agreement framework President Trump abandoned, Biden committed the United States to establishing national emissions-reduction goals in line with that agreement, thereby signaling that it embraces the agreement's overall goals of ensuring a safe global temperature, fostering climate resilience, and providing adequate climate financing for less affluent nations.

So, the thrust of current U.S. international climate diplomacy is to exercise U.S. leadership to significantly increase global climate ambition. The strategy involves pressing international bodies, such as the G7* and the G20,** to fully integrate climate concerns into their work and influencing other international bodies that work on relevant issues (climate finance, clean energy, aviation, shipping, the Arctic, oceans, sustainable development, and migration) to do likewise. In addition, the Biden administration intends to use its influence with the World Bank Group and the IMF to support the Paris goals through climate finance, economic stimulus, and debt-relief initiatives.

Penny-wise and Pound-Foolish

The funding provided for these purposes, however, is likely to be far short of the need. Currently, the United States spends only $2.5 billion a year on international efforts to address climate change. That's not even enough to meet the climate investment needs of a single major city, and it amounts to only 0.3 percent of the federal budget. (In 2022, Biden proposed spending $11.4 billion to help developing nations with climate change–related costs. Congress promptly slashed the amount to $1 billion, putting the administration in a tough corner.)

The jury is out at present on how much money the United States will be able to spend combating climate disruption in the years ahead, given the reluctance of some in Congress to fund global cli-

---

* The G7 is comprised of Canada, France, Germany, Italy, Japan, the United Kingdom, and the United States.

** The G20 is comprised of Argentina, Australia, Brazil, Canada, China, France, Germany, India, Indonesia, Italy, Japan, Republic of Korea, Mexico, Russia, Saudi Arabia, South Africa, Turkey, the United Kingdom, the United States, and the European Union. The 2020 G20 Summit was held virtually, under the presidency of Saudi Arabia, and little new specific action on climate change was proposed.

mate action. But the Biden administration has put its A-team on the job: Secretary of State Antony Blinken and Secretary of the Treasury Janet Yellen, in coordination with Special Envoy John Kerry and the heads of important domestic agencies, are focused on mobilizing climate finance to support sustainable development abroad.* Moreover, the administration has clearly signaled its intent to assist developing countries in reducing emissions, protecting critical ecosystems (like the Amazon rain forest), building climate resilience, and, possibly most significantly, "promoting the flow of capital toward climate-aligned investments and away from high-carbon investments." The size of these capital flows will be an acid test of the Biden administration's seriousness (and that of succeeding U.S. administrations) in confronting international climate change and of its determination and ability to get adequate funding through Congress. It bears repeating that the continued failure of the United States and other developed industrial nations to provide the capital promised to the developing world from 2009 to the present has been a major reason developing nations have not moved more quickly to arrest their growing emissions.

Unfortunately, the United States is not alone in lowballing international climate assistance. The United Nations' Green Climate Fund (GCF), the world's main instrument of climate-related financial assistance—which was to supply $100 billion per year by 2021—has actually raised less than $2 billion a year since its founding in 2010 and has dispersed even less.** By greatly increasing aggregate support for developing nations' climate mitigation and adaptation efforts, the United States and other developed nations would be

---

* Secretary Yellen is known to support a U.S. carbon-fee-and-dividend policy and efforts to force financial institutions to fully disclose their climate risks and to raise the standards they must meet to pass regulatory scrutiny from the Treasury Department's Financial Stability Oversight Council.
** The United States' last pledge to the GCF, under President Obama, was only $750 million a year for four years. Only $1 billion was ever delivered; the Trump administration ended all climate funding assistance and withheld the remaining $2 billion of the Obama pledge after Trump took office. See Joseph Aldy, Tim Profeta, and Himamauli Das, "Department of the Treasury," transition memo, Climate 21 Project, 2021, https://climate21.org/treasury/.

able to assist them in raising the emissions-reduction commitments they made at the global 2015 Paris climate conference. International funding vehicles should be doing some heavy lifting here, but many of them need replenishment—starting with the GCF and extending to the Adaptation Fund, the Global Environment Facility, the Least Developed Countries Fund, and the Multilateral Fund for the Implementation of the Montreal Protocol, in addition to the coffers of development banks (but targeted solely for climate purposes).

Quadrupling annual U.S. international climate assistance to $10 billion would seem to be a much-needed and de minimis down payment on supporting international climate efforts; given the overwhelming urgency of decreasing global emissions, this spending should rise quickly to far greater levels. The United States, after all, has contributed the most to the cumulative atmospheric carbon burden and has one of the world's biggest per capita carbon footprints. By being more generous with its international climate aid, it would be able to "buy down" the emissions of other nations on the cheap: emissions reductions in developing countries can cost as little as $5 per metric ton. Even at an average cost of $20 per metric ton, a $100 billion annual commitment could abate 5 billion metric tons of greenhouse gases, or about 10 percent of all global GHG emissions. This is a paltry sum given the $3 to $4 trillion the United Nations estimates is needed annually for the global transition to a low-carbon economy.

More generous spending would also increase U.S. influence with international funding entities and private financial institutions. According to one of its own guidelines, 35 percent of World Bank financing is to have climate "co-benefits." In practice, this means that, for the entire world, the bank provided only $16.6 billion for climate-related projects in 2019—an average of only $82 million per nation, of which half is for climate adaptation rather than emissions reductions. Multinational development banks collectively have a long history of financing oil and gas development, and as recently as 2018, they were not reporting how much those annual expenditures were. These organizations seem to be in urgent need of radical reform so that they focus primarily on helping developing nations achieve and enhance their emissions-reduction goals. (This is true also for the banks China uses to finance its international investments, such as its Belt and Road Initiative, a massive global infrastructure project.)

Both international and private funding institutions could be obliged to meet high standards of climate risk disclosure and management—which tends to reduce fossil fuel funding. So, whereas the United States needs to continue to fund international organizations that support fulfillment of the INDCs of the Paris Agreement, it also must increasingly act through supplementary bilateral and multilateral deals. To the extent possible, it should work with its allies within global powerhouses such as the G7, the G20, the World Bank Group, and the IMF to raise global climate assistance to developing nations and lower the emissions of high-emitting nations.

Zeroing In on the Super-Polluters

Since the United States is responsible for only 14 percent of current annual greenhouse gas emissions, the nation obviously needs the cooperation of China, India, and others if global emissions are to reach net zero and then go carbon negative. To figure out the most effective way to slash global emissions, we need to understand which nations and companies are most responsible and where within those nations the emissions are concentrated. The poster child for successful bilateral climate diplomacy is doubtless the 2014 climate agreement between the United States and China in the run-up to the 2015 Paris climate summit. After patient, high-level diplomatic contacts throughout 2014, the two nations finally agreed that the United States would cut its carbon emissions by 26 to 28 percent below its 2005 levels by 2025 and that China would peak its emissions by 2030. It was the first time China had ever agreed to any emissions restrictions. This deal between the world's two largest carbon emitters broke what had been a stalemate on global emissions reductions that likely would have handicapped or doomed the 2015 Paris conference.*

China is the world's largest emitter of greenhouse gases. At 30 percent of total global emissions, it produces more than twice the emissions of the United States, which is second in line, and four times those of India, the world's third largest emitter. China's emissions are growing rapidly, and in late 2022, the country announced that it would increase its coal production by 300 million tons, even though already, in 2020, it built three times the new coal capacity of the entire rest of the world. Not surprisingly, China's emissions rose 15 percent in the first quarter of 2021 and are approaching 12 billion metric tons of CO_2 a year. After the European Union as a whole, the world's fifth largest carbon polluter is Russia. So, China, the United States, India, and Russia must be induced to sharply reduce their emissions.** They also must be held to account for massive GHG emissions leaks and, in some cases, inaccurate emissions reporting. With the exception of Japan, at 3 percent of global emissions, the remaining nations of the world each produce 2 percent or less of annual global GHG emissions, with most under 1 percent.

Therefore, regardless of their venue or players, the most important future climate negotiations will be those that influence China and India. Whereas both nations have undertaken domestic clean-energy programs on their own timetables, both have also evinced great reluctance to be governed by any top-down international emissions-reduction agreement with compliance mechanisms. They prioritize their own economic growth and development, and their

---

* The stalemate was over how responsibility for reducing emissions should be shared between the developed and developing nations. Taylor Dimsdale, "U.S. Climate Diplomacy Leadership: The Evidence," briefing paper, E3G, November 1, 2016, https://www.jstor.org/stable/resrep17902.

** The European Union wants to be the world's first climate-neutral continent and is well on its way to that goal.

national sovereignty, far ahead of participation in any international scheme involving an emissions-reduction goal, any target date for that goal's achievement, or any legally binding enforcement provisions. In addition, while almost every other major country with a long-range energy plan has pledged to reach net-zero GHG emissions by 2050, China unilaterally has decided to give itself another decade to do so.

As noted in an excellent article by Lindsay Maizland for the Council on Foreign Relations, the problem of climate change, in China's view, was created by industrialized Western nations, and absent a willingness by the West to fund the costs of decarbonization for the developing world, China feels justified in pursuing independent, sovereign energy and climate policies based on its own goals and timetables. India has a similar outlook. Yet neither nation is impervious to world opinion, and both are aware of the impacts that climate change is having on them.

The China Conundrum

The impact of China's fossil fuel dependence and promotion is worse than might be imagined. As I've just pointed out, the central problem is its deadly romance with coal. In 2019, about two thirds of China's electricity was produced by burning coal—half the world's coal production and consumption—and tens of millions of jobs are dependent on its coal sector. China's coal production hit an all-time high in 2021, when five new major coal power plants and three billion-dollar coal mines were opened. Difficult and costly as it will be to pull the plug on coal in China, the country is making the problem worse by approving new mines at an increasing rate and by continuing to build new oil pipelines and coal power plants, including 240 of them in countries participating in its Belt and Road Initiative. That initiative *by itself* is likely to cause global warming to exceed 2.7°C. Furthermore, the customers for this high-carbon infrastructure no doubt expect it to operate for decades, locking in dangerously high emissions.

In my interviews with Jonathan Pershing, he did not offer a clear prescription for convincing China to reduce its emissions—except to say that doing so would require the United States to engage directly with Beijing. But bilateral engagement alone is not enough. Prime evidence of this is the United States–China agreement on the sidelines of the Paris conference: in a joint presidential statement, the two heads of state promised "to work towards strictly controlling public investment flowing into projects with high pollution and carbon emissions both domestically and internationally." It made for a great photo opportunity but changed nothing about China's heavy involvement in creating fossil fuel infrastructure overseas.

Pershing did suggest that one approach to deter China from this latter course would be for the United States to engage more actively diplomatically with Japan and South Korea, which together are building more coal-fired capacity than China. By exerting influence with these allies, with whom the United States has greater sway, to temper their coal use, Pershing felt that the United States might have some collateral influence in deterring China. The idea of using America's alliances in the service of getting China and India to reduce their emissions more quickly is similar to a proposal by former U.S. climate negotiator Todd Stern and Biden energy adviser John Podesta. The two suggested that if the United States marshaled allies such as the European Union, along with the United Kingdom, Canada, Japan, Mexico, and New Zealand, in a joint policy effort, their combined economic muscle could oblige China, India, and others to make concessions.

Former California governor Jerry Brown, who has been deeply engaged in international efforts to reduce greenhouse gas emissions for many years, believes that prolonged, serious, direct discussions between the United States and China are imperative for dealing with climate change and that threatening China or Russia bombastically in an age of nuclear weapons on hair-trigger alert puts the world in peril without helping to resolve the climate crisis. However, it is encouraging to remember that, despite the current strained relations between the United States and China, the United States, even during the Cold War with the Soviet Union, was still able to negotiate and sign the Strategic Arms Reduction Treaty in 1991 (START 1), to reduce the threat of nuclear war. It was also able, in 1987, to sign a treaty to collaborate with the USSR on space exploration, despite earlier Sputnik-era competition.

As with nuclear war, climate change presents an existential threat, and both "demand coordination and cooperation to avert an outcome that destroys everyone." Mutual self-interest may pave the way for greater cooperation between the United States and China on reducing emissions, even though, in the words of one expert, "climate change is a more complex problem [with] more actors, greater uncertainty, difficult trade-offs, and questions of equity between countries and generations."

In 2021, China indicated an openness to reengage with the United States over climate issues. However, it is likely that, in order to persuade China to reduce emissions more quickly than it otherwise would, the United States might need to make concessions on trade or even human rights. Currently, the two nations are at odds over China's theft of intellectual property, its moves to exert political control over Hong Kong, its long-standing domination of Tibet, its harsh repression of the Muslim Uyghurs in Xinjiang Province, and its belligerent behavior toward Taiwan. There is also the issue of China's territorial expansion in the South China Sea. Finding common ground will require walking some diplomatic tightropes.

Dealing with Russia—It's a Bear

It was chastening, but not unexpected, to learn that Pershing also seemed to have no easy prescription for dealing with Russia and seemed quite pessimistic that the United States had much near-term leverage. Whereas he believes that China and India see climate as only one part of a complex of domestic issues, the Russian economy is heavily dependent on its oil and gas exports; 45 percent of its revenue came from these sources in 2021. Unless Russia is able to both diversify its economy and increase its energy-efficient industrial base, it will face an increasing economic crisis as carbon cap-and-trade systems in the European Union and elsewhere, along with widespread antipathy toward its invasion of Ukraine, reduce the demand and prices for its fossil fuels. In addition, cross-border carbon adjustments—likely to be implemented soon in the European Union and elsewhere—will penalize imports of manufactured goods from energy-inefficient countries, including Russia. (Some of these concerns relating to the European Union are now moot, as Western economic sanctions against Russia have severely cut its energy exports to western Europe.)

Under President Vladimir Putin, who has ruled the nation since 1999, Russia faces domestic instability and has tried to deal with this through internal repression of dissidents and by trying to destabilize its neighbors and the United States in order to maintain power. These were the challenges *before* Russia's invasion of Ukraine, which, at the time of my dialogues with Pershing, hadn't yet taken place. Even then, Pershing said, "I don't see many [climate] strategies with respect to Russia except . . . to isolate them and to have them be in a minority, where they cannot disrupt other parts of global [climate] agreements."

One strategy to reduce the climate damage from Russia's fuel exports is to seek to permanently eliminate demand for them in Europe and elsewhere and to try to stave off future demand for those fuels from China (a goal that seems even less attainable). Currently, Russia seems able to function despite reduced fuel sales to Europe and has itself cut its natural-gas exports to the European Union by 80 percent, in an effort to force Europe to reduce its support for Ukraine.

Getting Our House in Order

Of course, the United States itself is still heavily addicted to fossil fuels and will have far greater influence on climate change with the rest of the world once its own economy becomes a model of low-carbon sustainability. In 2019, the United States got 80 percent of its energy supply from fossil fuels; it was the world's leading producer of natural gas (just ahead of Russia) and also the world's largest oil producer that year. Along with China and India, it uses more fossil fuels than the entire rest of the world combined.

The leverage the United States has over the emissions of other nations is therefore not unrelated to its own massive domestic fossil fuel production. Its deep engagement with fossil fuels weakens its credibility when it tries to jawbone other nations to cut their fossil fuel reliance. Alaska, Wyoming, and North Dakota, for example, are heavily dependent on fossil fuel revenue. In 2019, as much as 70 percent of Alaska's state revenue came from energy production, as did 52 percent of Wyoming's, while North Dakota has been enjoying an oil and gas fracking boom that began not long after the first well was fracked in the Bakken Formation in 2006.* Colorado, New Mexico, Oklahoma, Texas, and West Virginia are also significant fossil fuel producers.

Some of the domestic resistance to effective climate policies in the United States undoubtedly has its roots in U.S. dependence on fossil fuels. Thinking through how the most fossil fuel–dependent areas of the country can be weaned would be a good exercise for the United States in trying to tackle high-emission countries farther afield.**

---

* Underlying parts of North Dakota, Montana, and Saskatchewan, the Bakken Formation holds billions of barrels of oil and trillions of cubic feet of gas.

** Few if any major nations are coming to the climate negotiating table with clean hands. Germany is still burning copious amounts of lignite, the soft, brown, highly polluting form of coal. Canada is hell-bent on developing its tar sands resources, the world's dirtiest fossil fuels, Japan, which gets 87 percent of its energy from fossil fuel, is the world's largest importer of liquified natural gas and the third largest importer of coal.

Simpatico Nations

When it comes to climate policy, the United States is more in sync with the advanced industrialized nations of western Europe than with the emerging economies of the East or the Global South. The advanced nations understand the economic advantages of decarbonization and are not afraid to relinquish some of their autonomy to get them. Collectively, the twenty-seven nations of the European Union today account for only about 10 percent of the world's emissions, and those emissions are declining rapidly. While the group's economy has grown more than 60 percent since 1990, its emissions have fallen 25 percent below 1990 levels. Moreover, in 2020, the European Union raised its emissions-reduction targets to an ambitious 55 percent below 1990 levels by 2030.

Decarbonizing the Developing World: High Stakes, Huge Barriers

The major global diplomatic challenge on climate is to convince the top four emitting nations, and reluctant developing nations whose emissions are growing swiftly, to set ambitious emission-reduction targets. Engagement with emerging and developing nations is essential to prevent construction of emissions-generating fossil fuel infrastructure, such as coal or natural-gas power plants or pipelines, that would "lock in" emissions far into the future. In addition, to protect tropical forests, which are the global reservoirs of carbon and biodiversity, it is also an absolute top global priority to engage with Brazil specifically. Over a few decades, various Brazilian governments have countenanced the destruction of vast swaths of the Amazon rain forest; former Brazilian president Jair Bolsonaro, a climate denier, is merely the latest and most flagrant offender.

As can be seen in the relative success of the EU nations in reducing their emissions, the wealthier a nation is—such as the United States, Germany, the Netherlands, or Norway, with their advanced industrial and technological base—the easier it is likely to be for them to plan, manage, and afford major cuts in GHG emissions and a clean-energy transition. But other nations of the world, whose emissions may not be great individually, nonetheless are collectively significant, and so, ultimately, their emissions also must be eliminated.

Many such nations are in earlier stages of industrialization and economic development, and some are dealing with economic crises, war, civil unrest, or extreme poverty; for them, emerging from these dire conditions is a much higher priority than reducing their carbon emissions. These troubled nations include Afghanistan, Bangladesh, Ethiopia, Iran, Pakistan, Sudan, and Syria. Some nations, like Kuwait, Nigeria, Russia, Saudi Arabia, and Venezuela, have economies that are enormously dependent on fossil fuels, relying on these resources to provide government revenue for all domestic purposes, sometimes including vital food and water subsidies for poor people. (Some nations, like Iraq and Libya, suffer from both conditions.) Because global decarbonization will have profound geopolitical implications—altering the global balance of power, the flow of capital and resources, and migration—successful and cooperative climate diplomacy will become increasingly vital to facilitating a smooth, low-carbon energy transition and minimizing conflict in the decades ahead.

For the purpose of simplification, we can refer to the group of nations with sometimes overwhelming domestic issues apart from their fossil fuel dependency as "the overwhelmed," and to those whose economies are tightly bound with fossil fuels as "the dependent." The overwhelmed group needs economic and, in some cases, military help from wealthier and more advanced economies in order to deal with their various crises and, simultaneously, to both reduce emissions and prepare for the inevitable consequences of the climate change to which the world is already committed.

The dependent group will require major assistance in diversifying their economies away from fossil fuels, as well as other economic and climate change–adaptation assistance. Imagine how hard it would be to decarbonize the U.S. economy if, like Venezuela's, the government depended on oil for half its revenue! Government leaders would be racking their brains to figure out how to increase fossil fuel production and sales. "Petrostates" could lose $9 trillion in revenue by 2040 in a decarbonized world, according to Carbon Tracker, and $7 trillion dollars according to the International Energy Agency. As decarbonization takes hold, fossil fuel prices may fall at the same time as export volumes plummet. Nations and regions that don't prepare adequately for this extraordinary transition could experience a tectonic economic shock. The global decarbonization toward which climate diplomacy aims, and to which the United States and the European Union aspire, is therefore a profound threat to many nations and could easily lead to their political destabilization and to massive internal displacement and external refugee flows, even potentially culminating in the downfall of fragile governments.

Trade: Carrots and Sticks for Climate

The dependency of less developed petrostates on aid may give developed "climate champion" countries considerable leverage with these nations: in some cases, development assistance may be all that's needed; in others, a combination of assistance offers coupled with implied trade restrictions may be needed to overcome resistance to decarbonization. Developed nations can use carrots, such as partnership and cooperation agreements,* as well as free-trade agreements, to influence these and other nations to accelerate their decarbonization efforts. This quid pro quo can provide diplomats with leverage when negotiating the steep, slippery slope of achieving actual emis-

---

* Cooperation agreements can provide assistance with education, research, and training in energy transition policy development and technologies and in national security assistance for countries facing domestic and regional threats.

sions reductions, as opposed to orchestrating public events that make for good press releases but leave emissions unchanged.

When I asked Jonathan Pershing what could be done to make the United Nations' climate diplomacy process dramatically more efficient and productive under current political circumstances, his response was more cautious than encouraging. "The things that would be more efficient also tend to step on rights and authorities that countries hold as sovereigns." Yet he allowed that climate agreements could have a trade element to them, in which trade could be conditioned upon compliance with a global climate agreement.

To that end, the trade ministers of both France and the Netherlands have called for the Paris Agreement to be incorporated in future EU trade agreements. This would mean that each party to the trade agreement would need to live up to its commitment to upgrade its most ambitious emissions-reduction commitments every five years. Another way to incorporate trade in climate diplomacy, the trade ministers suggested, would be to include a cross-border carbon adjustment requirement in EU trade policy, to help reduce the European Union's carbon footprint. This would exert pressure on other nations that wanted to trade with the European Union by subjecting them to tariffs if they failed to also impose a domestic price on carbon through either a carbon tax or a cap-and-trade system.

In another effort to make sure that trade is used more effectively in the service of climate protection, New Zealand, together with Costa Rica, Fiji, Iceland, Norway, and Switzerland—nations with small, trade-dependent economies—announced an effort in 2019 to negotiate a new agreement on climate change, trade, and sustainability. Once the agreement is complete, the parties intend to open the treaty to members of the far larger and more powerful 164-member World Trade Organization. Initially, the signatories will agree to remove tariffs on environmental goods and services (those that pertain to environmental protection and management) to increase their sales. They also will take measures to reduce or eliminate fossil fuel subsidies and will develop eco-labeling programs.

The Lima-Paris Legacy: Universal Mobilization

At a time when emerging nations appear disinclined to relinquish their sovereignty to an international body or treaty, Pershing seemed resigned that major changes in how global-scale climate diplomacy is conducted were unlikely. He did, however, see hope that diplomats could advance the climate dialogue by amplifying ongoing efforts to increase the engagement and support of the business community and the banking sector in backing stronger emissions-reduction efforts.

That's the approach that former Peruvian environment minister Manuel Pulgar-Vidal took during his presidency of the global climate talks held in Lima a year before the Paris summit and a year after climate negotiators had turned in another lackluster performance at the 2013 UN Climate Change Conference in Warsaw, Poland. There, some eight hundred civil-society representatives had walked out toward the end of the meeting in protest, frustrated at the slow pace of progress and the deference shown at the conference to the coal industry.

A major aim of the 2014 Lima conference, Pulgar-Vidal told me, was to avoid a second global climate summit debacle after the 2009 Copenhagen global climate talks. He was eager to give people a hopeful, optimistic vision of what could be achieved in the coming Paris negotiations. Nonetheless, following the previous year's events in Warsaw, he had his work cut out for him in Lima.

The challenge didn't daunt him, however. "Sometimes failure is a good platform for success," he said. When he had envisioned the Lima talks, he saw their role as being not only to further negotiations but also to mobilize global climate action. Not coincidentally, one of the signature accomplishments of the meeting was the Lima-Paris Action Agenda. The underlying idea was to show that a global transformation to a low-carbon economy was under way and could be accelerated after Paris. Pulgar-Vidal also sought to broaden the whole climate dialogue to include issues related to forestry, ocean management, and sustainable agriculture. Pursuant to the Lima meeting, this has largely been accomplished.

Perhaps a bigger idea was to bring business and civil society back into the mainstream of climate negotiations, showcasing their accomplishments and giving civil society space within the conference venue to hold meetings—rather than forcing them into satellite hotels, as in the past. As Laurent Fabius, president of the Paris summit, commented, "To deeply transform our economic models and societies, we not only need universal awareness, but also universal mobilization."

Pulgar-Vidal wanted cities, regions, indigenous communities, rural areas, companies, investors, foundations, and universities all to become engaged in accelerating the transition to a low-carbon economy, while also exposing them to the diplomatic process. To this end, the Lima-Paris Action Agenda left another impressive legacy, in the form of a Global Climate Action Portal, an online space where the climate achievements and commitments of almost 31,000 "climate actors" are described. Most of the actions documented on the platform are those of "nonstate actors"—companies, investors, organizations, cities, states, and regions. Begun in 2014 by the UNFCCC, the portal was the first—or one of the first—to compile the achievements and commitments of these change agents and to show how formidable their combined contributions could be. The final product, a testament to Pulgar-Vidal's vision, includes a growing roster of 11,000 cities and 286 regions representing over 15 percent of the world's population and close to 14,000 companies.

A Radical Approach to Slashing Emissions

George Marshall, in *Don't Even Think About It*, has an interesting solution to the problem of inexorably rising emissions: rather than focusing exclusively on the emissions resulting from the use of fossil fuels, pay a lot more attention to curtailing those fuels' supply and production. Ignoring the political realities for a moment, conceptually, wouldn't it be a whole lot easier to just taper off the supply of fossil fuels, rather than allowing that supply to grow unconstrained while having to struggle after the fact to control emissions? As fossil fuel production declined, emissions would *automatically* fall.

A logical point of departure for applying this strategy would be, first, to set a ceiling on new fossil fuel exploration and development, then a complete moratorium on new production, followed by mandatory annual decreases. This concept is consistent with an ongoing NGO campaign for a "Fossil Fuel Non-Proliferation Treaty Initiative" to cap and then gradually shut down fossil fuel production. The campaign supports the creation of a global registry of fossil fuel production, to promote greater transparency on fossil fuel supply for purposes of multilateral nonproliferation treaty negotiations. Relatedly, the UN Environment Programme has called for a 6 percent annual reduction in fossil fuel production, in service to the Paris summit goal of keeping global temperature from rising more than 1.5°C above pre-industrial levels.

In a perfect world, these decrements would be enforced, perhaps by some powerful global organization—one might imagine, for instance, an International Atmospheric Authority with strong enforcement powers—but an institution of this nature is politically infeasible today. Theoretically, production reductions could also be applied in the context of national fossil fuel production quotas. This type of solution, however, suffers from the "pussycat syndrome"—as in "Who will bell the cat?" No credible political pathway for its adoption currently exists. Perhaps such an orderly exit from fossil fuels could at some point be voluntarily negotiated. Or, someday, climate conditions might get so bad that the global community is frightened enough to accept an international enforcement regime.

Solving the Diplomacy Conundrum

Many leading participants in the UNFCCC negotiating process are by now totally invested in the process and have gradually become true believers in the inevitability of its ultimate success. But, time and again, the UNFCCC, with its 195 members, has shown itself to be too big, unwieldly, and conflicted because of the participating nations' differing interests. In private negotiating assemblies and meeting rooms, oil-producing states and developing nations are often arrayed against advanced, developed nations. A coalition of the reluctant has often resisted a coalition of the willing, with the former doing its best to slow or, at times, overtly obstruct global negotiations. As a result, progress has often been reduced to the pace of the most reluctant participants.

The solution to this dilemma is for the United States to participate in the global negotiations but simultaneously to vigorously engage in bilateral and multilateral climate diplomacy, using all the tools in its toolbox—development aid, cooperation agreements, alliances, tariffs, and trade—to secure agreements to mitigate emissions. Aid and cooperation of all kinds must, in addition, be provided in return for tangible mitigation action, on a "pay-for-performance" basis. At the same time, the United States must clean its own house by more deeply reducing its own fossil fuel dependency so as to be a credible leader on the world stage in this effort.

Often, the confidence of participants in the UNFCCC process reposes on whether the work of the IPCC, the UNFCCC's authoritative scientific research and assessment arm, convinces them that the process is confronting the reality of climate change and putting the world on notice about it while using the best available diplomatic means to keep it in check. The IPCC, however, as a very conservative scientific body, moves so slowly that its reports are often already out of date in important respects by the time they are published—a big liability when it comes to confronting rapidly advancing climate disruption. These assessments also tend to err on the side of under-estimating the dangerous impacts and risks of climate change while stressing the scientific unknowns, offering detractors a field day.

Finally, before each quinquennial report is released, with thousands of pages of small print and citations, plainspoken summaries of the underlying scientific reports are prepared. In this final stage, politicians concerned with politics and public perception of climate change, rather than science, get an opportunity to edit the summaries and downplay the grave implications of the underlying scientific reports. Because those reports tend to be long and dense, the expurgated political summaries are what political leaders read, what the media reports on, and what the unsuspecting public believes.

Partly as a result of this, for more than thirty precious years now, nationalism and the short-term economic concerns of nations to raise their living standards have been allowed to take precedence over more urgent and transcendent, long-term, life-and-death, planetwide climate concerns. Major participant nations in the decades-long climate negotiations have tended to set very long-term targets—2050 for the world—to reach carbon neutrality, while evading difficult decisions in the shorter term. Chinese President Xi Jinping, however, has only agreed that China will "aim" for carbon neutrality by 2060.

This game of kicking the can down the road has been just about played out, to the point where achieving the UNFCCC's 1992 basic goals is no longer attainable. In 2023, the cumulative total of green-house gases released into the world's atmosphere had risen to 422 ppm, a level not seen for millions of years (long before modern humans walked the earth) and 50 percent higher than pre-industrial levels.

Meanwhile, the pernicious destruction of the world's rain forests, where carbon has been safely stored for millennia, continues. So does rapid global population growth, habitat destruction, and the loss of species and wildlife abundance, plus rising numbers of climate refugees. Extreme weather events are intensifying, and environmental impacts are worsening—to take just one example, vast areas of the Great Barrier Reef have now died. Clearly, this is an awesome agenda of global problems for any world leader or diplomat short of Superman or Wonder Woman.

When the UNFCCC was established in 1992, its primary purpose was to stabilize greenhouse gas emissions "at a level that would prevent dangerous anthropogenic [human-induced] interference with the climate system." Tragically, that ship has already sailed: dangerous global climate change *has arrived and can no longer be prevented*, though it can still be mitigated and prevented from roaring completely out of control.

The absence of a truly effective international agreement to cut global emissions is one reason many observers are beginning to be more receptive to the idea that various geoengineering schemes to prevent global overheating might be necessary on an emergency basis to slow the emissions whose impacts are now scourging the earth. The next chapter offers a description of the main geoengineering options, along with their pros and cons.

16.

Geoengineering: Friend or Foe?

Wouldn't it be terrific if we could just magically shunt sunlight away from the earth with mirrors, to make it cooler? Or make the atmosphere itself, or the oceans, shinier and more reflective, with similar results? At first blush, it all sounds appealing; reducing the sunlight reaching the earth by just 1 to 2 percent would be enough to cool the planet significantly.[*]

While we're cooling the earth anyway, how about rapidly planting a trillion trees around the world—a vast green canopy over much of the planet, especially on marginal and degraded land? As the trees grew, they could continuously incorporate more excess atmospheric greenhouse gases in their tissue for decades or even centuries.

Finally, what about just pumping excess CO_2 deep into the ocean, or dumping carbon-rich unwanted vegetation into the deep, or ingeniously fertilizing algae on the ocean's surface so that, when they died, they would take their embodied CO_2 into the depths with them? Or, why not churn up the ocean surface and create trillions of tiny bubbles to reflect the sun's rays; or alter clouds over the ocean to brighten them? Maybe send powerful jolts of electricity into the ocean to

---

[*] Computer models built by Harvard physics professor David Keith indicate that deflecting some of the sunlight striking the earth could reduce the global heating expected by 2100 by at least half. See James Temple, "Climate Hackers: One Man's Plan to Stop Global Warming by Shooting Particles into the Atmosphere," *The Verge*, August 24, 2016, https://tinyurl.com/456hz93r.

de-acidify it chemically, so it absorbs more CO_2 from the air? Or add finely ground, acid-absorbing rock to seawater to achieve the same results? Perhaps, in addition, millions of farmers and ranchers everywhere could manage their lands so that billions of tons of greenhouse gases were locked deep in the soils, simply and affordably.

Does all this seem far-fetched? Why are we even talking about this?

The reason people are already advocating these measures is that the world is currently far from being on track to keep global average temperature increases below 2°C, and emissions continue to climb rapidly. Consequently, so do GHG concentrations. Without some way of reducing them—even if the world achieves net-zero emissions by 2050—it would take a long time for temperatures to return to normal, depending upon how intense positive climate feedbacks have become by then and, subsequently, what we do to decarbonize the atmosphere. So, some scientists and engineers believe the answer lies in developing and applying new technologies now—and on a massive scale.

But what almost all these often innocuous-sounding proposals have in common is uncertainty as to their feasibility, scalability, cost, and potential to cause more harm than good. So, while the world still has a brief chance before being *forced* to embrace these potentially dangerous measures, let's examine the risks and projected benefits of the Hail Mary passes that leaders and nations might be tempted to gamble on in the throes of a climate catastrophe more obvious even than today's.

Until now, without naming it, we've been talking about various forms of geoengineering, which describes large-scale, intentional interventions in the climate system to alter the climate. Apart from solar radiation management (SRM)—the various means of deflecting sunlight away from the earth, including those just mentioned—and the direct air capture of GHGs, the universe of geoengineering techniques also includes large-scale enhancement of the carbon storage that occurs in natural biological systems such as oceans, forests, prairies, soils, and wetlands. The proponents of climate intervention, or "geoengineers," tend to coalesce into two camps: some favor SRM, while others advocate a crash program to accelerate the removal of CO_2 from the air, either by machine or through "nature-based" approaches.

Solar Radiation Management—A Flashy Fix

Although scientists have studied various SRM approaches for half a century, discussion of them has intensified in recent years. A number of techniques are on offer in this category, differing mainly in the material used to reflect the sunlight and the location in Earth's atmosphere where the reflecting takes place. Some scientists, for example, have proposed using mirrors in space to reflect sunlight away from Earth.

Other SRM options aim to mimic natural processes more than do giant space mirrors. When large volcanic eruptions occur in nature, in addition to releasing lava, they blast hot, sulfurous gases high into the stratosphere, where they circulate around the world. Chemical reactions there produce particles of shiny sulfur dioxide; these increase the reflectance of the atmosphere, sending solar energy back into space and measurably cooling the earth. SRM advocates have several prominent proposals that involve the dispersion of sulfur; the one most commonly discussed is to disperse chemicals into the stratosphere from balloons, planes, or rockets. This would make the upper atmosphere shinier, so it would reflect sunlight away from the earth, much as after a volcanic eruption.

Marine cloud brightening is yet another variation on the SRM theme. In this approach, ships or aircraft would spray seawater into the marine cloud layer, infusing the clouds with salt particles tainted with organic matter from the ocean's surface. This would increase the number of cloud condensation nuclei; the more cloud vapor particles, the more solar radiation the low marine cloud layer can reflect.

Some researchers have suggested that 1,500 GPS-guided, unmanned spray vessels traveling throughout the oceans could spray enough seawater into the clouds to return temperatures to pre-industrial levels, or at least to stabilize them.

Certainly, SRM is an intuitively appealing approach that would decrease the earth's temperature. Because the sun heats up the earth, why not just get rid of some extra sunlight? This "quick fix," however, would ultimately prove addictive, expensive, and morally hazardous. By blocking some of the sunlight reaching the earth's lower atmosphere, SRM would have the capacity to lower the earth's temperature *without* reducing the amount of CO_2 in the air—in effect, masking the full impact of climate change without addressing its underlying cause. Meanwhile, with atmospheric CO_2 still high or even rising, the oceans would continue to acidify. SRM would also bring changes to the world's weather systems. However, with the climate appearing to be improving, society might be even less motivated to reduce fossil fuel combustion than it is right now. That's the moral hazard associated with SRM and with some other forms of geoengineering.

SRM won't fundamentally address the cause of global warming. Yet, once we set off on the SRM pathway, we are locked into a cycle of *perpetual* SRM, as anytime we stopped, global heating would accelerate rapidly due to the massive amounts of CO_2 still in the air, the effects of which SRM would be hiding. So, instead of a path to a better world, SRM could become a highway to hell. Then there's another serious issue: reflecting sunlight away from Earth and back into space may significantly alter the planet's rainfall patterns and displace monsoons, so that their rain would fall over the ocean while leaving parts of Southeast Asia in drought. Alternatively, it could provoke an unusually wet monsoon season, with disastrous consequences.

Aside from these broad-based drawbacks to any SRM approach, each such technique has its own specific hazards. Mimicking volcanos by sowing chemicals in the stratosphere, for example, will no doubt cause the destruction of stratospheric ozone, which naturally occurs following a volcanic eruption. The ozone layer, of course, is critical to protecting all life from dangerous ultraviolet radiation, which causes cataracts, skin cancers, and damage to plants. But while a volcanic eruption prompts a relatively short-term release of sulfur dioxide, the proponents of sulfur dioxide injection want to discharge a more or less continuous supply of sulphates into the stratosphere in order to maintain the cooling—as if an artificial volcano were operating year after year. The constant annual addition of a million tons of sulfur to the atmosphere would cause an increase in acid rain—given that a sulfate plus moisture creates sulfuric acid. The sulfur dioxide would also change the color of the sky from blue to a whitish gray.

Although the costs would be in the tens of billions of dollars annually—cheap, by geoengineering standards—SRM's use of sulfurous particles presents another risk: the technology could be deployed relatively rapidly, even by rogue nations or private companies, without the broad consent of the eight billion people whom the decision would affect. This combination of low cost and easy deployment makes SRM's use dangerously tempting to those with ill intent or who are driven by an amoral pursuit of profits, heedless of the harmful consequences to others.*

According to some experts, sulfate injection into the stratosphere could lead to a 15 percent increase in sulfate particles in the troposphere, the lowest, densest part of Earth's atmosphere. "Studies have indicated that [this] could result in as many as 26,000 additional deaths associated with acid rain pollution," warns Professor Wil C. G. Burns, an expert on geoengineering. On the plus side, injection of particles of titanium oxide or limestone, instead of sulfur, could conceivably produce similar effects, without acidifying the rain. But this still would not negate the risk of fundamentally altering the

---

* A provocative start-up called Make Sunsets recently caused a furor by launching a balloon to release sulfur particles in an unannounced and unauthorized experimental test flight from a site in Mexico. The company plans further sulfur releases and is offering "carbon cooling credits" for sale. See James Temple, "A Startup Says It's Begun Releasing Particles into the Atmosphere, in an Effort to Tweak the Climate," *MIT Technology Review*, December 24, 2022, https://www.technologyreview.com/2022/12/24/1066041/a-startup-says-its-begun-releasing-particles-into-the-atmosphere-in-an-effort-to-tweak-the-climate/.

earth's rainfall patterns while also launching humanity into a perpetual cycle of climate intervention that does nothing to remove the original cause of climate change.

Some scientists have also proposed positioning 55,000 enormous mirrors, or countless tiny metal mesh mirrors, in space between Earth and the sun, at a gravitationally stable location known as a Lagrange point, to reflect sunlight away from Earth. Unfortunately, this strategy would carry some of the same risks as blasting sulfur into space. As one definitive review of geoengineering put it, "This option is widely considered unrealistic, as the expense is prohibitive, the potential of unintended consequences is huge and a rapid reversibility is not granted." A recent MIT study found that even if the mirror scheme were affordable and practical, the cooling would have major unintended climatic effects. Like other forms of SRM, it could decrease warming at the equator but increase it at the poles, thereby weakening intercontinental storm tracks. This, in turn, could reduce regional rainfall and produce stagnant air masses that could worsen heat waves and air pollution. It could also spell disaster in parts of Asia by disrupting the annual monsoon, putting millions of people at risk of starvation, and potentially causing massive flooding, displacing millions. "Solar geoengineering is not reversing climate change," the principal author of the MIT study stated, "but is substituting one unprecedented climate state for another."

As for marine cloud brightening, only one field experiment has been done on this technique; the rest of the inferences about cloud brightening are from modeling exercises. Because the behavior of clouds is complex and their properties are still not totally understood, theoretical studies have reached conflicting conclusions. Another problem is that stratocumulus clouds suitable for cloud brightening occur over only 10 percent of the earth's surface. The jury is thus very much still out on marine cloud brightening.*

---

* If this approach were shown to work over the poles, however, it could be a very useful means of slowing or bringing to a halt the abnormal melting of polar sea ice and ice caps.

Another geoengineering approach relies on a concept known as "earth radiation management" rather than solar radiation management. When sunlight strikes the earth, it is converted into infrared radiation (heat) that is reradiated skyward. A portion of this radiation is trapped by clouds, absorbed, and reradiated back toward the earth. Wispy cirrus clouds comprised of tiny ice crystals waft high above the earth, where they both reflect sunlight and absorb heat radiating from the earth—two opposing effects, but of unequal strength. Unlike with low marine clouds, however, the warming effect of the cirrus cloud layer is stronger than the cooling effect of reflecting sunlight. Therefore, if they could be thinned to allow more heat to escape from the earth, that would tend to cool the earth. The idea is to seed the clouds with ice-nucleating particles, causing water vapor in the cloud to condense on the particles to form tiny crystals of ice, increasing the cloud's transparency to heat. Many scientific questions remain unanswered about cirrus cloud thinning, however. It could, for example, be detrimental to the ozone layer if the seeding were done with sulfur and nitric oxides, as has been contemplated; paradoxically, some research has suggested that it could also potentially cause warming instead of the cooling intended.*

Decarbonizing the Air

Imagine, for a moment, millions of shiny, boxy sheet metal machines stacked in seemingly endless humming rows, one atop another. Whirring fans in the boxes inhale carbon dioxide from the air and onto chemically treated filters. Inside the boxes, the machinery performs reliably day in and day out, decade upon decade, its filters trapping CO_2. Then, with the addition of heat or liquid sprays,** the gases are stripped from the filters, concentrated, and then com-

---

* The use of dust particles or bromine triiodide has also been discussed.
** Some systems use amines or calcium hydroxide.

pressed at high pressure for transport as a liquid by pipeline (or other means) to a storage site. There, they are pumped underground forever, never to reemerge—one hopes. This process is known as "direct air capture" of carbon dioxide, or DAC.

Effective as many of the projects described in the preceding chapters may be for reducing carbon emissions, even if we are smart, determined, and lucky enough to zero-out global carbon emissions by 2050, the atmosphere will still likely be burdened with an extra trillion metric tons or more of CO_2 that humans have contributed to it since pre-industrial times. When I first began studying climate issues decades ago, few people had heard about the notion of mechanically removing massive amounts of CO_2 from the air, and even fewer took it very seriously. The process of removing CO_2 from a mixture of air and other gases was certainly not new—it's been used to purify the air on sealed submarines and spacecraft for decades (and by plants for millions of years). Yet its early industrial applications were small-scale challenges compared with the planetary-scale challenge of removing significant amounts of CO_2 from the entire atmosphere, at an affordable cost. The scale of the challenge—the vast amount of CO_2 in the atmosphere and its relatively low concentration (in the parts-per-million range)—seemed an insuperable obstacle. The more diffuse the CO_2, the more energy required to scavenge and remove it from the air.

Therefore, almost everyone assumed direct air capture would be impractical and impossibly expensive and that operating CO_2-removal machines would require so much energy that the idea didn't even merit serious examination. The technologies proposed and used today (at very small scales) suffer from the intrinsic limitations just mentioned, arising from fundamental laws of physics that cannot be repealed and that, frankly, preclude these particular technologies from ever inexpensively scaling up beyond niche or relatively small-scale applications to accomplish "deep decarbonization," meaning the removal of globally significant quantities of CO_2. But this hasn't stopped several very prestigious scientists from promoting and developing them, with millions of dollars from investors.

One of the scientists who took exception to the notion that large-scale CO_2 removal from the atmosphere with current technology is impractical was Dr. Klaus Lackner, a distinguished theoretical physicist who has held senior posts at Los Alamos National Laboratory and taught geophysics as a professor at Columbia University. Today, Lackner is professor of engineering at Arizona State University and director of the Center for Negative Carbon Emissions, which he founded to demonstrate that direct capture of CO_2 from ambient air is affordable.

He reasoned as follows: Only about half of all emissions are from large, stationary sources like power plants and industrial furnaces, where the CO_2 emerges from a flue in a concentrated form that is relatively easy to capture. Lackner was concerned about the other half of human emissions, which emanate from what policymakers call "nonpoint sources": hundreds of millions of vehicle tailpipes, gas furnaces and stoves, airplane engines, and ships. If humanity is ever to draw down atmospheric CO_2 to pre-industrial levels, Lackner concluded, it would have to be done partly by direct air capture.

Lackner therefore developed a device he calls an "artificial tree," with plastic "leaves" that originally looked like pieces of shaggy, filamentous white carpet. When dry, the plastic absorbs CO_2 a thousand times faster than natural trees, says Lackner; but when wet, the leaves release CO_2. As Lackner has continued to refine his technology, these shaggy leaves have been replaced by discs coated with a chemical resin and arranged in a vertical column about five feet wide.

The shaggy "leaves" originally had to be exposed to moving air for between forty-five minutes and an hour to become fully charged with CO_2. The newer, disc-shaped leaves need only about twenty minutes to become fully saturated; they are then allowed to sink into a sealable chamber into which steam and water are introduced. The moisture causes the discs to release their CO_2, creating a low-pressure mixture of CO_2 and steam. (The steam is produced by heating water with renewable energy.) CO_2 is then recovered from the gas and compressed, so it can be stored or reused to make chemicals or fuels. Then the process is repeated.*

In other direct-air processes, energy has to be used to power fans that force air through ceramic filters. By contrast, Lackner's "artificial trees" save energy by relying on the wind to move air across the surface of the "leaves." With his current technology, a boxy device roughly the size of a shipping container would be required to remove a ton of CO_2 per day, according to Lackner. At that rate, one hundred million of these devices would remove 36 billion metric tons of CO_2 a year (an amount roughly equal to the world's annual emissions from energy use).

This would be a huge, complicated, noisy, and expensive industrial undertaking, to say the least. But "we have the industrial capacity to get there if we choose," Lackner argues, pointing out that the world has more than a billion vehicles and builds eighty million new cars annually. He claims that, currently, the cost of the CO_2 captured by these devices would be about $100 per ton but could be reduced with innovation and mass production. He does not appear to be interested in engaging with the question of whether this is a sensible thing to do today, given the energy and materials required and the other available uses for these scarce resources that can actually reduce atmospheric carbon more quickly and cheaply. This is so even though Lackner says that the current cost of the raw materials, energy, and water required for each machine—just the operating costs, excluding land to put it on—would amount to $15 per ton of carbon removed.

Simply citing numbers like $100 or $120 a ton is easy, but it can be very misleading. The true cost of the entire enterprise, from CO_2

---

* One possible use of the captured CO_2 would be to raise algae, which would then be processed into synthetic fuels. Currently, making synfuels from algae is prohibitively expensive and would not result in a commercially viable product. And when the fuels were burned, the CO_2 would return to the atmosphere.

capture to disposal, would be *much* higher: after capture, the CO_2 would need to be compressed for transportation and, ultimately, injection under pressure into underground repositories, where it would need to be monitored indefinitely. A vast CO_2 transportation pipeline system would also need to be constructed at great expense, unless the CO_2 were trucked or sent in rail tank cars to disposal sites. But that option would be even more expensive and would require the creation of a large electric truck fleet or electric railways, to avoid using fossil fuels that would release yet more CO_2. There is also the cost of creating new renewable-energy capacity to provide the power to heat and otherwise operate the machines—unless one were to use fossil fuels to operate the DAC system, which would be counterproductive. (I have posed some of these issues to Lackner's Center for Negative Emissions but haven't received a response.)*

Almost any way you slice it, direct air capture of diffuse CO_2 is the most expensive way to remove the greenhouse gases from the atmosphere.** Even if you assume, for argument's sake, that the cost of removing the CO_2 is a nominal $100 a ton, *and* if there were sufficient available water and renewable energy to operate the machines in vast numbers, the annual cost to remove current annual human-caused emissions would be on the order of $4 to $5 trillion globally. That's about 4 to 5 percent of global GDP, and it would come on top of the rest of the costs of creating a clean-energy economy.

Natural carbon-removal methods are often an order of magnitude cheaper than direct air capture. So, investing in them, or using the

---

* If this renewable capacity were created to run DAC machines before all fossil fuel power plants and industries were phased out, then each kilowatt of renewable power built and dedicated to DAC would be a kilowatt that couldn't be devoted to replacing dirty fossil fuel power, which would provide greater benefit to the atmosphere. Still, if DAC were powered by fossil fuels, it would be like removing carbon with one hand while releasing it with the other. DAC therefore can be of no net benefit until all other, more cost-effective ways of removing or avoiding atmospheric carbon are exhausted, or until the entire grid is powered by clean energy and no cheaper methods of carbon removal are available.

** It is much easier and more cost-effective to remove CO_2 from the stacks of power plants and industrial facilities, such as cement kilns or steel mills (in which it is far more concentrated), thereby preventing its release into the ambient atmosphere. After the gas is captured, it can be cleaned, compressed, and resold. More than twenty such facilities currently operate worldwide, but the process is not cheap, and that's one reason they are rare.

money to reduce fossil fuel demand through investments in energy efficiency and new clean-energy sources, would eliminate far more carbon from the atmosphere than DAC. One shorthand way of understanding this is to realize that it is always going to be easier and cheaper to keep a ton of CO_2 out of the atmosphere than to strain it out once it's already there. Even if removing a ton of CO_2 via DAC cost only as much as avoiding a ton of releases by substituting, say, a new wind or solar farm for a coal or natural-gas generator, the investment in avoidance would still be preferable. That's because the avoidance investment is a net *producer* of clean energy in normal operation. Removing CO_2 produces no energy, and to do so, the DAC device, a net energy consumer, would require the burning of fossil fuel if it were connected to today's grid (which is 60 percent fossil fuel–powered). It would be as if DAC were on a carbon-removal treadmill on which each couple of steps forward required a step or more back.

Moreover, those who have advocated direct air capture using chemicals to remove atmospheric CO_2, at times, either have not been fully transparent about all the costs of DAC or, possibly, have not done their energy accounting rigorously enough. When the energy required just to produce all the chemical reagents for DAC and then for their regeneration is tallied, it would annually require *more than the world's entire electricity generation capacity* just to remove the world's annual CO_2 emissions. And this would hardly make a noticeable impact in removing the total extra CO_2 that would need to be purged to return the atmosphere to its pre-industrial state.

That's why people who have done detailed studies on the energy and material requirements of large-scale DAC call it "an energetically and financially costly distraction in effective mitigation of climate changes at a meaningful scale before we achieve the status of a significant surplus of carbon-neutral/low-carbon energy." In plain English: deploying DAC today would actually be counterproductive. In addition, there's no significant commercial market today in which corporations are willing to pay even one hundred dollars to remove or store a ton of CO_2. And governments are not willing to provide the trillions of dollars annually this system would require to be deployed and operated.

Some analysts have proposed that the captured carbon be put to industrial use to make plastics, methanol, or other biofuels, but the vast scale at which the carbon would have to be captured is far in excess of any conceivable industrial demand for it. (Current industrial demand for CO_2 as a commodity is small worldwide, and that demand is fully met with far cheaper, industrially available CO_2.) Also, goods and fuels made from captured carbon by DAC machines would not be cost-competitive with fossil fuels or products made from them unless a whopping carbon tax—far more than one hundred dollars a ton—were in place, which is hardly imminent.

Nonetheless, astonishingly, President Biden's Infrastructure Investment and Jobs Act appropriates about $3.5 billion for the construction of four DAC facilities around the country. Even Dr. Lackner knows this is a bad idea. "DOE [the U.S. Department of Energy] is scaring me," he says, "because they make it sound like the technology is already ready. . . . We have to assume this is a nascent technology." Dr. Lackner has also said, with surprising candor, "Don't expect (these trees) to help climate. It's too little, too late. But it will give us startup money and an early way of becoming profitable." Yet Lackner has his own ready (and unsound) answer for how to begin deploying his machines. He thinks that DAC technology could be jump-started by volunteers who would be willing to pay an extra charge whenever they bought a gallon of gasoline or used any other fossil fuel product formulated from DAC carbon.*

The bottom line: DAC technology will not appear on street corners or soccer fields anytime soon. But can its costs be driven down farther, in case we as a society get to the point at which we decide it's time to deploy even very expensive technologies?

---

* Recycling, too, Lackner points out, initially was done on a voluntary basis, until it won policy support and became ubiquitous. (For an indication of how well that has worked, just visit your local beach or landfill, or read up on the enormous floating gyre of plastic swirling mid-ocean.)

ASU's Center for Negative Carbon Emissions, along with Silicon Kingdom Holdings, a commercial partner, either think so or see a way, somehow, eventually to commercialize the university's costly technology, as they have begun a joint venture, perhaps to sell CO_2 for niche applications. At any rate, Lackner's strategy calls for providing his DAC technology to Arizona State for deployment and testing on a university scale.

Inside the Box

However skeptical you may be about the role of volunteers in launching direct air capture, a Swiss company called Climeworks (which counts Bill Gates among its funders) has already enrolled nearly five thousand people who voluntarily pay up to $55 a month to remove 1,320 pounds of CO_2 from the air per year. (That's $1,000 a ton.) Climeworks says that the company offers a technology to "reverse climate change." "Together," they claim, "we can inspire 1 billion people to remove CO_2 from the air." How that is to be done is left to the imagination.

Climeworks's founders are mechanical engineers Jan Wurzbacher and Christoph Gebald, avid skiers who met at ETH Zürich (the Swiss Federal Institute of Technology), where both got master's degrees and then PhDs. Skiing in the Swiss Alps, they were shocked at the shrinkage of mountain glaciers, and they began studying direct removal of CO_2 from the atmosphere as part of their research. Today, Climeworks is the largest direct air capture company in the world, removing four thousand tons of CO_2 a year, or about one ten-millionth of the world's annual emissions.

At present, Climeworks has only fifteen DAC machines operating in Europe and a production facility able to produce a hundred such machines a year. But Wurzbacher and Gebald are dream weavers, and they have been good enough at selling their dream to raise over one hundred million dollars. The ad on their website sounds sweet:

"This safe, sustainable solution is available for everybody! Act now. Remove unavoidable emissions."

Would that it were all so simple.

Climeworks's "Orca" plant is located at the Hellisheidi Geothermal Power Plant, in Iceland. The company's partner, Carbfix, mixes the CO_2 captured there with water and pumps the mixture into an underground basaltic formation for permanent storage. There, chemical reactions with the basalt turn the captured CO_2 into stone, removing it from circulation for geological time. As every business owner knows, you can't overestimate the importance of "location, location, location," and in this case, the presence of a co-located geothermal energy source and basalt formation is ideal, allowing the Orca plant to cleanly remove and store the CO_2 using renewable energy. But continental geothermally active sites with deep basaltic rock formations suitable for carbon storage are globally limited in number.

Unlike Lackner's passive DAC collectors, Climeworks's machines use fans to draw ambient air into stackable, modular metal boxes where filters remove the CO_2 from the airstream. Instead of spraying the filters with water to desorb the adsorbed CO_2, Climeworks closes the vents once the filter in a machine is saturated with CO_2, then heats the box with geothermal energy to between 80 and 100°C (212°F) to drive off a stream of high-purity CO_2, which is then collected.

Another major difference between Climeworks's technology and Lackner's passive mechanical tree might be cost. Climeworks's real-world costs have been reported at between $600 and $800 a ton to extract, compress, transport, and store CO_2. The whole-system costs for Lackner's technology are unclear. We've discussed the problem of high costs before, but consider the implications: At, say, $700 a ton, removing just a tenth of annual global emissions (estimated at over fifty billion metric tons from all human causes, including land use change) would cost around $3.5 trillion annually. And if we reached net-zero annual emissions and then wanted to remove *all* the excess

CO_2 now in the atmosphere using Climeworks's technology, that would be a $766 trillion one-time proposition.

Igniting an Industry

Another clever scientist who likes big challenges is Professor David Keith of Harvard University. Dr. Keith directs the Keith Group (a research team funded in part by the Bill and Melinda Gates Foundation) at Harvard and founded Carbon Engineering, which operates a small and controversial prototype carbon-removal plant in Alberta, Canada. Carbon Engineering says its raison d'être is to have "the greatest impact on the huge climate challenge" by deploying its DAC technology globally. "This proven technology," the company claims, "can help any company, country or individual achieve critical net-zero targets by delivering solutions for permanent carbon removal and ultra-low carbon synthetic fuels."

Keith, a self-described "avid mountaineer [and] Arctic explorer," works at the confluence of experimental physics, hardware engineering, and public policy. Not content to stay in academe or the lab, he also became an entrepreneur, founding Carbon Engineering in 2009. The company is currently soliciting and accepting small deposits from people who want carbon removed on their behalf. "If you're interested in addressing your carbon footprint with safe, permanent and verifiable carbon dioxide removal," the company says, "we'd like to talk to you."

In 2021, Carbon Engineering was doing the planning for the construction of a million-ton-per-day CO_2-removal plant in the oil-rich Permian Basin, in the Southwest United States, which it says will be the world's largest carbon-removal and synfuels facility when it opens in 2024. Electricity is used to power its fans, and heat—up to 900°C—is delivered to its porous ceramic filters to remove the captured CO_2. Once the CO_2 is captured, Carbon Engineering says it wants to reform it into synthetic fuels, so it can be burned as carbon-neutral fuel in hard-to-decarbonize sectors such as aviation or provide low-carbon fuel to meet California's Low Carbon Fuel Standards (LCFS).

The business case for these applications, however, has not been proven. Using engineering studies of its tiny (one-metric-ton-per-day) prototype plant, the company estimates that its cost of removing a metric ton of CO_2 from the atmosphere is between $94 and $232. On its website, however, it claims it can deliver CO_2 at large scale for just "$100 to $150 a ton." This remains to be seen and will depend on the efficiencies of Carbon Engineering's complex chemical engineering operations. If, instead, the CO_2 were destined for burial, its transportation under pressure to a suitable leak-proof repository site and its injection into the ground would entail significant additional costs.

The biggest market by far for captured CO_2 is, ironically, the petroleum industry—for its use in forcing more oil out of the ground. The industry needs CO_2 for its "enhanced-oil-recovery" operations, and some of the CO_2 injected in that process—no one really knows how much—remains permanently trapped underground.* This is the only significant demand for expensively captured CO_2 today, and Carbon Engineering is angling for this market. The company says its plants "can be co-located with an oilfield or industrial operation to deliver pipeline purity compressed CO_2 at point of demand."

So, if DAC companies were ever really successful at large-scale removal of CO_2 from the atmosphere—say, in the hundreds of millions or billions of tons—there is no other significant commercial demand for it, and that CO_2 would have to be compressed, transported, and stored at great expense. (Carbon Engineering is prospecting for future customers who might someday want to use carbon from direct air capture in producing industrial products such as plastics, carbon fiber, industrial chemicals, fertilizers, and carbonates—a use for which no business case currently exists on account of the cost of capture.)

---

* Drilling sites are not designed as permanent CO_2 repositories and could leak CO_2.

Thus, the biggest logical customer nowadays for captured CO_2 is the oil industry. Three major energy companies—Occidental, BHP, and Chevron—recently invested $68 million in Carbon Engineering. Perhaps it was a token strategic investment in support of carbon capture technology that they see as capable of prolonging the carbon fuel age. It may also have been for another reason: "It gives them a nice big marketing win in headlines that they are saving the planet from global warming," says CleanTechnica writer Michael Barnard. It also gives talking points to those in coal-producing states who profess that society can continue fueling power plants with coal because "technology is available" to capture CO_2, even though it doesn't capture all the CO_2 produced and is prohibitively expensive. Adding this technology to coal and gas power plants that are already more expensive to operate than wind and solar plants would make the fossil fuel power even less competitive than it already is, further stranding these assets in the marketplace.

In any case, because Big Oil helped finance the company, it stands to reason that they may ultimately be customers for its CO_2. If so, the rationale for a Carbon Engineering product as a climate-friendly technology would evaporate. "There is zero net removal of CO_2 from the atmosphere if air carbon capture CO_2 is used for enhanced oil recovery," Barnard writes. "And that's at a [bespoke] cost of $94 to $232 for the air carbon capture portion alone. All of the negative externalities of fossil fuels persist indefinitely."* In enhanced oil recovery, the oil industry pumps CO_2 (among other things) into older, low-producing oil wells where sludgy residual oil remains lodged in rock pores. Like a solvent degreasing a pan, the acidic CO_2 mixes with the oil and reduces its viscosity, enabling it to flow toward the well bore. Then, when the oil is burned, whatever climate benefit there may have been in coaxing the CO_2 out of the air is negated by

---

* The fact that the price of bulk industrial CO_2 is currently $380 per metric ton raises questions about how Carbon Engineering's new process would produce it so much more cheaply. See "Liquid Carbon Dioxide Price Trend and Forecast," quarter ending September 2022, ChemAnalyst, accessed February 15, 2023, https://www.chemanalyst.com/Pricing-data/liquid-carbon-dioxide-1090.

the production of additional fossil fuel CO_2 at two to three times the rate of CO_2 previously captured.

To make matters worse, in the case of Carbon Engineering, the company *burns* natural gas to power its machines, producing half a ton of *additional* CO_2 for each ton captured from its filters. The CO_2 from that plant therefore *also* has to be captured. The company says it can do this with 90 percent efficiency, but independent analysts concluded that more than a third of the CO_2 from the gas combustion escapes into the atmosphere. So, the company is, in a sense, running just to stay in place.

Of nineteen larger carbon capture facilities around the world, only four are not supplying CO_2 solely to the oil industry. In the United States, the oil industry is even given a 15 percent tax credit for enhanced-oil-recovery costs. Another credit, known as Section 45Q, gives the oil industry a $35-a-ton tax credit for each ton of CO_2 used in enhanced oil recovery. These are taxpayer-funded subsidies that serve to make fossil fuel production more profitable, a result that's not in the public interest.

No direct air capture of CO_2 is happening on a large scale today. "Basically, all CCS [carbon capture and storage] is a rounding error on the actual solution: just stop emitting CO_2," writes Barnard. The new plant that Carbon Engineering is constructing will remove only a tiny fraction of U.S. CO_2 emissions a year. Carbon Engineering's direct air capture is unlikely *ever* to be done on a large scale at a competitive price, despite the company's claims that California's Low Carbon Fuel Standards make its fuel technology economically viable today and that the technology will find a large global market. Yet the company maintains that it will roll out a large number of commercial plants by 2030. "It's our goal," it says, "to ignite a critical new Direct Air Capture industry." Even a million-ton carbon-removal plant like Carbon Engineering's, however, would require a wall of huge fans sixty-six feet high, twenty-six feet thick, and stretching for a mile and a quarter around a natural-gas-burning power plant to provide heat and electricity for the DAC plant. Two thousand such plants would have to be built just to remove 2.5 percent of annual global emissions, even if that removal were done with no new CO_2 creation in the process, which isn't currently the case.

Given the extravagant costs and logistical problems, it is nearly impossible to imagine that Carbon Engineering's DAC technology can make any appreciable dent in reducing global emissions or staving off climate change. In fact, it's very hard to imagine that the CO_2 resulting from all the phases of this operation, including the "upstream" energy used to produce the fossil fuels consumed, is not in fact *greater* than the amount of CO_2 it removes. In any case, as Michael Barnard writes, "it's not a rational choice for humanity to dig up ... sequestered carbon [fossil fuel], recapture it and re-sequester it at great expense where there are alternatives."

The best, most cost-effective way to protect the climate is to go full speed ahead in eliminating carbon emissions today with renewable energy sources and energy efficiency. These are orders of magnitude more cost-effective at reducing atmospheric CO_2, as it's much cheaper and easier to keep CO_2 out of the air than to strain it out once it's there. If strain we must, however, it behooves us to pick the low-hanging fruit. This means first deploying massive amounts of the most cost-effective CO_2-removal processes—those originally designed by Mother Nature.

Natural Geoengineering—with a Little Help

Nature's biological carbon-removal systems are the planet's oceans, forests, prairies, soils, algae, bacteria, and wetlands. These naturally cycle carbon and gently coax it out of the atmosphere—as they have for hundreds of millions of years, in the case of trees, and over a billion years in the case of algae.

Readers will recall that I devoted appreciative chapters to naturalistic carbon dioxide removal on farms and ranches and to the protection of forests from destruction. The enhancement of natural carbon-removal processes, however, has challenges, limitations, and uncertainties. Many unanswered questions remain even about carbon farming and carbon ranching on a global scale, especially about the ease of large-scale implementation. As just shown, even more issues remain to be resolved regarding the technological readiness, cost-effectiveness, and safety of engineered CO_2 removal for deployment at the enormous scale needed.

Natural carbon removal can be slow, challenging to scale up, and, in some cases, impermanent—as when a forest burns, or when carbon-enriched soil is abused so that it releases stored carbon instead of removing it. Plants grown to sequester carbon can be thirsty and can compete with other water uses. Land requirements for natural carbon solutions also tend to be large and can compete with food production and other land uses.

So, natural solutions are not a panacea; every approach that passes scrutiny must still be used judiciously. In addition, a simple bet-hedging strategy indicates that it would be a mistake to rely solely on enhancement of natural carbon removal, just as it would be foolish to rely on today's DAC technology for removal of carbon on a globally meaningful scale. So, what are some of these other bet-hedging technologies?

Professor Wil C. G. Burns has a PhD in international law and codirects the Institute for Carbon Removal Law and Policy, which he cofounded at American University. He has thought broadly about the philosophical, legal, and governmental aspects of geoengineering. I met him in Berkeley, California, for a deep dive into the subject. Dr. Burns teaches part time at UC Berkeley, though he boomerangs around the country to various institutions of higher learning. He is a proponent of emissions mitigation as a first resort and, only thereafter, of carbon dioxide removal to cool the earth, but he's firmly against solar radiation management.

My conversation with Dr. Burns soon turned to an approach to CO_2 removal that he's quite familiar with and that mimics a primordial geological process. The earth has been removing CO_2 naturally for billions of years, by a process known as mineral weathering. In it, silicate rocks dissolve in rainwater over long periods to form carbonate or bicarbonate compounds with atmospheric CO_2. This gets the CO_2 out of the atmosphere for the long haul. To enhance mineral weathering artificially, limestone or silicate rocks, such as common basalt (formed from the cooling of lava), would be mined and crushed to increase the surface area available for absorbing CO_2. The "rock flour" could then be spread on cropland, where it could significantly increase fertility and crop production—"by maybe twenty, twenty-five percent," Dr. Burns estimates. "With those co-benefits, you can provide incentives for farmers to allow that kind of spreading to happen on their lands, and you might be able to sequester one to two gigatons a year with that process." Thus, he believes, by cobbling mineral weathering together with a suite of other carbon-removal processes, the world could get a little closer to the carbon-removal targets needed.

The Rodale Institute of Kutztown, Pennsylvania, goes farther than that. In a white paper on regenerative agriculture, it concluded that if all the world's cropland and rangeland were to adopt regenerative practices, "that could drawdown [*sic*] more than 100% of annual CO_2 emissions, pulling carbon from the atmosphere and storing it in the soil." Easier said than done—aside from the challenges of fomenting mass participation, discussed in chapter 12, one difficulty is determining how much additional carbon is being added to the land by millions of participants, given the vast variability in soil type, crop type, and agricultural practices. Another challenge is to guarantee the persistence of the soil carbon over the long term; even when soil carbon is increased "as advertised," subsequent land mismanagement might once again release it in whole or part. For all these reasons, it would be unwise to rely only on agriculture to decarbonize the atmosphere. But carbon ranching and farming must be part of a suite of natural carbon-removal solutions.

The Limits to Growth—of Trees

Adding vast numbers of new trees to the earth is another natural approach to reducing CO_2 levels, but sometimes enthusiasts oversell it. In 2020, an appealing One Trillion Trees Initiative* of the World Economic Forum was publicized at their annual Davos-Klosters, Switzerland, meeting. Yet critics say the initiative is based on a 500 percent overestimate of the amount of land available for tree planting and that it neglects the important uses to which much of the suitable land is already being put, especially as cropland for growing food.

Simplistic, large-scale tree-planting campaigns raise numerous serious concerns and can even be counterproductive. For one thing, assisting natural forest regeneration and forest protection tends to produce better results than plantation-style reforestation, an approach in which monocultures of single-age trees are planted in uniform rows or blocs that are highly susceptible to pests and disease while capturing less carbon than natural forests do.

Where trees are planted is also critical. Planting them in inappropriate areas, such as high northern latitudes where the land surface is highly reflective and where no trees have grown before, can actually *increase* the absorption of solar radiation and contribute to global warming. In addition, whereas it is often fine to plant trees to assist in the recovery of a deforested or otherwise degraded area, advocates of vast tree-planting programs sometimes are actually seeking a rationale for logging old forests, which store far more carbon than the new plantations that would replace them. Dr. Burns calls tree planting "tricky." To make a large difference, he claims, would require "twenty to thirty percent of the land mass of the United States."

Planting trees in the Global South, however, raises concerns about social justice. Many areas that are classified as "abandoned" or "low-productivity" regions, Burns says, actually are not. People earn their livelihoods there, and they may be expelled because a lot of

---

* See 1t.org.

money could be made by planting trees on that land and claiming carbon credits. In addition, trees also need to be maintained. "One of the things that we found with tree planting often in the south is that within a few years, most of those trees have disappeared."*

Water availability is another big concern. "[Trees] may require a lot of water, and that may take water from crop-planting in the area or other needs." Finally, tree planters may covet prairie grasslands or savannah regions that are often high in biodiversity. "If you take out those areas and put a lot of monoculture plantation forests, you may significantly decrease biodiversity," said Burns.

If done correctly, with due respect for all the constraints just mentioned, tree planting would sequester only about a gigaton of carbon a year on a sustainable basis, according to Burns. He also worries that many of the trees in the One Trillion Trees Initiative might be planted in African countries, or elsewhere in the Global South, where regimes might not protect the interests of the most vulnerable. "We want to ensure that there are some protections in place, so that people aren't expelled from their land so elites can obtain carbon credits in the south, or [so] companies like Microsoft can avoid taking measures [at home] to aggressively reduce their own emissions by simply purchasing cheap credits."

Seedling survival and the creation of tree plantations is another big concern with massive tree-planting campaigns. Such large-scale "reforestation" efforts often end up producing monocultures or cultures of only a few species; it's simpler and cheaper to do so than to plant multiple species with different requirements. In addition, once planted, trees are susceptible to hazards, such as drought, flooding, hurricanes, mudslides, diseases, and blights. Moreover, as climate change accelerates, extreme weather and other such threats to trees will intensify. Therefore, one might need to plant a multiple of the number of trees actually desired. That's all the more important

---

* Some may have died of drought or disease, while others may have been cut for firewood or used in construction.

because trees grown in an existing forest are protected and provided with nutrients by other, neighboring trees through a web of mycorrhizal fungi that exist in healthy soil but that can become depleted in abused and eroding soils. These mycorrhizae are also critically important for conversion of the carbon pumped into the soil by trees into stable chemical forms and soil clumps, known as aggregates, that are less easily washed out of the soil.

Trees grown in forests also tend to grow more slowly than trees grown in open fields, because, in the forest, the trees need to compete for light and nutrients. This leads to denser wood that is more resistant to pests and pathogens and, thus, to hardier, longer-lived trees. By contrast, trees grown under suboptimal conditions on nutrient-poor soil that may not have hosted trees for centuries are more likely to succumb more easily to their predators and to environmental stresses, such as drought. In addition to their lower survival rates, these trees will probably be less capable of funneling atmospheric carbon into the soil through their roots, which is the main raison d'être of the tree-planting campaign.

Finally, if the goal is to plant a trillion trees, a vast land area would be needed, and trees by the hundreds of millions would need to be planted every week for decades—a scale never before undertaken—to arrive at the goal of a trillion healthy, living trees. If "only" fifty million trees a week survived from a sustained tree-planting effort, and if they were long-lived, it would nonetheless take *four hundred years* for a trillion trees to become established. That would be roughly sufficient to remove, annually, humanity's current CO_2 emissions of about 36 billion metric tons a year from energy use, but it would not be sufficient to remove humanity's accumulated emissions. However, if humanity reduced its annual emissions to net zero, then those new trees and other carbon sinks, notably the ocean, would gradually draw down the atmosphere's legacy CO_2.

To succeed, this effort would need to be carefully planned, financed, and managed by some competent, well-funded international coordinating body—even more so if the intended end result were the creation of diverse forests of healthy, thriving trees: simply scattering seeds from airplanes or planting millions of tiny and subsequently untended seedlings would not be a recipe for success. By contrast, properly planting billions of trees would be a more expensive, labor-intensive effort, albeit one with many co-benefits, including job creation, watershed protection, and biodiversity protection. So, if wisely executed, the planting of billions of trees would be well worth doing, as it could cost-effectively sequester large amounts of carbon, potentially for hundreds of years.

Putting the Carbon Genie Back into the Bottle

Geoengineering approaches alone cannot remove the threat of climate change from the earth. Dr. Burns has said it well: "What they could do is buy us time or help us to come back below critical temperature thresholds. But unless they are coupled with aggressive efforts to decarbonize the economy, they would all fail in the longer term." Sadly, though realistically, Burns views geoengineering both as inevitable and as largely a technology of despair.

"[If deployed], it acknowledges the fact that we failed to get our act together and reduce our greenhouse gas emissions sufficiently." Yet, if we're going to hold temperatures to below two degrees Celsius, "we're probably going to have to remove somewhere between ten [and] twenty gigatons of carbon dioxide annually" by 2040 and beyond.

Despite all the reservations expressed in this chapter, it is important to do more research on geoengineering. Society also needs to do more to educate and engage the public about the risks and benefits of geoengineering in a world where most people have not yet even heard of it. In addition, in order to prevent one or more countries from proceeding unilaterally and recklessly with a rogue geoengineering program, international governance mechanisms will need to be developed to control it, and the safeguards need to be put in place now. Finally, social justice concerns and trade-offs need to be discussed now, too, so that they can be properly factored into decisions about where (or whether) trees or other plants should be planted to promote carbon storage, and at what scale.

Given that no one geoengineering method will compensate for all of humanity's global carbon emissions, geoengineering needs to be seen as a portfolio of possible options ranging from plausible to wacky. Those found safe, effective, and cost-efficient could at some point be helpful to supplement—not replace—global emissions-reduction methods.

17.

A Just, Clean-Energy Transition for All

Hurricane Katrina is still what many of us picture when we think of an extreme weather event pummeling a city's poorest residents. Katrina took 1,833 lives when it hit New Orleans in 2005 and displaced a million people, many of whom lacked the money to get out of harm's way. That's not unusual: Floodwaters rose six feet deep along the coast of Puerto Rico in 2017 just thirty minutes after Hurricane Maria hit land with 155 mph winds. Most of the displaced there also were poor and lacked the resources to flee their homes or rebuild. Floods, storms, extreme heat, and dangerous air quality caused by climate change threaten low-income communities daily, especially in parts of cities where low-income people live and work.

The Hardest Hit

Cities in the United States tend to be hotter by as much as 7°F than the surrounding countryside. But during a heat wave, low-income neighborhoods with little shade can be 20 to 22°F hotter than the leafier parts of town. In rural communities, wildfires—made more frequent and extreme due to climate change–induced drought and heat waves—cause dangerous air pollution, a serious health risk that climate disruption only compounds. Higher temperatures increase the rate of photochemical reactions in the atmosphere, which create ground-level ozone, a potent air pollutant. Poor air quality (which is exacerbated by global warming) is also part of life near busy freeways, ports, or other polluting industrial sites. And many low-wage jobs, including manual labor, like harvesting crops, painting houses, or repairing roads—especially jobs where masks and protective clothing aren't provided—put workers at risk for respiratory disease.

Climate disruption poses a greater threat to low-income communities for two other reasons. First, homes in these neighborhoods are likely to be structurally less resilient to natural disasters and to lack air-conditioning or proper insulation. Low-income people are also more likely than average to suffer from preexisting health conditions, which makes them more vulnerable than average to additional health challenges caused by heat, displacement, flooding, smoke, and disease vectors such as ticks and mosquitos. Residents in disadvantaged communities suffer higher rates of cancer, asthma, heart disease, stroke, and systemic ailments (like diabetes), and are further challenged by having less access to medical care.

While low-income communities need more assistance to deal with the impacts of climate disruption, this population, across the nation, often gets *less* government help than the more affluent. It's clear that more funding is needed to make sure that lower-income communities receive not just their statistically equal share of clean-energy dollars but additional resources to mitigate the disproportionate climate impacts they face.

A Champion for Change

No one is more aware of this than Vien Truong, a leader in the environmental justice and climate-protection movements. Truong has spearheaded the passage of major state legislation that is funneling billions of dollars to low-income and disadvantaged communities for programs that reduce carbon emissions and improve local residents' lives and health. She is an attorney, political activist, coalition builder, and nonprofit organization leader and a driving force behind climate legislation that benefits the communities most affected by climate change, whose welfare is all too often ignored.

Truong sees environmental and climate policy "as a bridge to address other issues," including economic inequality. She has devoted much of her career to advancing policies to create an equitable and inclusive green economy. The coalitions of community and social service organizations she helped create and lead are responsible for the drafting and passage of two landmark pieces of California statutes known as Senate Bill 535 and Senate Bill 1275. Both reduce carbon emissions and improve local residents' lives and health; the former statute created the largest fund in the state's history for California's poorest and most polluted communities.

Truong is the youngest of eleven children whose family came to the United States as refugees from Vietnam. Her family first worked as migrant laborers, picking strawberries and snow peas in Oregon. Then they moved to California to work in garment sweatshops, from which they also brought piecework home. Truong grew up during the 1980s—the "crack years"—in a poor neighborhood of West Oakland between Twentieth and Twenty-Ninth Avenues, along International Boulevard, known for its high murder rate. Instead of lullabies, she went to sleep every night hearing gunshots and sirens, though she nonetheless sensed that this was not normal.

The neighborhood public schools she attended were also violent, with fighting in the classrooms and on school grounds. Community services were underfunded, leaving students with few opportunities. But, to make sure that at least one person in the family went to college, Truong's older siblings tutored her. Seizing every advantage, she resolved not only to escape West Oakland but also to create programs to help communities like hers end poverty and environmental pollution. "Throughout my life," she said, "I always had the sense that things could be better in my community."

Truong's family members continued working in sweatshops until she was in college. She sold shoes at Sears and worked in Chinese restaurants to make ends meet. Later, she attended college, graduate school, and law school. Once she became an attorney, she clerked for a federal judge and served on the staff of a California state senator before holding leadership posts at nonprofit organizations, such as the then-named Dream Corps (now Dream.org), Green for All, and the Greenlining Institute. At Greenlining, which was set up in 1993 in reaction to the discriminatory practice of "redlining" neighborhoods, Truong, as director of environmental equity, worked with the state legislature, the California Public Utilities Commission, and in localities around the state to create solutions for poverty and pollution. The Greenlining Institute was designed to "proactively bring investments and opportunities into these communities."*

Leveling the Playing Field

No matter what job title Truong held, however, her goal was always to improve the quality of life in disadvantaged communities, lower the cost of living there, and bring change through environmental policies. Working at the intersection of social justice and climate justice, she used her education and understanding of the law to create programs and policies to bring new investment to communities like the ones she grew up in. Her work culminated in the creation of the largest fund in history to "green up" some of the state's poorest and most polluted areas, using money from the state's biggest polluters. But all this did not happen overnight.

At the Dream Corps,** where she was president and CEO, Truong worked for climate justice and criminal justice, successfully creating

---

* The institute is devoted instead to proactively building a future where communities of color thrive by enabling them to meet the challenges posed by climate change while building wealth and living in healthy places with ample economic opportunities. See "A Future Where Communities of Color Can Thrive," Greenlining Institute, accessed May 10, 2023, https://greenlining.org/about/.
** The Dream Corps was cofounded by activist and political commentator Van Jones.

policies that "close prison doors and open doors of opportunity for all." She was also national director for Green for All, one of the Dream Corps's social impact initiatives. It was indeed a perfect fit: Green for All's goal is to build a green economy while lifting people out of poverty.

"The two biggest problems facing our country—and the world— are runaway climate change and rising inequality,"Truong said. "We can't fully solve one without addressing the other. If we eliminate racism and poverty but do nothing to address climate pollution, we'll all be equal on a planet plagued by floods, drought, disease, and disasters. And we can't win on climate unless we build a bigger and broader movement that meets the needs of the hardest-hit Americans. We're working to turn the tide on both fronts."

After holding a number of positions and founding her own law firm, Truong subsequently moved on and assumed her current post as director of global engagement and sustainability for Nike, where she oversees the sustainability practices of the world's largest manufacturer of athletic footwear, apparel, and sports equipment and its supply chain. Her prior grassroots community-organizing efforts, which ultimately brought her to public attention, have by now influenced not only local but state and national environmental and economic policies. She was also part of two presidential campaigns, helping to write the presidential platform for candidate Tom Steyer and serving as a member of the Biden-Harris campaign's Clean Energy Advisory Council.

She also is a recipient of a White House Champion of Change Award for her work on climate equity. One of her passions has been to help direct clean-tech investment to ethnic and minority communities. Her work may have played a role in convincing President Biden to stipulate during his campaign that 40 percent of his proposed climate investments would go to programs benefiting disadvantaged communities.

Targeted Assistance—For People and the Climate

When I first met Truong in 2015, she was directing the Greenlining Institute's Environmental Equity program. Her goal was to simultaneously improve the quality of people's lives and their environment. To Truong, this meant making homes more affordable and transportation systems more efficient and reliable and providing good jobs with career ladders. She saw the linkage between attaining environmental goals and community improvement. Reducing greenhouse gas emissions on roads and in factories reduces neighborhood pollution and helps reduce the cost of living, as more energy-efficient housing requires less energy, lowering emissions along with family energy bills.

In 2022, programs that Truong was instrumental in designing began providing free and low-cost energy efficiency and free and low-cost solar panels to low-income families. The low-income communities Truong worked with most are often close to roads with heavy traffic. One of the laws she helped write while at the Greenlining Institute, California Senate Bill 1275, provides money for electrifying trucks and buses. That, she says, is a huge issue in "impacted environmental justice communities . . . especially West Oakland." Those communities are often more affordable in part because they're located in hazardous areas with poor air quality near factories, oil refineries, coal power plants, ports, incinerators, and toxic waste sites. Sometimes they're located in flood-prone coastal areas, yet residents may be unable to afford or obtain flood insurance. When rivers overflow into flood zones or rising seas flood toxic waste sites, low-income communities are disproportionately affected. "We have to figure out a way to start thinking of environmentalism as not just the John Muirs of the world, but beginning to see it as an equity and [environmental justice] issue," Truong said.

"The programs that we want to get funded are the ones that have triple-bottom-line benefits," she told me during my visit to the

Greenlining Institute.* In her role there, she advocated for programs that created well-paying jobs with a career ladder, that provided "the most bang for our buck in terms of reducing greenhouse gases," and that planted trees to help cool overheated, unshaded neighborhoods. She sees such programs as vehicles that could help address income inequality. "We have a bigger income divide now than for a long time, so we want to begin bridging that."

Environmental Restitution

By working on behalf of Green for All and the Greenlining Institute with community groups, legislators, and broad coalitions she helped build, Truong and her allies, in 2012, got a "Disadvantaged Communities Designation" provision added to SB 535, which sets aside 25 percent of the revenue from the state's Greenhouse Gas Reduction Fund for disadvantaged communities.** Then, in 2014, they were instrumental in the drafting and passage of California Senate Bill 1275.

The earmarked proceeds from the Greenhouse Gas Reduction Fund, which were later raised to 35 percent of revenues, are generated by carbon cap-and-trade allowance auctions. The revenues are then invested in low-income and disadvantaged communities; the money can be spent to improve public health, quality of life, and economic opportunity while reducing the fossil fuel pollution that causes climate change. As of 2021, SB 535, with its "polluter pays" pot of money, had directed $5.5 billion to low-income communities and communities of color to improve both the environment and people's daily lives. "This is the biggest pot of money ever for environmental justice communities," Truong said, "and it's from big polluters, not taxpayers!"

---

* The Greenlining Institute instead was designed to proactively bring investments and opportunities into communities that had been redlined in the past. The "redlining" of neighborhoods by banks and other institutions was part of a process of denying communities of color access to mortgages and to capital for the development of local businesses.

** The designation was originally 10 percent but was amended in 2016.

These funds, she pointed out, went to programs such as free and low-cost energy efficiency and free and low-cost solar panels for low-income communities and families. You can see these investments play out in concrete ways for people, she said. "It's getting money for electrifying the trucks and buses in impacted environmental justice communities, which is a huge issue in places like the Bay Area, Long Beach, [and] especially for West Oakland. It went to thousands and thousands of free trees in urban jungles and for affordable housing in transit-oriented development." One family in California's Central Valley whom Truong's work benefited previously had a monthly energy bill of $200 in the summer. "Now it's $1.50, because they have solar panels," Truong said. "That's real. That's concrete, and that's a significant savings." Truong, the daughter of migrant farmworkers, has also helped bring pilot van-sharing programs to rural farmworkers without drivers' licenses through rural ride-sharing programs supported by SB 535.

"[Senate Bill 535] began looking at investments into our transit systems, our bus systems, our affordable housing, and those are things that low-income communities/people of color care deeply about," Truong noted. "For us to have national movement around climate change, we have to begin looking at climate changes as a way that addresses those fundamental community issues and [as] not divorced from [them]."

The other legislation Truong, for years, lobbied for and helped draft, SB 1275, known as the Charge Ahead California Initiative, accelerates the transformation of California's vehicle fleet to electric vehicles and has a target of getting a million electric vehicles on California's roads by 2033. "That includes passenger vehicles all the way to heavy-duty trucks and buses," Truong said. Since SB 1275 was passed, however, California has forged ahead and adopted the goal of putting five million zero-emission vehicles on the road by 2030. SB 1275 also has an interesting "retire-and-replace" program, a "cash-for-clunkers" provision, to help get polluting vehicles off the road. The law includes a financing program for low-income families to receive subsidies for the purchase of used electric cars—with some families able to acquire one for as little as five hundred dollars. Since so much low-income housing is located near busy, polluted traffic arteries, the air pollution reduction that electric cars bring locally would be particularly welcome.

In terms of an overarching strategy to serve disadvantaged communities, Truong argues that people need to work together on specific issues that not only bring environmentalists together but also appeal more broadly to labor groups, educators, low-income communities of color, civil rights leaders, and health leaders. For her, the call to action needed to unite everyone is embodied in these simple basic questions: "What kind of world do you want to live in? What does it look like? Wouldn't it be great if we [had a] world that had clean air, in which kids of all skin colors played together and in which we all had access to a decent job and never had to go to a gas station again? Wouldn't it be great if we didn't have twice as many people dying from traffic pollution as from traffic accidents?"

Truong sees an important intersection between poverty work and climate work. "For me, it's creating a community that's more peaceful, more equitable, and [fairer]; that has clean air for our kids, clean water, the basics, and affordable housing, basic fundamentals—food and shelter. It's not just looking at the fastest way to reduce greenhouse gases or to get [back] to three hundred fifty ppm [CO_2]. It's 'How do we get at the best society possible?'"

Conclusion: What We Can Do to Restabilize the Climate

When I first moved to California sixty years ago, the rainy season was long, and the winters were chilly and wet. I enjoyed hiking in the Sierra Nevada, rafting the wild rivers, and skiing in the winters. Many feet of snow covered the state's high Sierra Nevada through late spring and early summer; the snowpack often provided 30 percent of the state's water supply. In soaking into the soil and flowing off the mountains, the meltwater set meadows abloom with wildflowers as they recharged underground aquifers and replenished rivers, lakes, streams, and reservoirs. The same thaw sent water to cities and filled irrigation canals for the state's farmers, who grow much of the nation's fruits and vegetables.

But the snowpack has dwindled as the state has become hotter and drier. California in recent years has experienced its worst drought in more than 1,200 years. Drought's impacts have destroyed communities, taken human lives and livelihoods, caused illnesses, ravaged western forests, polluted the air, deprived the state of clean hydropower, set back its emission-reduction efforts, and cost billions to address. With these and other impacts expected to worsen, it is clear we are in a climate emergency.

It's no surprise that the source of the drought is heat trapped by greenhouse gases in the atmosphere. But the magnitude of the heat we've taken on board Spaceship Earth is stupendous. In the past twenty-five years, due primarily to human reliance on fossil fuels and the blanket of atmospheric greenhouse gases their combustion has produced, the earth has trapped as much extra solar heat as billions of exploding atomic bombs. Just one exploding nuclear weapon creates temperatures of a hundred million degrees Celsius, equal to temperatures in the sun's interior.

More than 90 percent of this extra heat has now been absorbed into the oceans, warming them. And, of course, warm water expands. The oceans will therefore continue rising—for a thousand years or more—even after greenhouse gas emissions peak and begin to decline. These rising seas will be lifted even higher by inflow from melting polar ice and glaciers. The ocean is indifferent that eight or ten of the world's largest cities are on the coast; nor does it care that 40 percent of the U.S. population—130 million people—is coastal. The ocean speaks the language of heat absorption, heat flow, and aquatic chemistry. In the United States, cities like New Orleans, New York, and Miami are among those at greatest risk and are already seeing increasing damage from tidal flooding and storm surges. Abroad, cities like Tokyo, Kolkata, Jakarta, Bangkok, and Shanghai are also especially vulnerable.

When we started tampering with the heat balance of the earth and oceans by using the atmosphere as a sewer for our industrial waste gases, we triggered a 1.2°C rise in global average surface temperature. Temperature has risen even faster on land. And in the Arctic, to the surprise of climate scientists, temperatures rose four times as fast as the global average in forty years.

We're already seeing climate catastrophes striking almost daily, and we're growing dangerously accustomed to it. Yet the worst may still be to come. The best science is now telling us that with current climate policies in place, under a medium- to high-emission scenario, global surface temperature is expected to soar still higher, reaching between 2.8°C (5°F) and 4.6°C (8.2°F) or more by 2100.* Then, beyond 2100, further increases would be likely.

Today's climate models, however, may not yet be capable of telling us exactly when these quickly rising temperatures will trigger dangerous, irreversible climate system feedbacks. Some have already occurred: portions of the Amazon tropical rain forest, for example, are now producing greenhouse gas rather than absorbing it, as they have for millennia. Other feedbacks, such as those connected to the thawing of Arctic permafrost (which holds roughly twice as much carbon as the atmosphere) and the combustion of vast boreal forests, are capable of triggering large, rapid, and unstoppable releases of greenhouse gases. Meanwhile, those tipping points connected to the melting of polar ice sheets can initiate many meters of sea level rise.

The ongoing, accelerating warming we're now experiencing is unquestionably bringing us closer to a point at which self-reinforcing and uncontrollable climate heating will begin. This kind of "runaway" global heating would occur if a positive climate feedback, such as the thawing of permafrost, began to release so much greenhouse gas that humans, no matter how much we reduced our emissions or tried to extract CO_2 from the atmosphere, would be unable to keep up. Global warming would then be out of human control, and the planet would continue heating inexorably until, eventually, far in the future, natural forces intervened once again to cool it.

Just because we are somewhat in the dark as to precisely where this fatal precipice lies, we cannot afford to be complacent as we draw near it. This is especially so in that scientists on whose forecasts we have relied to date have significantly underestimated the speed and extent of some important climate impacts we're already seeing, including sea level rise, Arctic temperature rise, polar ice sheet melting, and sea ice disappearance.

The climate crisis is also not unrelated to the intensifying catastrophe

---

* The IPCC's *Sixth Assessment Report* also found the likelihood of other emissions scenarios, including the estimate that, under a plausible intermediate GHG-emission scenario, global temperature increase might reach between 2.1 and 3.5°C by 2100.

that has been overtaking the natural world most dramatically over the past fifty years. A million plant and animal species are now threatened with extinction as humans obliterate natural habitats and drench farms and fields with toxic agricultural chemicals. Climate change is an additional force multiplier in this process. Thus, as habitat quantity and quality have declined, the abundance and diversity of virtually all wildlife have declined at a shocking pace. According to the World Wildlife Fund, vertebrate populations—birds, mammals, fish—are all in free fall, declining by almost 70 percent just since 1970, with insect populations not far behind.

Most dramatically—and preventably—the world is continuing to destroy its priceless old-growth tropical forests at breakneck speed. In 2021, 9.6 million acres of primary forest were deforested, causing the emission of *2.5 billion metric tons* of CO_2, an amount equal to the emissions of all U.S. cars and trucks. Not only does the destruction of the forest in itself lead to further global heating, but the water cycle and soil changes it sets in motion produce other positive climate feedbacks that cause still more climate disturbance.

As the world now witnesses the accelerating impacts of climate change, the future will be studded with a wide array of compounding and disastrous climate impacts, many of them, for all practical purposes, irreversible. They will include significantly lower global food production, worse extreme weather, an expansion of areas too hot for human habitation, more climate refugees, and a runaway increase in the number of species extinctions and collapsing ecosystems. The social impacts of all these interrelated and often mutually amplifying developments will render societies increasingly difficult or impossible to govern; this is what people mean when they refer to "an end to civilization as we know it."

The longer that governments around the world fail to reduce these threats by slashing their national greenhouse gas emissions, the more immense the danger becomes. Unfortunately, far from shrinking, these emissions hit their highest peak in history in 2021, swelling by more than two billion metric tons—6 percent. That's like throwing gasoline on a raging fire and assuming the flames will go out. Moreover, by a conservative estimate, the world spent nearly $700 billion on subsidizing the sale of coal, oil, and natural gas in 2021, almost double the amount spent in 2020.*

The Road Out of Hell

I hope I've shown in this book that pursuing policies that have already wreaked havoc on the planet and then doubling down on them won't work. What's needed instead is a national climate mobilization, an all-hands-on-deck, all-of-society effort to rapidly scale up the technologies and policies that the climate savers portrayed in this book have pioneered. Ordinary citizens can join the growing ranks of public-interest organizations committed to this approach.

The fossil fuel industry and its allies have in the past suggested that we can't afford a quick clean-energy transition, or that it's not workable or cost-effective. The climate savers' work, however, reaffirms what experts have been telling us for a long time: not only can we afford a rapid transition, but the sooner we do it, the more money we'll save, the more jobs we'll create, and the more climate disruption and damages we'll avoid. To do all this in time, though, will require a great deal of new investment.

We know exactly how to transform our economy to rely on clean energy. We have the technology required, although more R&D would indeed enable us to do it more cost-effectively. Now we must coordinate and rapidly scale up the kind of pioneering work described in this book. Doing so will require overcoming numerous obstacles that fall into three broad categories: resistance from "bad

---

* Using a different methodology that took account of environmental and climate damages, other researchers found that fossil fuel subsidies were projected to be $5.3 trillion (6.5 percent of global GDP) in 2015. See David Coady et al., "How Large Are Global Energy Subsidies?" working paper, International Monetary Fund, May 18, 2015, https://www.imf.org/en/Publications/WP/Issues/2016/12/31/How-Large-Are-Global-Energy-Subsidies-42940.

actors" who want to keep profiting from fossil fuels, logistical challenges to deploying new hardware, and lack of political power.

Powerful interest groups have dangerous energy resource development plans. The major oil companies currently plan to invest on the order of $5 trillion by 2030 to develop new oil, gas, and coal assets. The International Energy Agency, however, has warned that *no* new oil or gas fields or coal mines can be developed if the world is to remain below the 2°C limit agreed to in Paris in 2015. These new trillions of dollars in lethal fossil fuel investment would torpedo climate-protection efforts. A 2022 *Guardian* investigation revealed that "these plans include 195 carbon bombs, gigantic oil and gas projects that would *each* result in at least a billion tonnes of CO_2 emissions over their lifetimes, in total equivalent to about 18 years of current global CO_2 emissions. About 60 percent of these have already started pumping."

The outlook for halting these projects does not look good. Despite giving lip service to the 2015 Paris climate goals, the world's sixty largest commercial and investment banks have invested $3.8 trillion in fossil fuels since the Paris Agreement; household names like JPMorgan Chase, Citibank, Wells Fargo, Bank of America, Royal Bank of Canada, Barclays, and Bank of China were prominent among them. Some large banks say they will get to "net-zero" financed fossil fuel emissions by 2050. But why not now? Their distant climate goals and pieties leave current profit taking undisturbed for nearly another thirty years.

Most likely only the U.S. government—the executive branch, Congress, and the U.S. Treasury—is powerful enough to force these banks to abort these projects. Instead of allowing these companies to undermine the world's climate goals, the United States and its allies could embrace a proposed Fossil Fuel Non-Proliferation Treaty, inspired by the Nuclear Non-Proliferation Treaty. Such a treaty has already been endorsed by more than a thousand civil-society organizations and sixty-three cities and subnational governments on five continents. It applies not just to new reserves of fossil fuel but also to fossil fuel reserves at existing properties containing more carbon than can be burned without exceeding the 2°C target.

Adherence to this treaty should be coupled with the passage of a steadily rising national fee on carbon, to oblige fossil fuels to bear a share of the immense environmental, climate, and public health damage they cause. If fossil fuels are made progressively more expensive than competing energy sources, and less available, their use will decline.

Scaling Up

But to make clean energy the rule rather than the exception, a number of logistical obstacles will need to be overcome. In the United States alone, millions of new solar and wind energy devices and batteries will quickly need to be manufactured, permitted, and deployed. Fortunately, the United States already gets almost a quarter of its power from clean, nonnuclear renewables (which is more than it gets from coal or nuclear power). And the Department of Energy recently concluded that the nation could get 40 percent of its electricity just from solar power by 2035. That alone would be enough to power all the homes in the nation while creating 1.5 million new jobs. In addition, however, millions of new wind turbines will need to be sited in new power plants. Existing dams will need to be repowered or outfitted with generators, and new geothermal wells will need to be developed. More short-term and long-term (seasonal) energy storage will be needed as well. With a clean-energy transition, rural areas starved for infrastructure will receive new investment in renewable powerplants and all the ensuing economic activity.

Energy systems of all kinds will need to be made far more efficient across the economy. Fortunately, by applying "deep energy retrofits" and other examples of integrated design to buildings, equipment (such as appliances and vehicles), and industrial processes, deep and highly cost-effective reductions in overall energy needs are possible.

These integrated energy designs are generally far more efficient and cost-effective than a sequence of smaller, less ambitious, piecemeal energy efficiency fixes. More than one hundred million buildings will need energy efficiency retrofits—new, advanced energy-efficient windows, heat pumps, electrical appliances, and fans. These retrofits will create a cornucopia of installation, construction, operation and maintenance, and supply-chain jobs that cannot be outsourced.

Because the energy transition will require increased electrification throughout the economy, our aging transmission grid will need a major expansion and modernization. There will be lots of new jobs for engineers, technicians, and installers in implementing these upgrades and for utility and construction workers of all kinds. New transmission lines will need to be permitted and then built, from rural areas replete with renewable power to urban load centers, and regional grids will need to be interconnected for power sharing. Thousands of miles of new high-capacity transmission lines will therefore be needed every year, along with upgrades to the power distribution system.

In addition, the electrical grid needs to be made "smarter." Instead of just funneling power in one direction, from power plants to users, a smart grid will need to receive and manage power from hundreds of millions of rooftop solar systems, electric vehicles, other generating sources, and storage nodes. The smart grid will also depend heavily on price-sensitive demand management strategies. Householders and businesses will be paid to allow voluntary service curtailments at times of high usage, to balance supply and demand on the grid without the need for costly new power plant capacity, thus making the most efficient and inexpensive use of existing generating equipment. This modernization will ultimately cost hundreds of billions of dollars—far more than we have currently budgeted for modifying the grid. Greater use of hydrogen for energy storage, and for clean industrial and transportation fuel, will also require a network of electrolyzers and hydrogen storage facilities. All told, we will see a major expansion of the hydrogen industry, which is already beginning in the United States and Europe.

New Jobs to Do, Old Ones to Undo

I have often referred to the millions of jobs that will be created by moving toward clean energy, but there are also important jobs to be done in moving *away* from fossil fuels. Each leaky old oil and gas well, abandoned mine and tailings pile (a type of rock waste), or pond or stream contaminated with acid mine drainage is both an environmental blight and a job-creation opportunity; a properly handled transition can be a win-win for the environment and the economy. Fossil fuel industry workers who have lost their jobs due to the transition should be offered free, publicly funded training and new jobs relevant to their skill sets: Former coal miners could be employed reclaiming some of the five hundred thousand abandoned hard-rock mines in the United States and the tens of thousands of abandoned coal mines. Power plant workers at coal and natural-gas plants could be hired to help convert obsolete fossil fuel power facilities to solar and wind farms and transmission facilities; others could be trained to install and maintain solar, wind, and other renewable generating plants. Oil and gas workers could be hired to help cap wells among the nation's 57,000 confirmed abandoned oil and gas wells, reducing methane emissions. (The 2022 Inflation Reduction Act [IRA] contains funding for methane leak reduction.) Pipe fitters and pipeline workers might one day build pipelines and pumping stations to transport and inject captured industrial CO_2 into the ground. The IRA and its $369 billion in climate-related spending is expected to create some 9 million new jobs. The coal industry today only employs about 40,000 workers, while oil and gas extraction employ about 320,000. There will be plenty of clean transition work for these fossil fuel workers, and the nation can afford to pay to retrain them. Overall, fossil fuel industry job losses will be overshadowed by clean-energy transition job gains.

A commitment to environmental justice and to ensuring that no community is left behind will be essential to the successful implementation of the clean-energy transition. Under the IRA, qualifying clean-energy or climate-related projects built in communities dependent on fossil fuel industries—such as those with now-shuttered coal mines or coal power plants—will be eligible for more generous incentive payments than projects sited elsewhere. So will projects built in census areas that contain "brownfield properties"—that is, previously used industrial sites now contaminated by pollution or hazardous materials. Some of these communities have already suffered health-related impacts of refineries and mines from their noise, air and water pollution, and traffic.

Transportation now produces more greenhouse gas in the United States than any other economic sector. We must electrify it, with ever-tighter fossil fuel efficiency standards paving the way. With 100 percent EV sales and a 90 percent clean-power grid, consumers would save *$2.7 trillion* over thirty years—$1,000 per household annually. This would also prevent 150,000 premature air pollution deaths and $1.3 trillion in environmental and health costs through 2050. A delay of as little as five years would leave $400 billion in benefits on the table.

More than a billion new electric vehicles will need to be built globally, along with their charging stations. A huge demand will ensue for vehicular and utility batteries and for computer chips and the raw materials with which to build them. In a farsighted move to stimulate a domestic computer chip manufacturing industry while alleviating supply chain issues arising from the pandemic, Congress passed the CHIPS and Science Act of 2022* with the support of the Biden administration. The act appropriates $54.2 billion in subsidies to build chip plants in the United States and to support U.S. chip R&D, while also providing a 25 percent tax credit for building and equipping the plants.

Factories, of course, will also need to be built to produce the new solar, wind, energy efficiency, and energy storage equipment required, as well as the millions of new zero-emission vehicles, air-

---

* CHIPS stands for "Creating Helpful Incentives to Produce Semiconductors."

planes, and ships. In addition, the power and transportation sectors will require enormous numbers of electronic and microelectronic components and a flood of raw materials. Electric motor magnets for wind turbines or EVs will need rare earth metals; EV batteries will be hungry for lithium, cobalt, and nickel until battery technologies using cheaper materials are developed.

Abundant Financing for a Clean Transition

Trillions of dollars of investment capital will be needed up front to create this clean-energy economy, though the investments will eventually pay for themselves and produce millions of jobs. Public-private partnerships, however, along with other possible vast new revenue streams, are more than capable of supplying the needed capital. A securities transaction tax, the end of tax cuts and loopholes for the wealthy (like the carried interest exemption benefiting wealthy executives and hedge fund managers), and many other financial policies could help enable the clean-energy transition without inflicting hardships or raising taxes on ordinary Americans. The IRA provides for some needed tax reforms, including a corporate minimum tax of 15 percent—essential to prevent legal tax evasion by big corporations. Amazon, for example, now earns billions in profits but, in recent years, paid at single-digit tax rates—when it paid at all. According to the Roosevelt Institute, fifty-five other wealthy companies—including Salesforce, Nike, and Archer Daniels Midland—paid no taxes in 2020, depriving the public of billions of dollars in tax revenue.

You might think that the federal government could easily act to retrieve even more of the revenue slipping through the sieve of federal and state tax codes. But government's hands have been zip-tied: At least since the early 1980s, politicians beholden to corporate patrons have portrayed government as the enemy of the people and all taxation as theft of taxpayers' hard-earned money. Most of these taxpayers do not fully realize who the real beneficiaries of the current tax code are, nor that low- and middle-income taxpayers like them could pay far less in taxes and receive better services if the superrich and large corporations paid their fair share. Often lost in the discussion is the fact that government is essential to the defense of liberty, and taxes are the means of ensuring it remains capable of doing so and of funding other public goods.

Building a Climate Coalition

Protecting the climate will require good ideas from not only scientists and engineers, but also organizers able to build the political power nationally and globally to implement climate-saving measures. Yet citizens have more power than they think, even in the face of legal decisions that give corporations unlimited political spending power. The Supreme Court's 2010 *Citizens United v. FEC* decision is an example: By ruling that corporations were legal "persons" whose political donations were a form of "free speech," the court allowed corporations to make virtually unlimited political donations through political action committees (PACs), which shield donors' identities. Since that decision, though, Public Citizen, a nonprofit advocacy group, has spearheaded a national campaign for a constitutional amendment to undo *Citizens United*. So far, the amendment has been endorsed by twenty-two states. (Thirty-eight are needed for ratification.)

Public Citizen has also called for PACs to be obliged to disclose the names of these "dark money" donors and for a presidential executive order that would require all corporate federal contractors to disclose their political donations. Another approach would be new campaign finance regulations to free politicians from their dependence on corporate contributions. Whereas all these goals cannot be accomplished at once, they all must be put on the agenda and pursued.

How, then, can sufficient power be acquired to secure the passage of stronger climate legislation? A powerful, broad-based, national coalition of interest groups is needed. The partners must include business, labor, civil rights, environmental, climate, youth, and social justice groups as well as public officials. They also need to include women's organizations, public health groups, scientific and engineering organizations, academics, media representatives, and farsighted military leaders. Individually, these groups are already sympathetic to the need for bold climate action. They now must be forged into a single, massive coalition strong enough to dominate the political stage and set the agenda on energy and climate.

To win the public's support, this coalition must project an audacious and inspiring vision, along with credible policy proposals to deliver it. Coalition proponents must be mindful of smart-messaging ground rules, speaking of "investments," not "spending," of "fees" rather than "taxes." If the coalition's agenda weaves climate and economic fairness tightly together, support for one will pull the other along. The clean-energy transformation must be recognized as a unique historic opportunity for restorative justice and be conducted so as to particularly benefit low-income, marginalized, and disadvantaged communities, to begin remedying past injustices. Other coalition messages that could elicit broad support might focus on the need to keep people and property safe from extreme weather, avoid the risks of food shortages and hunger, and protect the wonders of nature.* Focusing on bread-and-butter issues such as jobs and prosperity could excite a national groundswell of activism powerful enough to sweep aside congressional roadblocks and win over the "99 percent."

As public support grows for strong climate action, Congress or the president could declare a national climate emergency, thereby gaining emergency powers to fast-track the clean-energy transition. Once the emergency is declared, the president could call for a national climate mobilization, much as Franklin D. Roosevelt did in

---

* Some voters may respond well to messages about how effective climate policies will dampen the elevated climate risks from wildfires, droughts, heat waves, and floods. These will include higher food prices linked to crop failures and increased threats to national security from risks of war, social turbulence, and unrest as climate change displaces residents of vulnerable areas within the United States and abroad. That will create domestic climate refugees even as waves of climate refugees may arrive at U.S. borders from devastated regions abroad.

the 1930s and '40s to rally the nation during the Great Depression and then to fight World War II. Millions of national environmental group members undoubtedly would rally to the cause.

Public Education and Democracy

The ability to understand climate policies depends directly on knowledge about climate and energy. Many people, even today, do not understand what CO_2 is, and in a 2016 survey, a third didn't realize that burning fossil fuels produced it. A 2012 survey of 2,200 Americans by the National Science Foundation also showed that about a quarter of Americans believe that the sun revolves around Earth, although this belief was debunked nearly five hundred years ago by Copernicus. About three quarters of Americans don't know what the atmosphere is made of and are unable to name its largest component, nitrogen. Thus, it isn't surprising that many people find it hard to understand how our use of fossil fuels is affecting the earth's climate.* For all the policies described in this book to stand a chance of success, a long-term national campaign devoted to educating the public about climate and democracy would be very valuable.

Effective climate education could help people understand the causes of and solutions to the climate crisis. It could also help them see that unmitigated climate change is a nightmare that will get a lot worse without an all-out effort to combat it. Climate education must stress that climate change is a scientific issue, not a political or ideological one, and should not be cynically portrayed as such. If this message is coupled with the idea that ordinary people will also benefit economically from a rapid clean-energy transition, there remains

---

* A recent Ipsos poll found that about half the American public today in 2023 still does not believe that humans are the main cause of climate change. (An earlier poll estimated the number at about 40 percent.) Seventy percent of Americans, however, do agree that climate change is occurring. See Daniel de Visé, "Do humans cause climate change? Even now, only half of Americans say yes," *The Hill* republished in *Yahoo News*, May 25, 2023, https://thehill.com/policy/energy-environment/4019474-do-humans-cause-climate-change-even-now-only-half-of-americans-say-yes.

a chance these people will rise to the climate's defense. A national educational program could show that a swift clean-energy transition offers greater well-being and prosperity for the nation than the status quo.

Public education efforts are also needed to address the issue of preserving our democratic way of life. If democracy no longer functions properly, a broad national climate movement will be much harder to create, and meaningful action to combat climate change of many kinds will become more difficult or impossible. A danger to democracy is a danger to the clean-energy transition. An early goal of the climate movement therefore must be safeguarding elections, the bedrock of democracy, and making sure that the public fully understands the value of democracy and the price that has been paid to preserve it.

Granted, this will be a tall order on top of all the other challenging tasks that have to be undertaken as soon as possible, particularly in a nation deeply divided along political lines. But because speedy and effective national action, not just on climate but on virtually anything, depends on strong and unified public support, it is of paramount importance to overcome these divisions. A national climate education program could and should be crafted also to reinforce basic, unifying democratic values.

Individual Action and Choices

Many people still believe that individual consumer actions—myriad small, "green" personal decisions—are the solution to climate change: hanging laundry on a line to dry, taking a cloth bag to the grocery store, riding a bike or walking more. But little, well-intentioned green gestures will never save the climate if societal decisions beyond the individual's control aren't themselves well aligned with climate goals. The large, infrequent energy decisions in our lives, however— what to drive, how to heat and cool our homes—*do* make a major difference when multiplied millions of times, for they largely determine an individual's or family's carbon footprint. The same is true when millions of companies make energy-conscious business decisions, even when energy costs may be a small part of their operating budget.

Another vital individual action that can have a major climate impact when aggregated is registering to vote, actually voting, and urging your elected representatives to vigorously support the ambitious climate policies outlined in this book.* Let them know your vote depends on their support for a quick clean transition and get them on the record publicly on it. Demand that they clearly declare their positions on specific critical climate issues and, if necessary, hold demonstrations to focus media attention on them.

The number of children per family is also a major determinant of each family's carbon footprint. Because it is an emotionally charged issue for some, population stabilization is often left out of discussions of climate policy for fear of giving offense. I believe it is such an important topic that it is essential both to include it and to discuss it frankly in any dialogue on how to solve the climate crisis. "If we want to stop climate change," says John Seager, president and CEO of the Population Connection in Washington, DC, "we must stop population growth."

Whether you agree with that conclusion or not, population stabilization should be an explicit component of climate policy. This is a call not for draconian population-control measures, but, instead, for expanded funding nationally and globally for contraceptive and family-planning services and for expanded access to education for girls in developing countries. Their educational attainments are

---

* Several environmental organizations have addressed what individuals can do. In the Sierra Club's "12 Ways to Halt Climate Change," for example, the individual actions recommended include growing a garden, purchasing green power for the home, using efficient appliances and light bulbs, turning off unneeded lights, shifting power use to off-peak hours, using energy-efficient windows, insulating walls and attics, driving an EV, adjusting the thermostat "to a comfortable temperature when needed and [to] save energy [at] other times," reducing resource consumption, going paperless, printing on both sides of the page, reusing goods, and recycling materials.

inversely correlated with fertility rates, and according to a study by Project Drawdown, educating girls and increasing the global availability of family planning services could avoid about seventy billion metric tons of CO_2 emissions by 2050, far more than electric vehicles, recycling, and composting combined.

As with the population issue, the subject of our dietary choices and how they affect the climate is a hot-button issue for some. Yet food loss, food waste, and dietary choices have surprisingly large climate consequences, and they need to be made a high strategic priority by the food industry and government. A recent study found that each year, somewhere between 33 and 40 percent of all food grown in the world is lost or wasted. By some estimates, producing that lost or wasted edible food releases 8 percent of the world's greenhouse gases and uses a quarter of the world's freshwater supplies. About half the total—two billion tons of food a year, worth $600 billion—is lost on the farm or in processing or transport. Roughly the same amount of food is then wasted by retailers and consumers, with people in Western countries not uncommonly wasting a quarter of food purchased, although one in nine people in the world suffers from hunger.

Some knowledgeable observers believe that 50 to 70 percent of this loss and waste could be eliminated. Doing so is complicated, however, and requires strong leadership by food manufacturers and retailers able to bring about cooperation across the food supply chain. Food efficiency is a bit like energy and water efficiency, in the sense that increasing the efficiency in how one resource input is provided opens up interlocking stores of value and, potentially, enormous financial savings. For example, when less water is required, less energy is needed to move that water; when less food is squandered, less water and energy are needed to produce enough food for all. In general, losses and waste will begin declining once smart executives and others begin realizing that by eliminating inefficiencies in the food system, they'll be able to tap into a vast and barely explored reservoir of value potentially worth billions to companies who make food efficiency a strategic priority. Wise public policy, educational efforts, and targeted incentives can accelerate this process. It's something we can do if we put our minds to it.

Our current food system's massive environmental and carbon footprint absolutely needs to be carefully considered in any comprehensive strategy to combat climate change. The food system occupies billions of acres of land, uses vast amounts of water, and produces large amounts of greenhouse gas. Livestock production alone may conservatively account for nearly 15 percent of total GHG releases. Thus, the food system is a highly relevant part of the climate problem. It is a large driver of deforestation, for example, which in turn releases stored carbon. Deforestation then triggers other serious environmental problems, sometimes causing serious water shortages while also polluting rivers that then create dead zones in coastal waters.

If we attempt to scale up our current food system just to keep pace with growing global population but without major changes, the environmental and climate impacts will be even less sustainable than they are today. Technological improvements can increase water use efficiency and improve fertilizer use on farms, but they alone will not suffice. What foods we as a society choose to produce and consume is enormously important.

Food production, however, is ultimately driven by consumer demand, and the notion of telling people what they ought to eat is probably as unpopular as trying to tell them how many children to have. Nonetheless, significant dietary changes will be necessary. We currently kill more than fifty billion animals for food every year, just in the United States. The pigs in Iowa alone produce as much manure as eighty million humans; manure releases methane gas and nitrous oxide. Raising, transporting, fertilizing, and processing the crops grown to feed these animals releases both CO_2 and nitrous oxide. It wouldn't be a bad idea if we reduced our meat consumption.

For generations, we have known that raising livestock takes more water and energy than producing an equivalent amount of food by raising grains, pulses, fruits, vegetables, and nuts. Almost 40 percent of all corn grown in the United States is fed to animals. That's partly because it takes about six pounds of grain used in fattening beef in a feedlot to produce a single pound of edible beef. We also know that, among livestock, beef cattle and pork have the greatest environmental impacts in terms of their water and grain requirements and in releasing methane gas and manure. During its life, each head of cattle will drink about four thousand gallons of water.

Researchers who have done authoritative work on the environmental impacts of global agriculture have concluded that we will not be able to keep global temperature from rising 2°C without reducing beef, pork, and lamb consumption while eating more protein-rich plant substitutes. Although most people are not ready to give up eating meat entirely, many would be willing to reduce the amount they consumed. Fortunately, people can get sufficient protein and virtually all other needed essential nutrients from plant-based sources, save for vitamin B12.* And a vegetarian diet or one lower in meat offers major health benefits; aggregated on a global scale, these translate into savings of a trillion dollars in annual health care costs.

A National Climate Action Plan

It may seem that in outlining what needs to be done to combat climate change, I've discussed everything but the kitchen sink. That makes it a bit difficult to understand what the priorities ought to be among all these urgent tasks: who is going to do what, when, where, how, and with funding from whom. Here is where thoughtful, scientifically informed planning comes in.

The critical mission of extending the climate savers' exemplary work can't be left to haphazard "free-market" forces, nor to scattershot efforts improvised on a year-to-year basis. A systematic, scientifically based, comprehensive long-range national climate

---

* Although B12 does exist in mushrooms and nutritional yeast, vegans are advised to consume fortified cereals, a B12 vitamin capsule, or small amounts of animal protein to get enough of it.

action plan must be developed to guide all sectors of the economy in the energy transition. The plan needs to be prepared by the best and brightest scientists and engineers the nation has; these experts should be truly independent and not financially compromised by ties to the fossil fuel industry or its think tanks or to agribusiness.

An effective national climate action plan must have a diplomatic component as well as a domestic one. After three decades of international negotiations, countless international meetings, and voluntary declarations of national commitments, both the greenhouse gas concentrations in the atmosphere and global temperature have continued rising dangerously. Even well-intentioned nations cannot solve this global problem alone.

It makes no more sense to have each nation set its own emissions targets independently than it would for each state in the United States to set its own clean-air and clean-water standards. That's why we have a federal Environmental Protection Agency and a federal judiciary, to enforce uniform national environmental standards. Fresh ideas are needed now in the diplomatic realm to break the international diplomatic logjam. The United States ought to seek the establishment of a powerful International Atmospheric Authority (IAA) or International Climate Authority with the power to set global climate targets and, from them, to derive emissions quotas for all 195 nations. An IAA would need to have the authority to encourage cooperation by leveraging a large financial investment endowment and the funding of international financial organizations, such as the World Bank—and to enforce its decisions with powerful sanctions. Otherwise, rogue nations and others will be able to continue undermining global climate goals.

Establishing such an organization would be extraordinarily challenging; politically, the odds are stacked against it. Seeking to create it by the unanimous consent of 195 nations committed to a common goal and an enforcement mechanism would represent a race to the bottom, propelled by the most obstructive nations. Therefore, the IAA and its global targets initially would need to be established by a relatively small group of advanced industrialized nations with progressive climate policies, including the European Union, the United States, the United Kingdom, Norway, Canada, Japan, and a few other like-minded nations. They would set an ambitious long-term target of returning the atmosphere to 325 parts per million of CO_2—a level 16 percent above the historic pre-industrial level but still consistent with the existence of polar ice and other climate-stabilizing conditions. The IAA would establish emission quotas and timetables for each nation, so that the entire world would be able to achieve the authority's global atmospheric targets.

The IAA would be generously capitalized with a fund that would begin in the hundreds of billions of dollars and grow from there. It would assist less affluent nations who joined and accepted its authority. Larger nations that refused to cooperate with the IAA would be subjected to stiff trade and other economic sanctions, to encourage them to comply with IAA quotas; members accepting the IAA's global targets would be relieved of sanctions and would receive scientific and technical assistance as well as means-tested financial help in achieving emission reductions and in adapting to climate change. The political pushback against a powerful global authority would be enormous, but it is nonetheless badly needed and worth trying.

Once the Battle's Won

During and after the clean-energy transition, some fossil fuel facilities, such as former coal and gas plants and former coal mine sites, will be suitable for siting new solar or wind facilities or for transmitting renewable power to the grid over existing utility connections.* But a complete clean-energy transition will, ultimately, render trillions

---

* The Nature Conservancy, for example, plans to build a large solar power plant on six reclaimed coal mine sites in West Virginia. See Steve Hanley, "Nature Conservancy To Build Solar Farms At Abandoned Coal Mines In Virginia," CleanTechnica, March 7, 2022, https://cleantechnica.com/2022/03/07/nature-conservancy-to-build-solar-farms-at-abandoned-coal-mines-in-virginia/.

of dollars of fossil fuel infrastructure and undevelopable resources worthless. When that time comes, the industry will without doubt seek a federal bailout. Should we, the taxpayers, grant it?

The fossil fuel industry and its investors have for generations made vast profits from their fossil fuel assets. They have taken enormous public subsidies in the form of tax breaks, discounts on federal oil and gas leases, and royalty payments. Oil and gas companies often lease federal lands for as little as two dollars an acre, and sometimes pay nothing. Then they remit only 12.5 cents to taxpayers for each dollar they retrieve from these publicly owned resources. Meanwhile, they have created vast public health and environmental damage, while destabilizing the climate, after fooling the public for decades about both climate science and climate legislation. More recently, when the opportunity arose during the COVID-19 pandemic and the Russian invasion of Ukraine, the industry charged previously unheard-of prices at the gas pump, remorselessly price-gouging the public. Although fossil fuel investors and executives had no scruples about profiteering, some friends of the industry have already floated trial balloons regarding a public bailout, citing job losses, pain for shareholders, and economic calamity if the industry were forced to fold.

I have just explained how the energy transition will be a boon to energy consumers and a jobs bonanza for workers, and how its impact can be cushioned for fossil fuel industry workers. Investors in fossil fuel stocks have made hay while the sun shined. Perhaps those were not the wisest or most ethical investment decisions they could have made; maybe it's finally time to unwind those positions. As to the idea that a financial catastrophe will ensue if the industry is obliged to absorb losses due to stranded assets, do not be misled. The sky will not fall if the industry doesn't get another free ride, and no multibillion-dollar company is entitled to a free ride at public expense *forever*. In the market economy, one profits when prices go up and accepts losses when they decline. Money spent bailing out the fossil fuel industry could be better spent accelerating the clean transition and making some of the industry's victims whole.

Precedent-Setting Climate Legislation

The 2022 Infrastructure Investment and Jobs Act (P.L. 117-58) included significant funding for some climate-related investments, and the Inflation Reduction Act (IRA), also in 2022, included $369 billion in energy security and climate change funding over a decade. Whereas this is the largest climate bill ever passed by Congress, it amounts to only a very modest outlay of under $37 billion a year—sixteen ten-thousandths of our GDP. Nonetheless, that relatively small sum is expected (by Congress and various research organizations) to be enough to enable U.S. greenhouse gas emissions to fall to 40 percent below 2005 levels.

While this measure is clearly a highly important step in the right direction, the net impact will barely be noticed globally and will have a less-than-game-changing impact in the United States. Citing a preliminary study by the Princeton-based REPEAT Project, climate scientist Dr. James Hansen points out that the IRA will diminish U.S. emissions by only an additional 15 percent if one takes into account the current downward trend in U.S. emissions, the reductions already achieved since 2005, and those expected from the Infrastructure Investment and Jobs Act.* Hansen also raises legitimate doubts about whether the projected 15 percent decline will actually be achieved; he is concerned about opposition often encountered in siting power lines and the law's reliance on carbon capture and storage technology that hasn't been proven at scale. Yet, if the IRA's $369 billion investment does result in an additional 15 percent reduction in emissions, think what investment on the trillion-dollar scale could do! At that rate, $2.5 trillion spread over a decade or more could bring us to carbon neutrality.

Even at its modest size, however, the IRA will move the United States from the "climate laggard" to the "climate leader" column

---

* U.S. emissions are 14 percent of global emissions, and reducing them by 15 percent will reduce global emissions only by 2 percent, if everything goes as planned.

internationally and will help get the country four fifths of the way to President Biden's announced goal of cutting U.S. emissions 50 to 52 percent by 2030. This target is now still feasible if the federal efforts in the bill give momentum to independent state, local, and corporate emissions-reduction efforts. The new IRA also takes on the issue of environmental justice, offering low-income households energy rebates for buying new, efficient home appliances and to pay for home efficiency retrofits and offering tax credits for the purchase of used clean vehicles. The act also provides $27 billion to the EPA for public-private partnerships that leverage federal dollars to be spent on any technologies that help combat climate change; more than 40 percent of that money is earmarked for low-income and disadvantaged communities. These households will also be eligible for another $3 billion in environmental and climate justice block grants to pay for low- and zero-emissions and resilient technologies, as well as for mitigating climate and health risks. All these are steps in the right direction, yet much more will be needed from legislators—and from all Americans—in the future.

Rising to the Occasion

This country is capable of extraordinary effort in crisis and of remarkable achievement when united and committed to a cause. The transformation of the United States of America to a clean-energy nation now calls for that commitment and is worthy of our utmost effort.

The climate savers in this book have shown us that we have the engineering technologies, ecological knowledge, and entrepreneurial spirit to confront the climate emergency effectively. Yet, currently, these technological, ecological, and political solutions are still not being deployed quickly, fully, or widely enough to arrest the most extreme and dangerous existential emergency humanity has ever faced. We are therefore currently on track to double the tragic climate disruption we have already had to date.

The highest levels of the world's governments have known about the nature and gravity of the climate crisis for decades, yet have failed to take adequate action, though the solutions have been repeatedly spelled out for them by scientists and other public figures. So, we cannot assume that government, left on autopilot, will act decisively enough to head off this crisis. They have often said it isn't politically possible. Our role now is to make it not only politically possible but politically *necessary*. We who are concerned about the climate emergency must unite and coalesce into a broad and effective movement to oblige government to greatly accelerate the clean transition.

A global and national climate movement already exists. We all need to join it and make it a great deal stronger, so it can call the shots. Everyone has a part to play, and we all must show up. We must *act*—educate, organize, and mobilize ourselves and others. It is now up to us to ensure that the resources are made available and the structures are put in place to scale up the climate savers' pioneering work and deploy it in the service of climate protection and the clean-energy transition. That transformation must become nothing less than the nation and the world's top priority. Only a society forever wedded to energy efficiency and renewable energy has any hope of saving itself from a global climate catastrophe.

The United States has the scientific knowledge, avid entrepreneurship, and availability of wise leadership to reinvent itself as a model clean-energy society. We could now show the rest of the climate-buffeted world what a modern, prosperous, and emissions-free society looks like. We could also prove that we're not only a thriving democracy but a generous people, willing to extend a helping hand to other, less affluent nations to help them accomplish their own clean-energy transition. So, let's now bestir ourselves, take control of our own destiny, and inspire other nations with our vision and success to accelerate the whole world's transition to a clean-energy economy. The future of the world really does depend on it.

Coda: Recommended Climate Policies and Strategies—The To-Do List

The policies proposed here are intended to accelerate the United States' transition to a clean-energy economy while protecting, restoring, and enhancing natural and managed ecosystems to increase their ability to extract and store atmospheric greenhouse gases. They also are intended to facilitate the phasing out of fossil fuels and to prevent further unsustainable investment in fossil fuel infrastructure, including power plants, and in resource development that would increase rather than diminish fossil fuel emissions. These policies are grouped by general type and, imperfectly, in a subjective order of priority. Prioritization is difficult and somewhat arbitrary because these activities are deeply interrelated and will need to occur simultaneously rather than sequentially.

National Clean-Energy Transition Plan

* As outlined in chapter 14, Congress should direct the National Academy of Sciences or another suitable government body to prepare a systematic, scientifically based, comprehensive long-range National Climate Action Plan

to guide all sectors of the economy in the energy transition. The plan should be approved and authorized by Congress and should set specific and ambitious national goals for a complete clean-energy transition sooner than the current 2050 national net-zero goal. The plan should therefore provide a new timetable, milestone targets, and a process for regular progress evaluation. It should strongly emphasize maximum feasible implementation of cost-effective energy efficiency investments and similar targets for the protection, restoration, and enhancement of natural and managed ecosystems to facilitate their absorption and storage of atmospheric carbon in soil and organic matter.

Emergency Declaration/National Mobilization

* To underscore the existential threat we face and to make emergency powers available to the executive branch, the president should issue a formal climate emergency proclamation. (There are legal safeguards in place against the abuse of such powers: Emergency proclamations must conform to invoked statutes and are subject to congressional review and revocation. They are legal under the National Emergencies Act; more than seventy are currently in place.)

Fossil Fuel Phaseout

* Public utilities and federal power authorities should put fossil fuel power plants on glide paths to shutdown, phasing them out as soon as renewable capacity and storage can be provided in each plant's service area or when replacement power can otherwise be ensured. The phaseout should begin with coal plants.

* The U.S. Department of the Interior should end the leasing of federal land for new coal, oil, and gas fields; not extend leases beyond their current term; and attempt to negotiate a phaseout of existing leases.

* As far as possible under existing statutes and regulations that oblige agencies to grant properly submitted permits, all permitting agencies should cease approving permits for all new fossil fuel infrastructure. Where agencies lack authority to deny permits, further applications should be discouraged while means are sought to grant broader permit denial authority.

* All federal, state, and local fossil fuel subsidies should be terminated as soon as legally possible—to the industry itself first and, ultimately, to consumers as it becomes practical.

* Federal, state, and local funding should be provided for the retraining or retirement of fossil fuel industry workers.

Financing

* Congress should create a new public-private federal clean-energy finance corporation to help finance investments in clean-energy supply, energy efficiency, energy storage, and clean transportation at low interest.

* Congress should establish a clean-energy bank to issue low-interest clean-energy transition loans.

* Congress should require the U.S. Treasury to issue national clean-energy-conversion bonds to raise clean-energy capital.

✳ Congress should generously fund the U.S. Department of Energy to offer grants, loans, and loan guarantees and to perform other financial services necessary to help finance the transition to a clean, sustainable economy.

✳ Congress should generously fund the U.S. Department of Housing and Urban Development to provide low-interest loans for energy efficiency retrofits of the U.S. housing stock.

✳ The U.S. Treasury, the Securities and Exchange Commission, and the federal Financial Stability Oversight Council should be directed by Congress to formalize a consistent set of rules requiring full climate risk disclosure by fossil fuel companies. (This will raise the cost of capital for them, reducing profitability and forcing some operations to be canceled, given that investors shy away from risk.)

Taxation

✳ Congress should approve a national carbon-tax system to gradually drive up fossil fuel costs and make fossil fuels progressively less cost-competitive, with tax rebates for lower- and middle-income consumers and cross-border adjustments to ensure fair trade.

✳ Congress should impose a minimum 20 percent income tax on individuals earning over $10 million a year to generate hundreds of billions of dollars for the clean-energy transition. (A much more modest "Billionaire Minimum Tax" plan proposed by the Biden administration for the 2023 federal budget would impose a 20 percent minimum tax on people earning over $100 million a year. Even that low a rate would raise $360 billion over ten years—virtually equal to the cost of *all* the climate provisions in the 2022 Inflation Reduction Act.)

* To free up additional billions of dollars for the clean-energy transition, Congress should impose a minimum wealth tax of 2 percent on very-high-net-worth individuals with over $100 million in assets.

* Congress should instruct and fund the U.S. Treasury to retain adequate staff to collect all backlogged unpaid taxes, starting with taxes owed by upper-income Americans and corporations. (Unpaid taxes cost the U.S. Treasury $7 trillion over a decade. Those funds—3 percent of GDP—could go a very long way toward financing the clean-energy transition and meeting social needs while lowering energy bills for ordinary people.)

* Congress should modify the tax code to ensure that accelerated depreciation for renewable-energy and energy efficiency investments be accessible to qualified users on terms at least as generous as now exist for fossil fuel investments. It should simultaneously end accelerated depreciation allowances for fossil fuel–related investments.

Energy Efficiency

* Congress should require a national building energy efficiency retrofit program with the goal of retrofitting 100 percent of U.S. multifamily, commercial and industrial buildings with all deliberate speed, while also setting minimum regional efficiency standards.

* Congress should provide cities and other jurisdictions with funds to encourage the use of energy-efficient local district heating-and-cooling systems and for co-generation (the combined production of heat and power) in industry, commerce, and residential communities.

* Congress should provide leadership and ample, predictable financing for an ambitious national building retrofitting campaign, offering low-interest loans and tax credits to encourage commercial and industrial building owners to make their buildings energy efficient.

* Congress should require that all cooking, heating, and hot water systems in new residential and commercial buildings be electric and that all heating systems be electric or use electric heat pumps, which are highly efficient.

* The Inflation Reduction Act already provides tax credits for personal energy efficiency and clean-energy investments, including EVs, energy-efficient appliances, rooftop solar panels, geothermal heating, heat pumps, home batteries, stand-alone energy storage, and biogas systems. However, the implementation of these credits is so important that they should be objectively reviewed at frequent intervals to gauge their impact. Funding must be augmented as needed to ensure that the credits adequately incentivize rapid, massive adoption of clean and efficient technologies.

* Cities should revise building codes to require highly energy-efficient designs and should require each building to obtain and make available to the public an energy efficiency rating when built or, in the case of existing buildings, sold.

* Congress should require that all utilities implement demand management programs to conserve energy and avoid construction of unnecessary generating capacity.

Energy Supply and Power Generation—Utility and Grid Issues

* Congress should provide ample support for a national smart electrical supergrid, so that renewable energy can be efficiently transmitted from rural and frontier areas of abundance to urban load centers over high-voltage AC or, where needed, DC lines.

* Public utility commissions and other pertinent regulatory agencies should direct public utilities, co-ops, and independent power providers to adopt brisk timetables for upgrading their grids to 100 percent clean power and shutting down their coal and natural-gas fleets while making maximal use of energy efficiency programs and strengthening their power grids. The U.S. DOE should similarly direct the federally owned Tennessee Valley Authority to accelerate the replacement of its fossil generation with renewable capacity.

* Congress should create a national net-metering policy and uniform interconnection standards for renewable-energy systems.* This would reduce disincentives to connect renewable-energy systems in states where traditional utilities are hampering the expansion of residential and commercial solar.

---

* In a net-metering system, customers who generate their own electricity, typically by rooftop solar, are allowed to feed back their excess production into the grid and receive a credit for the power provided, reducing their net usage and thus their utility bills.

* Congress should create minimum standards to facilitate the rapid introduction of vehicle-to-grid charging programs that use electric vehicles to provide decentralized energy storage to the grid and that repurpose used EV batteries for stationary power storage.

* Congress has already provided investment tax credits and production tax credits for clean wind, solar, hydrogen, and hydro projects. Permitting and interconnection to the grid for such renewable energy projects should be expedited by all relevant federal, state, regional, and local governmental entities, along with public utilities. Currently, the average interconnection time has almost doubled to four years, and more than eight thousand proposed projects consisting of 950 gigawatts (GW) of energy supply and 225 GW of storage, by one estimate, are waiting for interconnection approval—more than the grid's entire current capacity. Funding needs to be provided to increase the staffing at the overwhelmed regional transmission organizations responsible for reviewing interconnection applications.

* Congress should incentivize the construction of utility- and community-scale seasonal heat and cold storage and building-scale energy storage and the use of electrolytic hydrogen to store surplus renewably generated electricity.

Transportation

* Congress should require that sales of gasoline and diesel highway vehicles be phased out in the United States according to the timetable set in 2022 by the state of California, so that 100 percent of light-duty vehicle sales are electric, plug-in hybrid, or otherwise cleanly fueled by 2035,

with medium- and heavy-duty vehicles obliged to be in full compliance by 2040. (According to a recent study, this would result in consumer savings of $2.7 trillion over thirty years—$1,000 per household annually.) All federal service vehicles, including postal and conventional military service vehicles, should be replaced exclusively by electric- and hydrogen-powered vehicles at the end of their service life or sooner, if feasible.

* Congress should adequately fund the rapid construction of a national EV charging infrastructure. Completion of this network should be pursued with the same focus and intensity that created the National Highway System in the 1950s.

* Congress should provide additional federal funding to accelerate the construction of electrified public transit systems, from buses and trolleys to light commuter rail and heavy-duty high-speed rail.

* Congress and other legislative bodies should adopt and fund programs to encourage the use of zero- and hybrid-emission clean transit and school buses.

* Given that about 80 percent of light-vehicle trips are under ten miles and half are under three miles, federal, state, and municipal tax credits and other subsidies should be provided to encourage the purchase of electric bikes, scooters, and motorcycles.

* Congress should direct the U.S. Department of Transportation and the Federal Aviation Administration to set standards and a compliance timetable for a net-carbon-neutral U.S. aviation industry, with appropriate tax credits to speed adoption by smaller airlines and individuals.

* The U.S. Environmental Protection Agency currently regulates emissions from marine diesel engines and recreational boats. Congress needs to create a new maritime rule book for the EPA and set timetables for, and strongly incentivize the conversion of, working and recreational watercraft to electric propulsion and clean fuels. Similar effort needs to be applied to the regulation and conversion of heavy equipment and of small engines such as lawn mowers and leaf blowers.

* Working with other nations and with the International Maritime Organization, the U.S. Department of Transportation should set standards and a compliance timetable to create a net-carbon-neutral shipping industry.

* The Department of Transportation should provide additional funding for bike and pedestrian paths and infrastructure.

Minimizing Industrial Carbon Use

* The U.S. Department of Energy and the U.S. Environmental Protection Agency should be empowered, directed, and funded by Congress to require medium and large firms to adopt zero-emissions plans and to require high-carbon-output industries, such as cement and steel, to adopt non–fossil fuel energy inputs and clean production technologies, or to use carbon capture and storage technologies.

* The U.S. Department of Energy should be funded by Congress to require and incentivize the use of biobased, non–fossil fuel inputs in manufacturing and chemical industries in lieu of fossil fuel–based feedstocks, and should

increase support for the adoption of technologies consistent with the principles of a "circular economy."

* Congress should direct the U.S. EPA to phase out the use of any remaining powerful greenhouse gas refrigerants.

Methane Emissions

* Methane emissions leaking from oil and gas wells, coal mines, and landfills, and being released by the agricultural sector are responsible for a quarter of global warming. Congress should instruct federal and relevant private entities to eliminate methane leaks at natural-gas wellheads, gas pipelines, and within urban distribution systems and should pass additional legislation as needed to accomplish this goal, while providing adequate funding and financial incentives to motivate compliance.

* Congress should require the U.S. Department of Agriculture to prepare regulations requiring the use of safe feed supplements that reduce methane emissions by cattle and sealed manure digesters to capture methane from them.

Natural-Carbon Solutions and Carbon Credits

* The U.S. Department of Agriculture and its Natural Resources Conservation Service should remunerate farmers and land managers globally with direct carbon-credit cash payments in return for investing in the storage of carbon in their agricultural soils and rangelands.

* The U.S. Department of Agriculture and its U.S. Forest

Service, in consultation with the U.S. Department of the Interior and other relevant federal agencies, should develop and implement a program to restore and enhance all federal forest lands to capture greenhouse gases.

* All state governments should implement similar plans to incentivize the protection, restoration, and enhancement of carbon storage capacity on state lands and on private forest land in their states.

* The U.S. Department of the Interior and the U.S. Army Corps of Engineers should implement a coherent, long-term natural resource restoration program to increase carbon storage in natural ecosystems such as wetlands and forests and on crop and grazing lands.

Protection and Restoration of Democracy

* Congress should modernize and revitalize the Voting Rights Act of 1965 to safeguard the integrity of elections so that democracy continues to function properly. (If it does not, meaningful action to combat climate change will be nearly impossible.)

* The U.S. Department of Justice should fully enforce all laws respecting free and fair elections, in particular to safeguard against voter suppression and election sabotage.

* States must adopt a constitutional amendment to overturn the 2010 Supreme Court decision in *Citizens United v. Federal Election Commission*, which granted corporations a right to virtually unlimited political contributions.

* Congress must adopt campaign finance reform legislation to eliminate the power of dark money in U.S. elections at every level.

* Congress should strengthen existing legislation to more fully protect voters, poll workers, and election officials from harassment and intimidation.

Social and Environmental Justice

* The Federal Emergency Management Agency and the U.S. Department of Homeland Security should create and manage expanded programs to address the needs of Americans adversely impacted by climate-related extreme weather, with an emphasis on preventive measures such as relocating homes and businesses out of flood-prone areas.

* Elected representatives, as well as cities and counties across the nation, should convene town hall meetings to discuss the climate crisis and what needs to be done in their locality, consistent with a swift and economically just, clean transition.

Public Education

* The White House, in direct addresses to the American people, should keep elevating the importance of an all-out national mobilization to reduce greenhouse gas emissions and increase natural carbon sequestration. These messages should be constantly and skillfully reinforced on social media by administration staff.

* The U.S. Department of Education, in consultation with state boards of education, should establish policies and programs to ensure that the U.S. educational system provides greater scientific and climate literacy, so the public can improve its understanding of the climate crisis and its remedies.

* The Department of Education should also increase the scale and improve the effectiveness of the federal government's public climate education through the National Oceanic and Atmospheric Administration or through a new Climate Information Service, to disseminate to the public reliable climate information that communicates the gravity of the climate crisis and the need for national mobilization.

* Congress should order the Federal Communications Commission to restore an updated version of its Fairness Doctrine, so that media can be held accountable for false and misleading information disseminated to the public as fact.

Reducing Food Waste and Our Carbon Footprint

* The U.S. Department of Agriculture should conduct intensive public education programs to discourage food waste and to encourage reductions in meat consumption, while also increasing support for regenerative agriculture, rangeland composting, and organic farming to increase carbon storage in soils. The department should also provide ample support for the development of meat substitutes and cultured meats to increase their availability and affordability.

Civilian Climate Corps

 * Congress should provide the full $30 billion requested by the Biden administration in the unsuccessful Build Back Better Bill for a Civilian Climate Corps to create a robust, federally funded jobs program guaranteeing employment to all who are willing and able to work as part of the corps. Its projects would be designed to speed the clean-energy transition or to manage or restore natural ecosystems to enhance their ability to remove atmospheric greenhouse gases.

International Coordination of Effort

 * The U.S. Department of State and the National Oceanic and Atmospheric Administration should serve as lead agencies in designing and establishing a new International Atmospheric Authority to set an ambitious global greenhouse gas–reduction timetable and an international system of national GHG-reduction quotas.

 * The National Oceanic and Atmospheric Administration should lead other federal agencies in validating and implementing safe, proven methods of enhancing oceanic and coastal ecological processes and ecosystems to better sequester carbon.

Military Energy Use and the Military Budget

 * Congress should require the Department of Defense to employ all cost-effective energy efficiency technologies

across all installations and vehicles and to procure its energy, wherever possible, from renewable energy sources.

* Congress should require the Department of Defense to give preference in supply procurement to contractors that demonstrate superior performance with respect to their carbon footprint and those of their supply chains and subcontractors.

Population Management

* Congress should expand funding for contraceptive and family planning services across the nation and for expanded access to education for girls in developing countries. USAID should ramp up funding for global population management to make access to contraception, reproductive health care services, and universal public education accessible to all, regardless of income.

Research and Development

* Congress should provide adequate funding through the National Science Foundation, the U.S. Department of Energy, and other relevant federal agencies to greatly accelerate research, development, and demonstration support for technologies pertaining to renewable energy, energy efficiency, and industrial (point-source) carbon-removal technologies.

Recommended Reading

Bond, Becky, and Zack Exley. *Rules for Revolutionaries: How Big Organizing Can Change Everything*. White River Junction, VT: Chelsea Green Publishing, 2016.

Griffith, Saul. *Electrify: An Optimist's Playbook for Our Clean Energy Future*. Cambridge, MA: MIT Press, 2021.

Harvey, Hal, with Robbie Orvis and Jeffrey Rissman. *Designing Climate Solutions: A Policy Guide for Low-Carbon Energy*. Washington, DC; Covelo, CA; and London: Island Press, 2018.

Hawken, Paul, ed. *Drawdown—The Most Comprehensive Plan Ever Proposed to Reverse Global Warming*. New York: Penguin Books, 2017.

IPCC. *Special Report: Global Warming of 1.5°C, Summary for Policymakers*, https://www.ipcc.ch/sr15/, and *Technical Summary*, https://www.ipcc.ch/sr15/resources/technicalsummary, both in *Global Warming of 1.5°C: An IPCC Special Report on the Impacts of Global Warming of 1.5°C Above Pre-industrial Levels and Related Global Greenhouse Gas Emission Pathways, in the Context of Strengthening the Global Response to the Threat of Climate Change, Sustainable Development, and Efforts to Eradicate Poverty*. Cambridge, UK: Cambridge University Press, 2018.

Jacobson, Mark Z. *No Miracles Needed: How Today's Technology Can Save Our Climate and Clean Our Air*. Cambridge, UK: Cambridge University Press, 2023.

Johnson, Ayana Elizabeth, and Katharine K. Wilkinson, eds. *All We Can Save: Truth, Courage, and Solutions for the Climate Crisis*. New York: One World, 2020.

Lakey, George. *How We Win: A Guide to Nonviolent Direct Action Campaigning*. Brooklyn, NY: Melville House Publishing, 2018.

McAlevey, Jane F. *No Shortcuts: Organizing for Power in the New Gilded Age*. New York: Oxford University Press, 2016.

Acknowledgments

I am immensely grateful for the many people who have provided invaluable assistance to me on this project since its inception in 2014. I've been blessed with every conceivable kind of support, ranging from generous research grants, to research assistance, to the patience of interview subjects. I've also enjoyed editorial input from dozens of readers and the guidance of scientific and technical experts, as well as invaluable letters of support from friendly organizations. Indispensable as all this assistance has been, I am solely responsible for any errors in the final work.

I have had the good fortune to have had a tireless crew of excellent research and administrative assistants; they include Costanza Gonzalo, Gina Cotos, Rashida Hanif, Amanda Hsiung, Joanna Jiang, Kris Lynch, Revathi Muralidharan, Michael Quiroz, Katie Rice, Sasan Saadat, Christine Vandevoorde, and Eric Willits. Each made unique and valuable contributions to the project, assisting in ways too numerous to recount.

I am deeply indebted to Dr. Kevin Trenberth of the National Center for Atmospheric Research, who has been the most stalwart and gracious of my scientific advisers, patiently and ever promptly responding to my climate science questions for more than a decade. The discussions of climate change in this book (and previous climate books of mine) have benefited enormously from his input. I have also received useful scientific input from Professor John Harte at UC Berkeley; Professor Michael Mann of the University of Pennsylvania; Drs. Ben Sanderson and Jeffrey Kiehl of the National Center for Atmospheric Research; Dr. Benjamin Santer, formerly of Lawrence Livermore National Laboratory; and Professor Mark Z. Jacobson of Stanford University. Nonetheless, as previously noted, any errors in the book are my sole responsibility.

This book went through countless revisions on its way to submission for publication. Family members as well as colleagues and friends read portions of the book, and I benefited mightily from their advice. I owe a tremendous debt to my dear friend Professor David K. Dunaway of the University of New Mexico who diligently went through multiple early drafts of the book over several years and provided extraordinary editorial input. I also received early and extremely valuable encouragement and editorial guidance from author Adam Hochschild. Freelance editor Jill Rothenberg greatly improved the draft and proposal. In addition, I received valuable editorial and commercial guidance and encouragement from literary agent Jessica Papin, vice president of Dystel, Goderich and Bourret LLC; and agent Deborah Grosvenor, of the Grosvenor Literary Agency, who handled contract negotiations. Seven Stories Press's phenomenal developmental editor, Molly Lindley Pisani, did wonders for the book. City Lights Books editor Greg Ruggiero (also an acquisition editor for Seven Stories Press) commended the book to Seven Stories Press. Without the help and guidance of all these people, this book might never have been commercially published. For that publication, I have to offer my deepest thanks to Dan Simon, founder and publisher of Seven Stories Press, who believed in the book and took a chance on it. Dan is a courageous, principled, and truly independent publisher, a consummate professional, and a pleasure to work with.

In addition to the writers, agents, and editors above, I also benefited greatly from the climate news brought to my attention by my wife, Dr. Nancy Gordon, and from her insightful editorial feedback, along with the astute editorial advice from my sons, Daniel and Michael Berger. I am similarly grateful to the following additional readers for their useful suggestions: Dr. David Lenderts; Dr. Peter Joseph; Michael Shuman, Esq.; Dr. Jeremy Lent; Lisa Ferguson; Amri Shahpuri; and Steve Jawetz, Esq. I am particularly grateful to my friend and colleague Benson Lee, an environmental planner, for the research he shared with me, particularly on the impacts of climate change on the polar regions, and for the many intensive discussions we have had about climate issues over the past twenty years. My thanks as well to my friend Ron Feldman, who brought many useful articles to my attention.

I greatly appreciate the generous publishing industry advice and introductions from freelance editor Elizabeth R. Bruce; Norman Solomon, national director, RootsAction Education Fund; and Denis Hayes, president of the Bullitt Foundation. I am also indebted to Michael Hajilou, publisher and editor of *Sustain Europe*, for his marketing assistance and enthusiastic support.

I am especially grateful for research financial support provided by the Roy A. Hunt Foundation; the Laird Norton Family Foundation and its director, Katie Briggs; and the Newton and Rochelle Becker Charitable Trust, directed by Dr. David Becker. I would particularly like to thank the individual donors who provided me with generous research support, especially Marion R. Weber, founder of the Flow Fund Circle (FFC), May Waldroup, Virginia Mudd, Betty Lo, the late George Helmholtz, and the late Nancy Kittle. I am also grateful to Peter Gleick and his colleagues at the Pacific Institute of Oakland, California, for administering these funds, and to Professor Dan Kammen, head of the Renewable and Appropriate Energy Laboratory at UC Berkeley, which hosted the project before the Pacific Institute. I would also like to thank the individuals and foundation who have provided support for future planned nonprofit educational and policy development efforts to be based on this book, including Marion R. Weber (FFC), May Waldroup, Kristin Waldroup, Michael Waldroup, Marc Waldroup, Benson Lee, Colin Wiel, Charles Popper, the Becker Trust, and the Creative Planning Donor Advised Fund.

I am grateful for the fund-raising advice and assistance of: fundraising consultant Judith Kunofsky, JMJ Consulting; Norman Solomon, RootsAction (and for his letter of endorsement); Cole Wilbur, trustee emeritus and past president, the David and Lucile Packard Foundation; Hollie Ainbinder, program and development director, Institute for Public Accuracy; Drummond Pike, former CEO of the Tides Foundation; Rafe Pomerance, Distinguished Senior Arctic Policy Fellow, Woodwell Climate Research Center; Nathan Gray, president and founder of Geoversity and cofounder of the Mamoní Valley Preserve; and administrative assistant Amanda Hsiung.

Equally crucial to the success of this project were early letters of support, for which I would like to thank John Adams, founding director and trustee, the Natural Resources Defense Council; U.S. secretary of energy Jennifer Granholm; Nathan McKenzie, program director, Strategic Energy Initiatives; Dan Raudebaugh, executive director, Center for Transportation and the Environment; Allen Brown, founder and treasurer, Climate Action Pathways for Schools; and Katie Rogers, communications specialist, Audubon Canyon Ranch. Letters of support were also generously provided by the following Sierra Club officials: Michael Brune, former executive director; Bruce Hamilton, national policy director; and Jodie Van Horn, director of strategic relations. I also very much appreciate the encouragement and support of Pradnya Leitner of Thinking Into Results.

I would like to express my deepest gratitude to the many individuals I interviewed for this book, who generously shared their time and knowledge with me. Their names and affiliations can be found in these pages. Because the scope and nature of the book evolved over the years in which it was written, some interviews, particularly those on the subject of democracy, didn't make it into the final manuscript. I therefore want to express my special gratitude to those people with whom I spoke on these issues and ask for their forbearance for their perspectives not being included in the final book. These experts include Derek Cressman, California-based political reform advocate and author; Professor Lawrence Lessig, Roy L. Furman Professor of Law and Leadership at Harvard Law School; Mary Creasman, California Environmental Voters' (EnviroVoters) chief executive officer; former presidential candidate, attorney, author, and activist Ralph Nader; the late U.S. senator Mike Gravel; and Getachew Kassa, the national field manager for the Democracy Initiative.

I also was obliged to omit distillations of a few other interviews, either because they didn't fit the schema of the book as it evolved or because the information obtained overlapped that of other interviews. Those interviewed in these groups include former professor Roger Falcone, director, Advanced Light Source, Lawrence Berkeley National Laboratory; U.S. energy secretary Jennifer Granholm; professor of energy and resources Daniel M. Kammen; Danny Kennedy, clean-tech entrepreneur, environmental activist, and author; U.S. energy secretary Ernest Moniz; Dr. David W. Orr, Paul Sears Distinguished Professor of Environmental Studies and Politics Emeritus at Oberlin College, and Professor of Practice at Arizona State University; Professor Ramesh Ramamoorthy, associate laboratory director for energy technologies, Lawrence Berkeley National Laboratory; and Purnendu Chatterjee, Endowed Chair in Energy Technologies.

This book would be very different—and far worse—were it not for the tireless and faithful support of my wife, Nancy P. Gordon. She and the family suffered gallantly through the many conversations about climate change I inflicted upon them.

Endnotes

AUTHOR'S INTRODUCTION

Texas . . . will add far more clean-energy projects: Peter Weber, "Texas Will Get More Electricity from Solar and Wind Power than Natural Gas Next Year, EIA Projects," *The Week*, December 9, 2022, https://theweek.com/energy/1019043/texas-will-get-more-electricity-from-solar-and-wind-power-than-natural-gas-next-year.

second largest contributor to global warming: Renee Cho, "The Damaging Effects of Black Carbon," *State of the Planet*, Columbia Climate School, March 22, 2016, https://news.climate.columbia.edu/2016/03/22/the-damaging-effects-of-black-carbon.

misinformation about climate science: Case in point: for decades, the powerful fossil fuel industry has used misinformation to mount successful political opposition to meaningful cuts in fossil fuel use and greater reliance on clean energy, despite knowing the devastating consequences of such policies. See John J. Berger, *Climate Myths: The Campaign Against Climate Science* (Berkeley, CA: Northbrae Books, 2013), and many other excellent, more recent sources, including the three-part *Frontline* documentary series "The Power of Big Oil" (Part 1: "Denial"; Part 2: "Doubt"; Part 3: "Delay"), directed by Michael Kirk and Rick Young, Episodes 10–12, Season 783–85, WGBH, aired April 22, 2022, on PBS, https://www.pbs.org/wgbh/frontline/documentary/the-power-of-big-oil/.

terrible humanitarian consequences: International Federation of Red Cross and Red Crescent Societies, *The Cost of Doing Nothing: The Humanitarian Price of Climate Change and How It Can Be Avoided* (Geneva, Switzerland: The International Federation of Red Cross and Red Crescent Societies, 2019).

between 6 and 29 percent: David Hrivnak, "Climate Change—The Cost of Doing Nothing," *Times News*, December 27, 2020, https://www.timesnews.net/opinion/blogs/climate-change-the-cost-of-doing-nothing/article_5b8dea88-478d-11eb-a0dc-8f25f9859e42.html.

1. DAUNTING CHALLENGES, RAYS OF HOPE

fleeing the impacts of extreme weather: John Podesta, "The Climate Crisis, Migration, and Refugees," policy report, Brookings Institution, July 25, 2019, https://www.brookings.edu/research/the-climate-crisis-migration-and-refugees.

largest river, the Yangtze, dried up: Samar Marwan, "Floods, Heat Waves, Drought, Wildfires, Tornadoes: Summer 2022 Will Be One for the Record Books," *Fast Company*, September 1, 2022, https://www.fastcompany.com/90785152/natural-disasters-of-summer-2022.

one and a half times as much: Rowan T. Sutton et al., "Land/Sea Warming Ratio in Response to Climate Change: IPCC AR4 Model Results and Comparison with Observations," *Geophysical Research Letters*, January 16, 2007.

forests to burn more frequently: Andrew Freedman, "Arctic's Boreal Forests Burning at 'Unprecedented Rate,'" Climate Central, July 22, 2013, https://www.climatecentral.org/news/arctics-boreal-forests-burning-at-unprecedented-rate-16278.

our most precious biodiverse regions: "FAQ 2: How Will Nature and the Benefits It Provides to People Be Affected by Higher Levels of Warming?" *IPCC Sixth Assessment Report: Impacts, Adaptation, and Vulnerability*, Intergovernmental Panel on Climate Change, accessed March 16, 2023, https://www.ipcc.ch/report/ar6/wg2/about/frequently-asked-questions/keyfaq2.

in 2022, a new draft paper: James Hansen et al., "Global Warming in the Pipeline," *Atmospheric and Oceanic Physics*, last revised December 12, 2022, https://doi.org/10.48550/arXiv.2212.04474.

the "wild card" of sea level rise: Julie Brigham-Grette and Andrea Dutton, "Antarctica Is Headed for a Climate Tipping Point by 2060, with Catastrophic Melting If Carbon Emissions Aren't Cut Quickly," The Conversation, May 17, 2021, https://theconversation.com/antarctica-is-headed-for-a-climate-tipping-point-by-2060-with-catastrophic-melting-if-carbon-emissions-arent-cut-quickly-160978.

sea level rise will then accelerate: Robert M. DeConto et al., "The Paris Climate Agreement and Future Sea Level Rise from Antarctica," *Nature* 593 (2021): 83–89, https://www.nature.com/articles/s41586-021-03427-0.

dying from saltwater intrusion: Brady Dennis, "The Swift March of Climate Change in North Carolina's 'Ghost Forests,'" *Washington Post*, May 12, 2022, https://www.washingtonpost.com/climate-environment/2022/05/12/ghost-forests-carolina-climate-change/.

multiplier of virtually all ecosystem threats: The White House, "The National Security Implications of a Changing Climate: Findings from Select Federal Reports," May 2015, https://obamawhitehouse.archives.gov/sites/default/files/docs/National_Security_Implications_of_Changing_Climate_Final_051915.pdf.

diminish the quality of our daily lives: John J. Berger, *Climate Peril: The Intelligent Reader's Guide to Understanding the Climate Crisis* (Berkeley, CA: Northbrae Books, 2014).

climate change as a major threat: Cary Funk and Brian Kennedy, "For Earth Day, How Americans See Climate Change and the Environment in 7 Charts," Pew Research Center, April 21, 2020, https://www.pewresearch.org/fact-tank/2020/04/21/how-americans-see-climate-change-and-the-environment-in-7-charts/.

doing too little about climate disruption: Funk and Kennedy, "For Earth Day, How Americans See Climate Change and the Environment in 7 Charts."

More than 2,287 American firms: The pledge was to reduce U.S. carbon emissions 26 to 28 percent below 2005 levels by 2025. "U.S.A. First NDC Submission," United Nations Climate Change, Nationally Determined Contributions Registry, https://www4.unfccc.int/sites/NDCStaging/Pages/Party.aspx?party=USA.

the "We Are Still In" coalition: "We Are Still In," home page, We Are Still In, accessed December 17, 2018, https://www.wearestillin.com/.

nearly 4,000 signatories: The signatories include but are not necessarily limited to members in the following coalitions: We Are Still In, the U.S. Climate Alliance, and Climate Mayors.

numerous mayors and governors: The nine states of the alliance are California, Connecticut, Hawaii, New York, North Carolina, Oregon, Rhode Island, Virginia, and Washington. If those nine states were a country, it would be the fifth richest and sixth highest polluter in the world. Former mayor of New York City Michael Bloomberg has been instrumental in coordinating this alliance and has pledged to donate $15 million from Bloomberg Philanthropies to the United Nations. Several U.S. states are now drawing large shares of their power from renewables. Vermont is the undisputed champ, with 100 percent renewable electric power ("Vermont—State Profile and Energy Estimates," U.S. Energy Information Administration, last updated October 20, 2022, https://www.eia.gov/state/?sid=VT#tabs-1). Maine is next, at 54 percent, followed by Idaho, at 47 percent; and Iowa, at 39 percent. California, Kansas, and South Dakota are all tied at 30 percent.

who represent 173 million Americans: *Fulfilling America's Pledge: How States, Cities, and Businesses Are Leading the United States to a Low-Carbon Future,* report, America's Pledge, September 2018, https://www.americaspledgeonclimate.com/reports/.

58 percent of U.S. GDP: "Who's In," We Are Still In, 2020, https://www.wearestillin.com/signatories.

Many Fortune 500 companies: "RE100 Members," RE100, 2020, https://www.there100.org/re100-members.

achieved net-zero carbon: Still other well-known companies in the 100 percent club include Autodesk, Clif Bar, Coca-Cola Europacific Partners (formerly Coca-Cola European Partners), Estée Lauder, Etsy, and many, many more. Facebook also committed to reaching net-zero carbon emissions across its "value chain" in 2030. See "Renewable Energy," Google Data Centers, 2020, https://www.google.com/about/datacenters/renewable/; "Apple Now Globally Powered by 100 Percent Renewable Energy," press release, Apple Newsroom, April 9, 2018, https://www.apple.com/newsroom/2018/04/apple-now-globally-powered-by-100-percent-renewable-energy/; "Sustainability," Facebook Sustainability, 2020, https://sustainability.fb.com; "Citi Announces New Five-Year Sustainable Progress Strategy to Finance Climate Solutions and Reduce Climate Risk," press release, Citi, July 29, 2020, https://www.citigroup.com/citi/news/2020/200729a.htm; Adele Peters, "Ikea Has Invested in Enough Clean Energy to Power All of Its Operations (Plus Extra)," *Fast Company,* September 19, 2019, https://www.fastcompany.com/90405756/ikea-now-generates-more-renewable-electricity-than-it-uses.

Google, American Express, Apple, Bank of America: Anheuser-Busch InBev, the world's biggest beer brewer, and many other massive firms have adopted the goal of increasing their reliance on renewable power to 100 percent by 2025. See "Climate Action: Championing Low Carbon Technology," AB InBev, 2020, https://www.ab-inbev.com/sustainability/2025-sustainability-goals/climate-action.html.

Citigroup, Facebook, and Ikea: "Jobs in Every Constituency: This Time It Must Be Different," Green New Deal Group, September 2018, https://www.greennewdealgroup.org/wp-content/uploads/2018/09/GNDJobsReport9-18.pdf; "RE100 Members," RE100, accessed February 23, 2023, https://www.there100.org/re100-members.

211 million passenger cars: "Big Business Sees the Promise of Clean Energy," *The Economist,* June 10, 2017, https://www.economist.com/business/2017/06/10/big-business-sees-the-promise-of-clean-energy.

Three million Americans and counting: "Clean Energy Now Employs 3 Million Americans," E2 (Environmental Entrepreneurs), April 19, 2021, https://e2.org/releases/clean-energy-now-employs-3-million-americans.

350 occupations will expand: Mark Muro et al., "Advancing Inclusion Through Clean Energy Jobs," Metropolitan Policy Program at Brookings, Washington, DC, April 2019.

one in every five new construction jobs: "Clean Energy Now Employs 3 Million Americans."

an investment of $11.4 billion: "Ford to Lead America's Shift to Electric Vehicles with New Mega Campus in Tennessee and Twin Battery Plants in Kentucky; $11.4B Investment to Create 11,000 Jobs and Power New Lineup of Advanced EVs," press release, Ford Motor Company, September 27, 2021, https://media.ford.com/content/fordmedia/fna/us/en/news/2021/09/27/ford-to-lead-americas-shift-to-electric-vehicles.html.

batteries continue their steep declines: Jonathan Lopez, "EV Battery Prices on the Rise, Says Report," GM Authority, December 7, 2022, https://gmauthority.com/blog/2022/12/ev-battery-prices-on-the-rise-says-report/.

More than a thousand companies: "About SEIA," Solar Energy Industries Association, accessed February 23, 2023, https://seia.org/about.

so much potential for clean energy: Quoted in "Clean Energy Now Employs 3 Million Americans."

"We were able to weather the pandemic": Quoted in "Clean Energy Now Employs 3 Million Americans."

carbon-neutral or zero-carbon propulsion: When burned, hydrogen produces no carbon emissions, but when it is produced from natural gas or other fossil fuels, carbon emissions do result from the process. When produced by electrolysis, with the use of electricity from clean power sources, however, hydrogen produces no emissions.

first oceangoing carbon-neutral container ship: "A.P. Moller-Maersk Joins Methanol Institute," press release, Maersk, March 24, 2021, https://www.maersk.com/news/articles/2021/03/24/ap-moller-maersk-joins-methanol-institute; Andy Corbley, "Seven Years Ahead of Schedule, Maersk Will Deploy World's First Carbon Neutral Shipping Liner by 2023," Good News Network, February 23, 2021, https://www.goodnewsnetwork.org/largest-shipping-company-maersk-makes-carbon-neutral-liner/.

a suitable and affordable fuel: For information on SunHydrogen, see "Clean Hydrogen from the Power of the Sun," home page, SunHydrogen, https://www.sunhydrogen.com/. For information on fuel cells under development for shipping and locomotives, see "On a Mission Towards Zero-Emission," home page, TECO2030, https://Teco2030.no.

larger global renewable-energy market: Bernd Radowitz, "'Warp Speed away from Russian Energy' as Germany Seals Canadian Green Hydrogen Pact," Recharge, August 24, 2022, https://www.rechargenews.com/energy-transition/warp-speed-away-from-russian-energy-as-germany-seals-canadian-green-hydrogen-pact/2-1-1283577; Jonathan Touriño Jacobo, "Denmark's CIP Reaches Close on US $3 Billion Fund for Industrial Green Hydrogen," PVTech, September 1, 2022.

$2.5 trillion in other direct investment: Callum Keown, "BP Launches Green Hydrogen Project with Danish Energy Giant Ørsted. The Stock Is Surging," *Barron's*, November 10, 2020, https://www.barrons.com/articles/bp-launches-green-hydrogen-project-with-danish-energy-giant-rsted-the-stock-is-surging-51605012886.

sodium . . . can be refined cheaply: Brian Westenhaus, "Sydney Based Researchers Announce Lithium Ion Battery Rival," Yahoo! News, December 15, 2022, https://www.yahoo.com/now/sydney-based-researchers-announce-lithium-190000574.html.

New aquaculture and aquaponics: A robotics start-up, IronOx, reportedly produces thirty times the yield per acre with 10 percent of the water required by traditional agriculture. But not all these systems are created equal, and some use large amounts of energy. See Helen Breewood, "Spotlight on Urban, Vertical, and Indoor Agriculture," Resilience, January 22, 2019, https://www.resilience.org/stories/2019-01-22/spotlight-on-urban-vertical-and-indoor-agriculture/.

a Beyond Meatball Marinara sandwich: Frank Bruni, "Beef Is Past Its Prime," *New York Times*, March 7, 2021, https://www.nytimes.com/2021/03/06/opinion/sunday/beef-meatless-burger.html.

Wildtype is using cultured salmon cells: Tanay Warerkar, "S.F. Is Getting the World's First 'Cultured Sushi' Tasting Room, Where Fish Is Grown in a Lab," *San Francisco Chronicle*, June 29, 2021.

cultured meats and plant-based substitutes: John Lynch and Raymond Pierrehumbert, "Climate Impacts of Cultured Meat and Beef Cattle," *Frontiers in Sustainable Food Systems* 3 (February 19, 2019), https://doi.org/10.3389/fsufs.2019.00005. For an earlier, simpler study that reached opposite conclusions, see Hanna L. Tuomisto and M. Joost Teixeira de Mattos, "Environmental Impacts of Cultured Meat Production," *Environmental Science and Technology* 45, no. 14 (June 17, 2011): https://doi.org/10.1021/es200130u.

growing four times faster: Advanced Energy Economy and Navigant Research, *Advanced Energy Now 2017 Market Report* (Washington, DC: Advanced Energy Economy, 2018), https://www.aee.net/reports.

exceeding that in oil and gas: Neither gas nor nuclear power is included as clean energy in these statistics.

to divest $11 trillion: Monica Tyler-Davis, "And the Momentum Is Getting Stronger: A New Fossil-Free Milestone, $11 Trillion Has Been Committed to Disinvest from Fossil Fuel," Go Fossil Free Campaign, 350.org, September 8, 2019, https://350.org/11-trillion-divested/.

increased financial regulation: Jess Shankleman and Rich Miller, "Yellen, Carney Back Central Bank-Like Councils on Climate Change," Bloomberg Green, October 8, 2020, https://www.bloomberg.com/news/articles/2020-10-08/yellen-carney-back-central-bank-like-councils-on-climate-change; Saleha Mohsin and Jennifer A. Dlouhy, "Yellen Gets a Shot to Put Treasury Clout into Climate Fight," Bloomberg News, December 11, 2020, https://www.bloomberg.com/news/articles/2020-12-11/yellen-gets-a-shot-to-throw-treasury-s-clout-into-climate-fight#xj4y7vzkg; Financial Stability Board, *The Implications of Climate Change for Financial Stability*, November 23, 2020, https://www.fsb.org/2020/11/the-implications-of-climate-change-for-financial-stability/.

Solar-electric panel costs have dropped: Jake Richardson, "Solar Costs Drop over 99% Since 1975, Thanks to R&D and Economies of Scale," Red, Green, and Blue, February 1, 2019, http://redgreenandblue.org/2019/02/01/solar-costs-drop-99-since-1975-thanks-rd-economies-scale/; "Quarterly Solar Photovoltaic Module Prices in the U.S. from Q1 2016 to Q1 2020 (in U.S. Dollars per Watt)," Statista, September 3, 2020, https://www.statista.com/statistics/216791/price-for-photovoltaic-cells-and-modules/; Paul Maycock, "The PV Module Experience Curve, 1976 to 2012," Bloomberg New Energy Finance, https://data.bloomberglp.com/bnef/sites/3/2013/12/2013-04-30-BNEF-University-renewable-energy-CCS-EST.pdf; Report no. 5108923, *Sustainable Energy in America Factbook 2018*, Bloomberg New Energy Finance, 2018, https://about.bnef.com/blog/sustainable-energy-america-factbook-2018/.

fastest-growing new power source: GlobalData Energy, "Renewable Energy to Reach 22.5% Share in Global Power Mix in 2020," Power Technology, July 16, 2018, https://www.power-technology.com/comment/renewable-energy-reach-22-5-share-global-power-mix-2020/.

are increasing seventeen times faster: Rabia Ferroukhi et al., *Renewable Energy and Jobs Annual Review 2017*, report, International Renewable Energy Agency (IRENA), 2017, https://www.irena.org/-/media/Files/IRENA/Agency/Publication/2017/May/IRENA_RE_Jobs_Annual_Review_2017.pdf.

They also pay more: William Lawhorn, "Solar and Wind Generation Occupations: A Look at the Next Decade," *Beyond the Numbers* (U.S. Bureau of Labor Statistics) 10, no. 4 (February 2021), https://

www.bls.gov/opub/btn/volume-10/solar-and-wind-generation-occupations-a-look-at-the-next-de-cade.htm.

growing by almost 42 percent annually: Jeff Cardello, "How Solar Drives the American Economy," Solar Informatics, *Alternative Energy Magazine*, February 23, 2022, https://www.altenergymag.com/article/2022/02/how-solar-drives-the-american-economy/36775.

enough power for every home: Valerie Volcovici, "Update 3—Solar Energy Can Account for 40% of U.S. Electricity by 2035-DOE," Reuters, September 8, 2021, https://www.reuters.com/article/usa-biden-solar-idAFL1N2QA1JW.

Wind power costs . . . are nosediving: Seb Henbest, "Power System Will Dance to Tune of Wind, Solar, Batteries," Bloomberg New Energy Finance, July 25, 2018, https://about.bnef.com/blog/hen-best-power-system-will-dance-tune-wind-solar-batteries/.

Onshore wind power fell 60 percent: Craig Richard, "Renewables 'Cheapest Option for Most of the World,'" *Wind Power Monthly*, April 29, 2020, https://www.windpowermonthly.com/article/1681740/renewables-cheapest-option-world.

turbines getting bigger: Economies of scale occur in construction of concrete foundations, cabling, transformers, and in turbine operation and maintenance costs. See Richard, "Renewables 'Cheapest Option for Most of the World.'" These developments are making offshore wind affordable without subsidies. In 2017, Dong/EnBW won a bid for the world's cheapest offshore wind power plant at just under five cents per kilowatt-hour, the first unsubsidized German offshore wind farm. Tino Andresen, "Offshore Wind Farms Offer Subsidy-Free Power for First Time," Bloomberg, April 13, 2017, https://www.bloomberg.com/news/articles/2017-04-13/germany-gets-bids-for-first-subsidy-free-offshore-wind-farms.

growth in renewable-energy capacity: Victoria Masterson, "5 Milestones in Green Energy," World Economic Forum, April 14, 2021, https://www.weforum.org/agenda/2021/04/renewables-record-ca-pacity-solar-wind-nuclear/.

for two thirds of the people on earth: Richard, "Renewables 'Cheapest Option for Most of the World.'"

roughly twice as much new U.S. capacity: Cara Marcy, "New Electric Generating Capacity in 2019 Will Come from Renewables and Natural Gas," *Today in Energy* (blog), U.S. Energy Information Administration, January 10, 2019, https://www.eia.gov/todayinenergy/detail.php?id=37952.

30 percent of the world's CO2: "Emissions," *Global Energy & CO2 Status Report 2019*, International Energy Agency, 2019, https://www.iea.org/reports/global-energy-co2-status-report-2019/emissions.

coal plants now cost more *just to operate*: "The Coal Cost Crossover 2.0," Energy Innovation, May 4, 2021, https://energyinnovation.org/publication/the-coal-cost-crossover-2021/.

fossil fuel plants are now closing: Scott Jell and Michelle Bowman, "Almost All Power Plants That Retired in the Past Decade Were Powered by Fossil Fuels," *Today in Energy* (blog), U.S. Energy Information Administration, December 19, 2018, https://www.eia.gov/todayinenergy/detail.php?id=37814.

More than 300 U.S. coal plants: "Preliminary Monthly Electric Generator Inventory," U.S. Energy Information Administration, November 27, 2018, https://www.eia.gov/electricity/data/eia860M/; "Bloomberg Philanthropies and Sierra Club's Beyond Coal Campaign Reaches Landmark Closure of 318th U.S. Coal Plant, on Track to Retire All Coal Plants by 2030," press release, Bloomberg Philan-thropies, September 15, 2020, https://www.bloomberg.org/press/releases/bloomberg-philanthropies-

and-sierra-clubs-beyond-coal-campaign-reaches-landmark-closure-of-318th-u-s-coal-plant-on-track-to-retire-all-coal-plants-by-2030/.

fossil fuel investments have been canceled: "Divestment Commitments," Go Fossil Free, 350.org, accessed January 28, 2019, https://gofossilfree.org/divestment/commitments/.

Renewable-power sources are also challenging: Report no. 5108923, *Sustainable Energy in America Factbook 2018*; "Levelized Cost and Levelized Avoided Cost of New Generation Resources in the *Annual Energy Outlook 2018*," U.S. Energy Information Administration, March 2018, http://large. stanford.edu/courses/2018/ph241/asperger2/docs/eia-mar18.pdf; *Renewable Power Generation Costs in 2017: Key Findings and Executive Summary*, International Renewable Energy Agency, January 2018, https://www.irena.org/-/media/Files/IRENA/Agency/Publication/2018/Jan/IRENA_2017_Power_Costs_2018_summary.pdf.

ninety-six nuclear power plants: "Nuclear Power Plants: Permanent Shut Downs 2005–2022," Statista, August 1, 2022, https://www.statista.com/statistics/513639/number-of-permanent-nuclear-reactor-shutdowns-worldwide/.

due to their advanced age: "Nuclear Power in a Clean Energy System," International Energy Agency, May 2019, https://www.iea.org/reports/nuclear-power-in-a-clean-energy-system.

currently increasing its renewable capacity: Liu Yuanyuan, "China's Renewable Energy Installed Capacity Grew 12 Percent Across All Sources in 2018," *Renewable Energy World*, March 6, 2019, https://www.renewableenergyworld.com/baseload/chinas-renewable-energy-installed-capacity-grew-12-percent-across-all-sources-in-2018/; "China: Overview," International Energy Agency, September 30, 2020, https://www.eia.gov/international/analysis/country/CHN. The International Energy Agency's Monthly OECD Electricity Statistics, however, state that China got 28 percent of its power from renewables in 2019. International Energy Agency, August 14, 2020, https://iea.blob.core.windows.net/assets/4979da92-959d-41b7-9bb5-948ff5cc0224/mes.pdf.

peaking its carbon emissions in 2026: "STATEMENT: China Commits to Stronger Climate Targets at Climate Ambition Summit," World Resources Institute, December 12, 2020, https://www.wri.org/news/statement-china-commits-stronger-climate-targets-climate-ambition-summit. "Energy" includes both fuels and electricity, so it is larger, and therefore renewable energy largely comprised of electricity is a smaller fraction of total energy supplied than of all electricity.

one soccer field every hour: Jing Yan and Lauri Myllyvirta, "China Has Already Surpassed Its 2020 Solar Target," *Unearthed*, Greenpeace, August 24, 2017, https://unearthed.greenpeace.org/2017/08/25/china-raises-solar-power-target/; "The Solar-Powered Future Is Being Assembled in China," Bloomberg Green, September 14, 2020, https://www.bloomberg.com/features/2020-china-solar-giant-longi/#xj4y7vzkg.

one wind turbine installation every hour: International Energy Association, *2017 IEA Wind TCP Annual Report*, Publications, IEA, https://iea-wind.org/wp-content/uploads/2020/12/Annual-Report-2017.pdf.

The Chinese invest almost twice as much: Michael Forsythe, "China Aims to Spend at Least $360 Billion on Renewable Energy by 2020," *New York Times*, January 5, 2017, https://www.nytimes.com/2017/01/05/world/asia/china-renewable-energy-investment.html.

$918 billion, compared to $528 billion: "Clean Energy Investment Trends, 1H 2020," Bloomberg New Energy Finance, July 13, 2020, https://data.bloomberglp.com/professional/sites/24/BNEF-Clean-Energy-Investment-Trends-1H-2020.pdf.

the world's largest manufacturer: Charlie Campbell, "China Is Bankrolling Green Energy Projects Around the World," *Time*, November 1, 2019, https://time.com/5714267/china-green-energy/.

still heavily dependent on fossil fuels: While constituting only 5 percent of the world's population, the United States is responsible for a fifth of daily world oil consumption. The nation is also a major natural gas exporter; in 2019, gas exports hit their highest level ever (4.7 trillion cubic feet). See U.S. Energy Information Administration, "U.S. Natural Gas Production, Consumption, and Exports Set New Records in 2019," *Today in Energy* (blog), U.S. Energy Information Administration, October 5, 2020.

doubling its renewable capacity: U.S. capacity in 2020 was 250 gigawatts.

solar capacity has grown eightyfold: Monique Hanis, "2020 'Factbook' Reveals a Decade's Shift to Advanced Energy," Advanced Energy Perspectives (blog), Advanced Energy United, February 19, 2020, https://blog.advancedenergyunited.org/2020-factbook-reveals-a-decades-shift-to-advanced-energy.

24 percent of U.S. electrical power: "In the first half of 2022, 24% of U.S. electricity generation came from renewable sources," *Today in Energy* (blog), U.S. Energy Information Administration, September 9, 2022, https://www.eia.gov/todayinenergy/detail.php?id=53779.

U.S. electricity-generating system decarbonized: "U.S. Pledges to Slash Solar Energy Costs by 60% in a Decade," Reuters, March 25, 2021, https://www.reuters.com/article/us-usa-solar/u-s-pledges-to-slash-solar-energy-costs-by-60-in-a-decade-idUSKBN2BH2ME.

two thirds of its power from non-carbon sources: "New Data Shows Nearly Two-Thirds of California's Electricity Came from Carbon-Free Sources in 2019," press release, California Energy Commission, July 16, 2020, https://www.energy.ca.gov/news/2020-07/new-data-shows-nearly-two-thirds-californias-electricity-came-carbon-free.

$38 billion state budget surplus: Dan Walters, "Newsom Budget Surplus Gets Reality Check," CalMatters, May 18, 2021, https://calmatters.org/commentary/2021/05/newsom-budget-surplus-lao/.

$102 billion in public and private investments: "California Climate Investments Drive Statewide Economic Development," E2 (Environmental Entrepreneurs), accessed February 1, 2023., https://e2.org/wp-content/uploads/2021/08/State_CA2021_081621.pdf.

taking 14 million cars off the road: "CARB Approves First-in-Nation ZEV Regulation That Will Clean the Air, Slash Climate Pollution, and Save Consumers Money," press release, California Air Resources Board, August 25, 2022, https://ww2.arb.ca.gov/news/california-moves-accelerate-100-new-zero-emission-vehicle-sales-2035; comments of Margo Oge, former director of U.S. EPA's Office of Transportation and Air Quality, on "Baby, You Can Drive My Electric Car," hosted by Mina Kim, *Forum*, KQED, August 31, 2022, https://www.kqed.org/forum/2010101890398/baby-you-can-drive-my-electric-car.

43 percent of the global economy: "Under2 Coalition," Climate Group, 2020, https://www.theclimategroup.org/under2-coalition.

more than 140 cities: "We Are Still In to Deliver on America's Pledge: A Retrospective," Bloomberg Philanthropies and We Are Still In, September 2020, https://assets.bbhub.io/dotorg/sites/28/2020/09/We-Are-Still-In-to-Deliver-on-Americas-Pledge_.pdf; "Jobs in Every Constituency: This Time It Must Be Different," Green New Deal Group, September 2018, https://www.greennewdealgroup.org/wp-content/uploads/2018/09/GNDJobsReport9-18.pdf."

Sweden gets essentially 100 percent of its power: Under Sweden's Climate Act of 2017, the nation is legally bound to attain net-zero emissions by 2045. That's five years earlier than previously planned, exceeding what Sweden pledged under the Paris Agreement. "Sweden Plans to Be Carbon Neutral by 2045," United Nations Climate Change, June 19, 2017, https://unfccc.int/news/sweden-plans-to-be-carbon-neutral-by-2045.

80 percent of Denmark's electric power: "Monthly OECD Electricity Statistics," International Energy Agency, August 14, 2020; "Pioneers in Clean Energy: Denmark," Denmark.dk, accessed February 23, 2023, https://denmark.dk/innovation-and-design/clean-energy; "Denmark's Energy and Climate Outlook: 2019," Danish Energy Agency, October 2019, https://ens.dk/sites/ens.dk/files/Analyser/deco19.pdf.

the United Kingdom . . . got 80 percent of its power: Masterson, "5 Milestones in Green Energy."

minus 12 metric tons of CO2: Dyani Lewis, "Energy Positive: How Denmark's Samsø Island Switched to Zero Carbon," *Guardian*, February 23, 2017, https://www.theguardian.com/sustainable-business/2017/feb/24/energy-positive-how-denmarks-sams-island-switched-to-zero-carbon.

2. SHOW ME THE MONEY: PAYING FOR THE CLEAN-ENERGY TRANSFORMATION

"a floor, not a ceiling": "How John Kerry Hopes to Combat Climate Change," U.S. Climate Envoy John Kerry, interview by Audie Cornish, *All Things Considered*, NPR, August 11, 2021, https://www.npr.org/2021/08/11/1026831081/how-john-kerry-hopes-to-combat-climate-change.

an explosion of growth in green bonds: Generation Investment Management, "The Impact Gap," *Insights* 07, November 17, 2021, https://www.generationim.com/our-thinking/insights/the-impact-gap/.

green bonds spread project risk: International Renewable Energy Agency, "Renewable Energy Finance: Green Bonds," January 30, 2020, https://www.irena.org/publications/2020/Jan/RE-finance-Green-bonds.

Investment in green bonds: Climate Bonds Initiative, "Green Bond Market Hits USD2TN Milestone at End of Q3 2022," November 9, 2022, https://www.climatebonds.net/2022/11/green-bond-market-hits-usd2tn-milestone-end-q3-2022.

"the rapid market growth to date": Green Bonds Initiative, "Five by Five Manifesto, Five Steps to Five Trillion," https://www.climatebonds.net/about/five-five-manifesto.

"taken the investing world by storm": Andrew Poreda, "Green Treasuries: Is Now the Time for the Federal Government to Jump on the Green Bond Bandwagon?" ETF Strategist Channel, June 29, 2021, https://www.etftrends.com/etf-strategist-channel/green-treasuries-is-now-the-time-for-the-federal-government-to-jump-on-the-green-bond-bandwagon/.

a five-billion-dollar first issue: James Langton, "Feds Plan $5-Billion Green Bond Issue," *IE Investment Executive* (Canada), April 19, 2021, https://www.investmentexecutive.com/news/industry-news/feds-plan-5-billion-green-bond-issue/.

global flood of capital: John Caramichael and Andreas Rapp, "The Green Corporate Bond Issuance Premium," International Finance Discussion Paper no. 1346 (Washington, DC: Board of Governors of the Federal Reserve System, June 2022), https://doi.org/10.17016/IFDP.2022.1346.

A "green Treasury bond": Poreda, "Green Treasuries: Is Now the Time?"

no more than 2 percent: Devashree Saha and Joel Jaeger, *America's New Climate Economy: A Comprehensive Guide to the Economic Benefits of Climate Policy in the United States* (Washington, DC: World Resources Institute, July 2020), https://www.wri.org/research/americas-new-climate-economy-comprehensive-guide-economic-benefits-climate-policy-united.

macroeconomic effects on the economy: Jean Pisani-Ferry, "Climate Policy Is Macroeconomic Policy, and the Implications Will Be Significant," Policy Briefs 21–20, Peterson Institute for International Economics, Washington, DC, August 2021, https://www.piie.com/publications/policy-briefs/climate-policy-macroeconomic-policy-and-implications-will-be-significant.

6 to 13 percent of GDP on energy: Saha and Jaeger, *America's New Climate Economy.*

moving 6.3 billion tons of fossil fuels: "Freight Shipments by Commodity," Bureau of Transportation Statistics, U.S. Department of Transportation, January 12, 2021, https://www.bts.gov/topics/freight-transportation/freight-shipments-commodity.

eight million people a year: Leah Burrows, "Deaths from Fossil Fuel Emissions Higher than Previously Thought," News and Events, Harvard University School of Engineering and Applied Sciences, February 9, 2021, https://seas.harvard.edu/news/2021/02/deaths-fossil-fuel-emissions-higher-previously-thought.

a human life is worth one million dollars: Sarah Gonzales, "How Government Agencies Determine the Dollar Value of Human Life," *All Things Considered*, NPR, April 23, 2020, https://www.npr.org/2020/04/23/843310123/how-government-agencies-determine-the-dollar-value-of-human-life.

"Unearthing, processing, and moving underground": Melissa Denchak, "Fossil Fuels: The Dirty Facts," National Resources Defense Council, June 29, 2018, https://www.nrdc.org/stories/fossil-fuels-dirty-facts.

57,000 abandoned oil and gas wells: David Manthos, "Mapping Abandoned Coal Mines," SkyTruth, October 16, 2015, https://skytruth.org/2015/10/mapping-abandoned-coal-mines/.

the public must pay the untallied tab: Mitch Bernard, president Natural Resources Defense Council, Letter to Members, New York City, August 2021.

the rate on the thirty-year U.S. Treasury: Interest Rate Statistics, "Daily Treasury Par Yield Curve Rates," U.S. Department of the Treasury, accessed February 6, 2023, https://home.treasury.gov/resource-center/data-chart-center/interest-rates/TextView?type=daily_treasury_yield_curve&field_tdr_date_value_month=202302.

inflation rate of 6.5 percent: "United States Inflation Rate," Trading Economics, January 2023 data, accessed February 6, 2023, https://tradingeconomics.com/united-states/inflation-cpi.

launched on the false assertion: Paulina Cachero, "US Taxpayers Have Reportedly Paid an Average of $8,000 Each and over $2 Trillion Total for the Iraq War Alone," Business Insider, February 6, 2020, https://www.businessinsider.com/us-taxpayers-spent-8000-each-2-trillion-iraq-war-study-2020-2.

the interest . . . exceeded $925 billion: Cachero, "US Taxpayers Have Reportedly Paid."

larger than those of the next nine countries: "The 15 Countries with the Highest Military Spending Worldwide in 2021," Statista, https://www.statista.com/statistics/262742/countries-with-the-highest-military-spending/.

less than 0.01 percent of all deaths: Hannah Ritchie et al., "Terrorism," Our World in Data, November 2019, revised November 2022, https://ourworldindata.org/terrorism.

the U.S. government provided net bailouts: John J. Berger, "Economic Peril," in his *Climate Peril*, 119–49; *Quarterly Report to Congress*, Special Inspector General for the Troubled Asset Relief Program (hereafter "SIGTARP"), October 26, 2010, https://www.sigtarp.gov/sites/sigtarp/files/Quarterly_Reports/October2010_Quarterly_Report_to_Congress.pdf; *Quarterly Report to Congress*, SIGTARP, July 21, 2010, https://www.sigtarp.gov/sites/sigtarp/files/Quarterly_Reports/July2010_Quarterly_Report_to_Congress.pdf; *Quarterly Report to Congress*, SIGTARP, April 20, 2010, https://www.sigtarp.gov/sites/sigtarp/files/Quarterly_Reports/April2010_Quarterly_Report_to_Congress.pdf; and *Quarterly Report to Congress*, SIGTARP, July 21, 2009, https://www.sigtarp.gov/sites/sigtarp/files/Quarterly_Reports/July2009_Quarterly_Report_to_Congress.pdf.

holds $25 trillion in wealth: Isabel V. Sawhill and Christopher Pulliam, "Six Facts About Wealth in the United States," Brookings Institution, June 25, 2019, https://www.brookings.edu/blog/up-front/2019/06/25/six-facts-about-wealth-in-the-united-states/.

***added* another $6.5 trillion:** Robert Frank, "The Wealthiest 10% of Americans Own a Record 89% of All U.S. Stocks," CNBC, October 18, 2021, https://www.cnbc.com/2021/10/18/the-wealthiest-1opercent-of-americans-own-a-record-89percent-of-all-us-stocks.html.

never paid tax on more than half: Laura Davison, "Wyden Unveils His Own Wealth Tax That Targets Investments Annually," Bloomberg News, September 12, 2019, https://www.bloomberg.com/news/articles/2019-09-12/senator-wyden-echoes-tax-the-rich-cries-with-new-mini-wealth-tax#x-j4y7vzkg.

$10 *trillion* for the clean-energy transition: William Rice et al., *Fair Taxes Now: Revenue Options for a Fair Tax System*, April 2019, Americans for Tax Fairness, Washington, DC.

perversely lead to a *loss* in tax revenue: John A. Tatom, "Biden's Capital Gains Tax Increase Is More Unproductive Misdirection," The Hill, July 6, 2021, https://thehill.com/opinion/finance/561274-bidens-capital-gains-tax-increase-is-more-unproductive-misdirection/.

would raise an extra $1.7 trillion: Grace Enda and William G. Gale, "How Could Changing Capital Gains Taxes Raise More Revenue?" Brookings Institution, January 14, 2020, https://www.brookings.edu/blog/up-front/2020/01/14/how-could-changing-capital-gains-taxes-raise-more-revenue/.

including well-known billionaires like: Jesse Eisinger et al., "The Secret IRS Files: Trove of Never-Before-Seen Records Reveal How the Wealthiest Avoid Income Tax," ProPublica, June 8, 2021, https://www.propublica.org/article/the-secret-irs-files-trove-of-never-before-seen-records-reveal-how-the-wealthiest-avoid-income-tax.

avoid more than $100 billion in taxation: Erik Sherman, "A New Report Claims Big Tech Companies Used Legal Loopholes to Avoid Over $100 Billion in Taxes. What Does That Mean for the Industry's Future?" *Fortune*, December 6, 2019, https://fortune.com/2019/12/06/big-tech-taxes-google-facebook-amazon-apple-netflix-microsoft/.

avoid $75 billion in taxes: Jesse Drucker and Danny Hakim, "Private Inequity: How a Powerful Industry Conquered the Tax System," *New York Times*, September 2021, https://www.nytimes.com/2021/06/12/business/private-equity-taxes.html.

$3.6 trillion in checking accounts: Federal Reserve Bank of St. Louis, "Total Checkable Deposits," FRED Economic Data, *Economic Research*, June 2020, https://fred.stlouisfed.org/series/TCDSL/25/Max.

$12 trillion in time deposits: U.S. Federal Reserve, Federal Reserve Statistical Release Z.1, *Financial Accounts of the United States: Flow of Funds, Balance Sheets, and Integrated Macroeconomic Accounts*, Table L.205: "Time and Savings Deposits," Line 1, Third Quarter 2018, page 115, https://www.federalreserve.gov/releases/z1/20181206/z1.pdf.

$3 trillion in money market funds: Board of Governors of the Federal Reserve, *Financial Accounts Guide*, Table F.206: "Money Market Fund Shares," First Quarter 2020, https://www.federalreserve.gov/apps/fof/DisplayTable.aspx?t=f.206.

even at very modest rates of return: For daily Treasury yield curve rates, see Interest Rate Statistics, "Daily Treasury Par Yield Curve Rates."

more than $2.5 trillion: "United States Corporate Profits," Trading Economics, accessed May 7, 2023, https://tradingeconomics.com/united-states/corporate-profits.

corporate deals in 2022 . . . totaled \$1.73 trillion: "Dealmaking Remains Active as Dark Clouds Form," 2020 M&A Report, Boston Consulting Group, October 20, 2022, https://www.bcg.com/publications/2022/the-2022-m-a-report-dealmaking-remains-active.

"failed to prevent" or directly financed: Oil Change International, *Unused Tools: How Central Banks Are Fueling the Climate Crisis*, Oil Change International and Cosponsors, August 24, 2021, https://priceofoil.org/2021/08/24/unused-tools-central-banks/.

"central banks are not using their influence": Theodore Whyte, "Central Banks Helping Funnel Trillions of Dollars into Fossil Fuels Despite Climate Pledges, Research Finds," *DeSmog* (International) (blog), August 25, 2021, https://www.desmog.com/2021/08/25/central-banks-helping-funnel-trillions-of-dollars-into-fossil-fuels-despite-climate-pledges-research-finds/.

to deter further fossil fuel industry financing: Oil Change International, *Unused Tools*.

\$3.3 trillion in subsidies: Damian Carrington, "'Reckless': G20 States Subsidized Fossil Fuels by \$3TN Since 2015, Says Report," *Guardian* (London), July 20, 2021, https://www.theguardian.com/environment/2021/jul/20/g20-states-subsidised-fossil-fuels-2015-coal-oil-gas-cliamte-crisis. More than half of the total subsidies came from China, India, Russia, and Saudi Arabia.

subsidies remain robust today: Carrington, "'Reckless.'"

5.5 billion metric tons of CO_2: Carrington, "'Reckless.'"

"lower[ing] global carbon emissions by 28 percent": David Coady et al., "Global Fossil Fuel Subsidies Remain Large: An Update Based on Country-Level Estimates," Working Paper No. 19/89, International Monetary Fund, May 2, 2019, https://www.imf.org/en/Publications/WP/Issues/2019/05/02/Global-Fossil-Fuel-Subsidies-Remain-Large-An-Update-Based-on-Country-Level-Estimates-46509.

"locking in future carbon emissions": Berger, *Climate Peril*, 142.

3. MORE SMART WAYS TO FINANCE THE TRANSFORMATION

digging our own graves: United Nations, "UN Secretary-General at COP26: 'Either We Stop It—Or It Stops Us,'" United Nations Regional Information Center for Western Europe, https://unric.org/en/un-secretary-general-at-cop26-either-we-stop-it-or-it-stops-us-2/.

in the past three million years: Nicola Jones, "How the World Passed a Climate Threshold and Why It Matters," YaleEnvironment360, January 26, 2017, https://e360.yale.edu/features/how-the-world-passed-a-carbon-threshold-400ppm-and-why-it-matters.

\$3.3 to \$5 trillion annually: The estimate is from the Glasgow Financial Alliance for Net Zero (GFANZ) and is consistent with estimates of \$3 trillion by Bloomberg New Energy Finance (see *Financial Institute Net-zero Transition Plans: Recommendations and Guidance*, GFANZ, June 2022, https://assets.bbhub.io/company/sites/63/2022/06/GFANZ_Recommendations-and-Guidance-on-Net-zero-Transition-Plans-for-the-Financial-Sector_June2022.pdf, 12) and \$3 to \$6 trillion annually by the International Monetary Fund (Ananthakrishnan Prasad et al., "Mobilizing Private Climate Financing in Emerging Market and Developing Economies," *Staff Climate Notes*, International Monetary Fund, July 27, 2022, https://www.imf.org/en/Publications/staff-climate-notes/Issues/2022/07/26/Mobilizing-Private-Climate-Financing-in-Emerging-Market-and-Developing-Economies-520585). See also Joseph Feyertag, Samantha Attridge, and Neha Kumar, "GFANZ: a watershed moment or a drop in the ocean of the Global South's climate finance needs?" ODI, November 9, 2022, https://odi.org/en/insights/gfanz-a-watershed-moment-or-a-drop-in-the-ocean-of-the-global-souths-climate-finance-needs/.

hit that mark by 2038: Michael Mann, *The New Climate War* (New York: Public Affairs Books, 2021), 213.

229 of the world's largest fossil fuel developers: Paddy McCulley, *Throwing Fuel on the Fire: GFANZ Financing of Fossil Fuel Expansion*, Reclaim Finance et al., January 2023, https://reclaimfinance.org/site/wp-content/uploads/2023/01/Throwing-fuel-on-the-fire-GFANZ-financing-of-fossil-fuel-expansion.pdf.

maximize their "risk-adjusted returns": Generation Investment Management, "The Impact Gap."

a new subsidiary, Just Climate: Instead, the company's researchers found that "capital flows are concentrated in a few technologies and geographies while other areas are starved of investment" (Generation Investment Management, "The Impact Gap"). This is not surprising for many reasons, not the least of which is the lack of a coherent national climate action plan, as described later in this book. Given the urgent need to massively scale up annual energy transition investments if the United Nations' Paris goals are to be met, Generation's new subsidiary, Just Climate, will focus on hard-to-decarbonize industries and on parts of the economy that are "often too capital-intensive, unproven at scale or in geographies too challenging for investors to consider." See David Blood, "In His Own Words: David Blood on Why Generation Is Launching Just Climate," press release, Generation Investment Management, October 27, 2021, https://www.generationim.com/our-thinking/news/in-his-own-words-david-blood-on-why-generation-is-launching-just-climate/.

operating the U.S. entirely on clean energy: Manish Ram et al., *Global Energy System Based on 100% Renewable Energy: Power, Heat, Transportation and Desalination Sectors* (Berlin: Lappeenranta University of Technology and Energy Watch Group, April 2019), http://www.energywatchgroup.org/wp-content/uploads/EWG_LUT_100RE_All_Sectors_Global_Report_2019.pdf.

will ultimately be profitable: Amory B. Lovins, *Profitably Decarbonizing Heavy Transport and Industrial Heat: Transforming These "Harder-to-Abate" Sectors Is Not Uniquely Hard and Can Be Lucrative*, Emeritus Insight Series, Rocky Mountain Institute, Basalt, CO, July 2021, https://www.rmi.org/profitable-decarb/.

tens of trillions of dollars: Ram et al., *Global Energy System Based on 100% Renewable Energy*.

three to seven times its costs: International Renewable Energy Agency, *Global Energy Transformation: A Roadmap to 2050*, 2019 Edition, https://www.irena.org/publications/2019/Apr/Global-energy-transformation-A-roadmap-to-2050-2019Edition.

"$26 trillion in economic benefits": New Climate Economy, *Unlocking the Inclusive Growth Story of the 21st Century: Accelerating Climate Action in Urgent Times*, Global Commission on the Economy and Climate, Washington, DC, 2018, https://www.newclimateeconomy.report/2018.

less than half the energy: Saul Griffith, *Electrify: An Optimist's Playbook for Our Clean Energy Future* (Cambridge, MA: MIT Press, 2021), 51.

"the world's biggest energy 'source'": Amory B. Lovins, "Twelve Energy and Climate Myths," Climate Voices, Climate and Capital Media, August 14, 2021, https://www.climateandcapitalmedia.com/twelve-energy-and-climate-myths/.

low-hanging fruit ripe for plucking: Joe Romm, "Energy Efficiency, The Low Hanging Fruit That Grows Back," *Think Progress* (blog), August 20, 2009, https://archive.thinkprogress.org/energy-efficiency-the-low-hanging-fruit-that-grows-back-4cad18617226/.

is also invariably cheaper: Amory B. Lovins, "Energy Strategy: The Road Not Taken?" *Foreign Affairs*, October 1976, https://www.foreignaffairs.com/articles/united-states/1976-10-01/energy-strategy-road-not-taken; Amory B. Lovins, *Soft Energy Paths, Toward a Durable Peace* (Cambridge, MA: Ballinger Publishing Company and Friends of the Earth International, 1977).

"deploying the entire efficiency resource": Amory B. Lovins, "How Big Is the Energy Efficiency Resource?" *Environmental Research Letters* 13, no. 9 (September 18, 2018).

40 percent of all end-use energy: The sectoral energy use is calculated by adding all the energy actually consumed in the buildings themselves to the electrical system losses incurred in producing and transmitting that power from primary energy sources, such as coal, oil, natural gas, nuclear, and renewable energy. See "How Much Energy Is Used in Buildings?" Frequently Asked Questions, U.S. Department of Energy, Energy Information Administration, last updated December 23, 2022, https://www.eia.gov/tools/faqs/faq.php?id=86&t=1.

half **that energy is being wasted:** MEETS Accelerator Coalition and the Bullitt Foundation, *Unlocking Deep Efficiency in Commercial Buildings: The Metered Energy Transaction Structure*, May 2016, https://www.meetscoalition.org/wp-content/uploads/docs_public/MEETS-Pilot-Interim-Report-Web-May-2016-rev-12-16-Public.pdf.

"Metered Energy Efficiency Transaction Structure": Information on MEETS and the consortium can be found at https://www.meetscoalition.org/.

analysis conducted for the Seattle 2030 District: Susan Wickwire (executive director, Seattle 2030 District) and Brett Phillips (board chairman, Seattle 2030 District), letter to Larry Weiss (General Manager & CEO, Seattle City Light) re: "City-wide Benefits of Large-Scale MEETS Implementation," August 8, 2016.

A massive *national* building retrofit program: Heidi Garrett-Peltier, *Employment Estimates for Energy Efficiency Retrofits of Commercial Buildings*, Political Economy Research Institute, University of Massachusetts, Amherst, June 2, 2011, https://www.peri.umass.edu/fileadmin/pdf/research_brief/PERI_USGBC_Research_Brief.pdf; "Build Back Better: Rebooting the US Economy After COVID-19," COVID-19 Expert Notes, World Resources Institute, April 2020, https://www.wri.org/publication-series/build-back-better-rebooting-us-economy-after-covid-19; Greg Carlock, "Building Energy Efficiency and Energy Assistance: Creating Jobs and Providing Relief to States Across the Country," COVID-19 Expert Notes, World Resources Institute, April 27, 2020, https://www.wri.org/research/building-energy-efficiency-and-energy-assistance-creating-jobs-and-providing-relief-states; Adele Peters, "Focusing the Recovery on Green Infrastructure Could Create Millions of Jobs," *Fast Company*, May 14, 2020, https://www.fastcompany.com/90504339/focusing-the-recovery-on-green-infrastructure-could-create-millions-of-jobs.

The dam also caused landslides: Nectar Gan, "China's Three Gorges Dam Is One of the Largest Ever Created. Was It Worth It?" CNN, July 31, 2020, https://www.cnn.com/style/article/china-three-gorges-dam-intl-hnk-dst/index.html.

the massive Pangue Dam: Jon Bowermaster, "Last Run Down the Bio-Bio," *Town and Country*, November 1992, https://earthriver.com/library/conservation/conservation/town-and-country.

a major new Stanford University study: Dan Reicher et al., *Derisking Decarbonization: Making Green Energy Investments Blue Chip*, Steyer-Taylor Center for Energy Policy and Finance, Precourt Institute for Energy, and the Hoover Institution, Stanford University, Palo Alto, CA, 2017.

"the single greatest economic opportunity": Reicher et al., *Derisking Decarbonization*.

$58 trillion in total by 2040: Reicher et al., *Derisking Decarbonization*.

jobs already in clean energy: E2 (Environmental Entrepreneurs), *Clean Jobs America 2021*, April 19, 2021, https://e2.org/reports/clean-jobs-america-2021/.

decarbonization investments would leverage: Reicher et al., *Derisking Decarbonization*.

$41 billion lay unused: The program to authorize loan guarantees for innovative technologies that avoid greenhouse gases was enacted as an amendment to the American Recovery and Reinvestment Act of 2009. See Section 406, Energy Policy Act of 2005, 109th Congress, PUBLIC LAW 109-58-AUG. 8, 2005, https://www.congress.gov/109/plaws/publ58/PLAW-109publ58.pdf.

the DOE's enacted budget: Congressional Research Service, "DOE Office of Energy Efficiency and Renewable Energy FY2022 Appropriations," *In Focus*, updated June 9, 2022, https://sgp.fas.org/crs/misc/IF11948.pdf.

plans to spend $360 billion: Dominic Chiu, "The East Is Green: China's Global Leadership in Renewable Energy," in *New Perspectives in Foreign Policy* 13 (Summer 2017), October 6, 2017, https://www.csis.org/east-green-chinas-global-leadership-renewable-energy.

creating 13 million new jobs: Dan W. Reicher, Testimony to Subcommittee on Energy, House Committee on Energy and Commerce, "Hearing on Alignment and Execution of DOE's Missions: Advancing National and Energy Security in an Era of Energy Abundance," January 9, 2018, https://law.stanford.edu/wp-content/uploads/2018/02/House-Committee-on-Energy-and-Commerce-Hearing-on-DOE-Mission-Reicher-Testimony-1-8-18-Final-V.2-1.pdf.

"removing 7.3 million cars from the road": Letter from Bobby L. Rush, ranking member, Energy Subcommittee, and Joe Barton, vice chairman, Energy and Commerce Committee, to Thad Cochran, chairman, U.S. Senate Appropriations Committee; Patrick Leahy, ranking member, U.S. Senate Appropriations Committee; Rodney Frelinghuysen, chairman, House Appropriations Committee; and Nita M. Lowey, ranking member, House Appropriations Committee, January 19, 2018, https://law.stanford.edu/wp-content/uploads/2018/02/LPO-Rush-Barton-Letter-to-Appropriators-Loan-Program-Office-1-19-18-1.pdf.

a substantial down payment: "Remarks by President Trump in Joint Address to Congress," White House, issued on February 28, 2017, https://trumpwhitehouse.archives.gov/briefings-statements/remarks-president-trump-joint-address-congress/.

pressing them to continue the program: Letter from Reps. Peter Welch and Paul Tonko to Rodney Frelinghuysen, chairman, House Appropriations Committee, and Nita M. Lowey, ranking member, House Appropriations Committee, January 24, 2018, https://law.stanford.edu/wp-content/uploads/2018/02/LPO-Welch-Tonko-Letter-to-Appropriators-1-25-18.pdf.

to the Senate Appropriations Committee: Letter from Sen. Joe Manchin II; Sen. Christopher A. Coons; Sen. Bill Cassidy, M.D.; and Sen. Shelley Moore Capito to Sen. Thad Cochran, chairman, Senate Appropriations Committee; Patrick Leahy, ranking member, Senate Appropriations Committee; Lamar Alexander, chairman, Subcommittee on Energy and Water Development, Senate Appropriations Committee; and Sen. Dianne Feinstein, ranking member, Subcommittee on Energy and Water Development, Senate Appropriations Committee, February 8, 2018, https://law.stanford.edu/wp-content/uploads/2018/05/Title-XVII-LPO-Letter-to-Appropriators.pdf.

4. "THE WORLD'S BIGGEST ENERGY 'SOURCE'"

the building came to exceed: Architects and engineers measure a building's energy use in British thermal units (BTUs) needed per square foot, called its "energy use index" (EUI). In Seattle, the EUI for an average commercial building is 60; design calculations indicated that the Bullitt Center had to reach an EUI of only 16 to be a net-zero-energy building. To everyone's delight, the building achieved an EUI of 10.

seven hundred net-zero-energy ... projects: Elsa Wenzel, "In the Business of Buildings, Net Zero Becomes a Towering Target," GreenBiz, November 1, 2021, https://www.greenbiz.com/article/business-buildings-net-zero-becomes-towering-target.

energy-positive ... home projects: "The 2022 EEBA Inventory of Zero Energy Homes Report Released," Energy and Environmental Building Alliance, Cision PR Newswire, December 21, 2022, https://www.prnewswire.com/news-releases/the-2022-eeba-inventory-of-zero-energy-homes-report-released-301708608.html. For earlier data, see Net Zero Energy Coalition, *To Zero and Beyond: 2017 Inventory of Residential Projects on the Path to Zero in the U.S. and Canada*, report, Net Zero Energy Coalition, April 2018, https://netzeroenergycoalition.com/zero-energy-inventory/.

to get all U.S. buildings certified: Brian Potter, "Every Building in America—An Analysis of the U.S. Building Stock," *Construction Physics*, November 2, 2020, https://www.construction-physics.com/p/every-building-in-america-an-analysis.

($1.3 trillion worth) is wasted: "About the Commercial Buildings Integration Program," U.S. Department of Energy, Office of Energy Efficiency and Renewable Energy, accessed February 8, 2023, https://www.energy.gov/eere/buildings/about-commercial-buildings-integration-program; "U.S.: This Chart Shows Just How Much Energy the US Is Wasting," World Economic Forum, May 25, 2018, https://www.weforum.org/agenda/2018/05/visualizing-u-s-energy-consumption-in-one-chart.

an in-depth conversation: For more of this interview, see John J. Berger, "Washington Governor Jay Inslee Calls for Carbon Tax, Champions Climate Protection and Clean Energy," *Sustain Europe*, Spring 2018, https://www.sustaineurope.com/u.s.-governor-jay-inslee-calls-for-carbon-tax-20180322.html.

potential for energy savings: John A. "Skip" Laitner et al., *The Long-Term Energy Efficiency Potential: What the Evidence Suggests*, American Council for an Energy-Efficient Economy, Washington, DC, 2012, https://www.aceee.org/research-report/e121.

while saving $12 to $16 trillion: Laitner et al., *The Long-Term Energy Efficiency Potential*, Table ES-1, viii.

a cumulative $3.6 trillion: Personal author communication with John A. "Skip" Laitner, senior fellow, American Council for an Energy-Efficient Economy, January 2019.

output of Grand Coulee Dam: Carol Winkel, "2021 Regional Conservation Progress Report Finds Lower Achievement," Northwest Power and Conservation Council, September 27, 2022, https://www.nwcouncil.org/news/2022/09/27/2021-regional-conservation-progress-report-finds-lower-achievement/.

seven times the current power consumption: Winkel, "2021 Regional Conservation Progress Report Finds Lower Achievement."

over two million energy efficiency jobs: *Energy Efficiency Jobs in America 2022*, E2 (Environmental Entrepreneurs), January 25, 2023, https://e2.org/reports/energy-efficiency-jobs-in-america-2022; *2018 U.S. Energy and Employment Report*, National Association of State Energy Officials and Energy Futures Initiative, Arlington, VA, May 2018, https://www.usenergyjobs.org/report. For comparison, the electric power generation sector employs fewer than nine hundred thousand workers.

5. 100 PERCENT CLEAN ENERGY FOR ALL

his conclusions were anathema: Mark Z. Jacobson, "The 7 Reasons Why Nuclear Energy Is Not the Answer to Solve Climate Change," Heinrich Böll Foundation, April 21, 2021, https://eu.boell.org/en/2021/04/26/7-reasons-why-nuclear-energy-not-answer-solve-climate-change.

the world could secure 100 percent: Mark Z. Jacobson et al., "A Low-Cost Solution to the Grid Reliability Problem with 100% Penetration of Intermittent Wind, Water, and Solar for All Purposes," *Proceedings of the National Academy of Sciences of the United States* 112, no. 49 (2015), https://www.pnas.org/doi/10.1073/pnas.1510028112. See also Mark Z. Jacobson et al., "100% Clean and Renewable Wind, Water, and Sunlight (WWS) All-Sector Energy Roadmaps for the 50 United States," *Energy and Environmental Sciences* 8 (2015): 2093–117; Mark Z. Jacobson et al., "100% Clean and Renewable Wind, Water, and Sunlight (WWS) All-Sector Energy Roadmaps for 139 Countries of the World," *Joule* 1, no. 1 (2017); and Mark Z. Jacobson et al., "100% Clean, and Renewable Wind, Water, and Sunlight (WWS) All-Sector Energy Roadmaps for 53 Towns and Cities in North America," *Sustainable Cities and Society* 42 (October 2018): 22–37.

intentional misrepresentations of his work: Email from Mark Z. Jacobson to *PNAS* editorial manager Etta Kavanagh, February 26, 2017.

the youth climate lawsuit: The lawsuit demands that the government take meaningful action against climate change and contends that the youthful plaintiffs' constitutional rights to life, liberty, and property are being violated by current government policies. See "Youth v. Gov: Juliana v. U.S.," Our Children's Trust, accessed January 26, 2023, https://www.ourchildrenstrust.org/juliana-v-us.

the critical paper appeared: Christopher T. M. Clack et al., "Evaluation of a Proposal for Reliable Low-Cost Grid Power with 100% Wind, Water, and Solar," *Proceedings of the National Academy of Sciences of the United States* 114, no. 26 (2017), https://www.pnas.org/doi/10.1073/pnas.1610381114.

articles critical of his work: Eduardo Porter, "Fisticuffs Over the Route to a Clean-Energy Future," *New York Times*, June 20, 2017.

rebuttal letter in *PNAS*: Jacobson wrote several responses. See Mark Z. Jacobson, "Note to National Review: A 100% Renewable Future Is Alive and Well," EcoWatch, July 7, 2017, https://www.ecowatch.com/national-review-mark-jacobson-2454398939.html; Mark Jacobson, "Response to Forbes: Stop Inaccuracies—100% Renewable Energy Is Possible," EcoWatch, July 6, 2017, https://www.ecowatch.com/mark-jacobson-pnas-2454280135.html; Mark Jacobson, "What New York Times Got Wrong on Assessment of Transition to 100% Renewables," EcoWatch, July 10, 2017, https://www.ecowatch.com/new-york-times-mark-jacobson-2455117917.html; and M. Z. Jacobson and M. A. Delucchi, "Line-by-Line Response by M. Z. Jacobson, M. A. Delucchi to 'Evaluation of a Proposal for Reliable Low-Cost Grid Power with 100% Wind, Water, and Solar,'" PDF, May 25, 2017, http://web.stanford.edu/group/efmh/jacobson/Articles/I/CombiningRenew/Line-by-line-Clack.pdf. For articles and interviews by others, see Stephen Lacey, "An Interview with Mark Jacobson on the 100% Renewables Debate," Greentech Media, July 12, 2017, https://www.greentechmedia.com/articles/read/an-interview-with-mark-jacobson-about-100-percent-renewable-energy; Karl Burkart, "100% Clean, Renewable Energy Is Possible, Practical, Logical—Setting the Record Straight," CleanTechnica, July 22, 2017, https://cleantechnica.com/2017/07/22/100-clean-renewable-energy-possible-setting-record-straight/; "Lying Is Not Okay," *Izuru* (blog), last updated November 11, 2017, https://www.hi-izuru.org/wp_blog/2017/11/lying-is-not-okay/.

With each doubling: Felix Creutzig et al., "The Underestimated Potential of Solar Energy to Mitigate Climate Change," *Nature Energy* 2, no. 17140 (2017).

annual volume increased: Finlay Colville, "PV Industry Production Hits 310 GW of Modules in 2022; What About 2023?" PVTech, November 17, 2022, https://www.pv-tech.org/pv-industry-production-hits-310gw-of-modules-in-2022-what-about-2023/; "Annual Solar Module Production Globally from 2000 to 2021 (in Megawatts)," Statista, accessed February 8, 2023, https://www.statista.com/statistics/668764/annual-solar-module-manufacturing-globally/.

preferable to current nuclear power: Gar Smith, "Nuclear Roulette: The Case Against a 'Nuclear Renaissance,'" ["No. 5 in the International Forum on Globalization series focused on false solutions to the global climate crisis"], International Forum on Globalization, San Francisco, CA, June 2011. See also Jacobson, "The 7 Reasons Why Nuclear Energy Is Not the Answer to Solve Climate Change," chapter 5, endnote 1.

until Jacobson's 2015 paper: Jacobson et al., "A Low-Cost Solution to the Grid Reliability Problem."

His hard work: Mark Z. Jacobson, "Curriculum Vitae," Stanford University, last updated February 7, 2023, http://web.stanford.edu/group/efmh/jacobson/vita/.

100 percent of their electricity: Byron Gudiel, "Saying Farewell to Ready for 100," Sierra Club, April 11, 2022, https://www.sierraclub.org/articles/2022/04/saying-farewell-ready-for-100.

Sixteen U.S. states: Mark Z. Jacobson, *No Miracles Needed: How Today's Technology Can Save Our Climate and Clean Our Air* (Cambridge, UK: Cambridge University Press, 2023), 378.

"A Path to Sustainable Energy by 2030": Mark Z. Jacobson and Mark Delucchi, "A Path to Sustainable Energy by 2030," *Scientific American*, November 1, 2009, https://www.scientificamerican.com/article/a-path-to-sustainable-energy-by-2030/.

materials required for their construction: Fossil fuel energy currently is used to produce the materials for renewable-energy systems, but in operation, these systems more than repay their energy debt by cleanly generating many times the energy required for their raw materials and construction. See Michaja Pehl et al., "Understanding Future Emissions from Low-Carbon Power Systems by Integration of Life-Cycle Assessment and Integrated Energy Modelling," *Nature Energy* 2 (2017): 939–45, https://doi.org/10.1038/s41560-017-0032-9.

templates to produce road maps: Jacobson et al., "100% Clean and Renewable Wind, Water, and Sunlight (WWS) All-Sector Energy Roadmaps for 139 Countries of the World;" Jacobson et al., "100% Clean and Renewable Wind, Water, and Sunlight (WWS) All-sector Energy Roadmaps for the 50 United States;" Jacobson et al., "100% Clean and Renewable Wind, Water, and Sunlight (WWS) All-Sector Energy Roadmaps for 53 Towns and Cities in North America."

work is being corroborated: Sven Teske, ed., *Achieving the Paris Climate Agreement Goals: Global and Regional 100% Renewable Energy Scenarios with Non-Energy GHG Pathways for +1.5°C and +2°C* (Springer Open, 2019), https://doi.org/10.1007/978-3-030-05843-2; *Exponential Climate Action Roadmap: Scaling 36 Solutions to Halve Emissions by 2030*, Future Earth, Sweden, September 19, 2019, revised January 2020, https://exponentialroadmap.org/wp-content/uploads/2020/03/Exponential-Roadmap_1.5.1_216x279_08_AW_Download_Singles_Small.pdf.

His energy road maps show: See Jacobson's website at Stanford University, https://web.stanford.edu/group/efmh/jacobson/Articles/I/susenergy2030.html, for links to specific road maps.

cut projected power demands: Jacobson, *No Miracles Needed*, 317.

By studying continuous energy demand: Jacobson, *No Miracles Needed*, 353.

storage costs have plunged: Whereas electricity once was very costly to store because batteries were so expensive, costs have been declining rapidly. Lithium-ion technology can now store a million watt-hours (1 MWh) of power for only $187, as opposed to almost $800 in 2012. See H. J. Mai, "Electricity Costs from Battery Storage Down 76% Since 2012: BNEF," UtilityDive, March 26, 2019, https://www.utilitydive.com/news/electricity-costs-from-battery-storage-down-76-since-2012-bnef/551337/. This now makes new renewable wind or solar plus storage (aka "storage-plus") competitive with fossil fuel–fired power plants and even with efficient combined-cycle natural-gas peaking plants. Thus, utilities bought nearly 9 gigawatts (GW) of energy storage capacity just from 2011 to

2018. (A gigawatt is 1 billion watts, or 1,000 megawatts, roughly equal to the power output of a large coal or nuclear power plant.) See *How Utilities Can Look Beyond Natural Gas with Cost-Effective Solar Plus Storage Strategies* (Boulder, CO: Navigant Research, 1Q 2019).

can now store surplus power affordably: Renewable power can also be stored by pumped hydro, a system in which surplus electricity is used to pump water up into a reservoir; the water is then released through a dam to produce power again when needed.

Solar panels on U.S. rooftops: Pieter Gagnon et al., *Rooftop Solar Photovoltaic Technical Potential in the United States: A Detailed Assessment*, Technical Report NREL/TP-6A20-65298, National Renewable Energy Laboratory, Golden, CO, January 2016, https://www.nrel.gov/docs/fy16osti/65298.pdf.

to form microgrid systems: With adequate storage, a grid connection may not even be necessary in many situations, especially when other renewables are connected to form a microgrid, such as geothermal heat pumps and biogas-powered district heating and cooling systems. For those who want to be independent of the grid and who require high reliability, a standby generator can be included in the system for rare occasions when backup is needed.

Are renewable resources really adequate: Austin Brown et al., *Estimating Renewable Energy Potential in the United States: Methodology and Initial Results*, Technical Report NREL/TP-6A20-64503, National Renewable Energy Laboratory, Golden, CO, 2015, revised 2016, https://www.nrel.gov/docs/fy15osti/64503.pdf; and U.S. Energy Information Administration, Table 1.2: "Summary Statistics for the United States, 2011–2021." For more information, see "Renewable Energy Technical Potential," National Renewable Energy Laboratory, n.d., https://www.nrel.gov/gis/re-potential.html; Anthony Lopez et al., *U.S. Renewable Energy Technical Potentials: A GIS-Based Analysis*, Technical Report NREL/TP-6A20-51946, National Renewable Energy Laboratory, Golden, CO, July 2012, https://www.nrel.gov/docs/fy12osti/51946.pdf; and "WindExchange," Office of Energy Efficiency and Renewable Energy, Department of Energy, n.d., https://windexchange.energy.gov/.

roughly eight times: Walt Musial et al., *2016 Offshore Wind Energy Resource Assessment for the United States*, Technical Report NREL/TP-5000-66599, National Renewable Energy Laboratory, Golden, CO, 2016, https://www.nrel.gov/docs/fy16osti/66599.pdf; and U.S. Energy Information Administration, Table 1.2: "Summary Statistics for the United States, 2011–2021." See also Cara Marcy and Philipp Beiter, *Quantifying the Opportunity Space for Future Electricity Generation: An Application to Offshore Wind Energy in the United States*, Technical Report NREL/TP-6A20-66522, National Renewable Energy Laboratory, Golden, CO, 2016, https://www.nrel.gov/docs/fy16osti/66522.pdf; Patrick Gilman et al., *National Offshore Wind Strategy: Facilitating the Development of the Offshore Wind Industry in the United States*, Report DOE/GO-102016-4866, U.S. Department of Energy and U.S. Department of the Interior, 2016, https://www.energy.gov/sites/prod/files/2016/09/f33/National-Offshore-Wind-Strategy-report-09082016.pdf; and U.S. Offshore Wind Industry, "Offshore Wind Energy Development in the U.S.," Status Update, American Wind Energy Association, https://www.awea.org/Awea/media/Resources/Fact%20Sheets/AWEA_Offshore-Wind-Industry-FINAL.pdf.

more than sixteen times: Xi Lu, Michael B. McElroy, and Juha Kiviluoma, "Global Potential for Wind-Generated Electricity," *Proceedings of the National Academy of Sciences of the United States* 106, no. 27 (July 7, 2009): 10933–38, https://doi.org/10.1073/pnas.0904101106.

total potential PV generation: Total U.S. PV generation potential—including rooftops—is 299,036 TWh/yr (terawatt-hours per year), excluding Alaska and Hawaii (Brown et al., *Estimating Renewable Energy Potential in the United States*). In 2016, U.S. electricity sales were 3,762 TWh (U.S. Energy Information Administration, Table 1.2: "Summary Statistics for the United States, 2011–2021").

delivers more energy in an hour: Nathan S. Lewis, "Toward Cost-Effective Solar Energy Use," *Science* 315, no. 5813 (2007): 798–801, https://science.sciencemag.org/content/315/5813/798.

a thousand times the energy needed: Joseph N. Moore and Stuart F. Simmons, "More Power from Below," *Science* 340, no. 6135 (May 24, 2013): 933–34, https://www.science.org/doi/abs/10.1126/science.1235640.

hard-to-electrify energy demands: Bioenergy Technologies Office, *2016 Billion-Ton Report*, Office of Renewable Energy and Energy Efficiency, U.S. Department of Energy, Washington, DC, https://www.energy.gov/eere/bioenergy/2016-billion-ton-report.

as a building block for synfuels: Hydrogen can be burned cleanly or converted into synfuels when it is combined with carbon from waste streams of carbon dioxide. Those fuels can then be used in manufacturing steel, cement, concrete, aggregates, and even asphalt without the combustion of fossil fuels.

others . . . opposed to fracking: Jacobson, *No Miracles Needed*, 356.

Jacobson at last began writing: Jacobson, *No Miracles Needed*, 357–58.

Cuomo was informed of the plan: Jacobson, *No Miracles Needed*, 361.

Governor Cuomo finally banned: Thomas Kaplan, "Citing Health Risks, Cuomo Bans Fracking in New York State," *New York Times*, December 17, 2014, https://www.nytimes.com/2014/12/18/nyregion/cuomo-to-ban-fracking-in-new-york-state-citing-health-risks.html.

clearing the way for: Mark Jacobson et al., "Examining the Feasibility of Converting New York State's All-Purpose Energy Infrastructure to One Using Wind, Water, and Sunlight," *Energy Policy* 57 (2013): 585–601.

the groundwork for large investments: Jacobson, *No Miracles Needed*, 362–63; author interview with Mark Z. Jacobson, January 31, 2019.

New York Public Service Commission adopted: Jacobson, *No Miracles Needed*, 362.

Climate Leadership and Community Protection Act: Andrew M. Cuomo, *2019 Justice Agenda: The Time Is Now*, Office of the Governor, Albany, NY, January 15, 2019, PDF, https://www.governor.ny.gov/sites/governor.ny.gov/files/atoms/files/2019StateoftheStateBook.pdf; New York State Senate Bill S6599 and Assembly Bill A8429, Climate Leadership and Community Protection Act, 2019–2020 Legislative Session, June 18, 2019, https://nyassembly.gov/leg/?default_fld=&leg_video=&bn=A08429&term=2019&Summary=Y&Actions=Y&Text=Y; and Jackson Morris and Miles Farmer, "Unpacking New York's Big New Climate Bill: A Primer," Natural Resources Defense Council, June 20, 2019, https://www.nrdc.org/experts/miles-farmer/unpacking-new-yorks-big-new-climate-bill-primer-0.

"to virtually eliminate poverty": New York Senate, Senate Bill 2878B, to Create a Green New Deal Task Force. The bill, introduced by Sen. James Sanders Jr. (D, 10th Senate District of New York) and cosponsors, is currently in the Senate's Environmental Conservation Committee. See "State of New York: 2878-B, 2019–2020 Regular Sessions, In Senate," January 30, 2019, https://legislation.nysenate.gov/pdf/bills/2019/S2878B.

100 percent clean energy for California: Mark Jacobson et al., "A Roadmap for Repowering California for All Purposes with Wind, Water, and Sunlight," *Energy* 73 (2014): 875–89.

the 2014 gubernatorial election: Mark Z. Jacobson, *100% Clean, Renewable Energy and Storage for Everything* (New York: Cambridge University Press, 2020), 368.

Clean Energy and Pollution Reduction Act: Governor Brown also proposed reducing petroleum use in cars and trucks by 50 percent by 2030, but that measure encountered stiff oil industry opposition. See California Senate Bill 350, "Clean Energy and Pollution Reduction Act, an Act to Amend Sections 365.1 and 380 of the California Public Utilities Code," introduced February 2, 2019, https://

www.energy.ca.gov/rules-and-regulations/energy-suppliers-reporting/clean-energy-and-pollu-tion-reduction-act-sb-350.

a law (Senate Bill 100): Senate Bill 100, "California Renewables Portfolio Standard Program: Emissions of Greenhouse Gases" (2017–2018), https://openstates.org/ca/bills/20172018/SB100/; California Public Utilities Code, chap. 312, secs. 399.11, 399.15, 399.30, and 454.53.

a goal of 100 percent renewable power: Governor Jerry Brown, "California Executive Order B-55-18 to Achieve Carbon Neutrality," PDF, September 10, 2018, https://www.ca.gov/archive/gov39/wp-content/uploads/2018/09/9.10.18-Executive-Order.pdf.

new homes to have solar panels: Title 24: "2019 Building Energy Efficiency Standards," California Building Standards Code, Part 6, California Energy Commission, n.d., https://www.energy.ca.gov/programs-and-topics/programs/building-energy-efficiency-standards/2019-building-energy-efficiency.

codifying an executive order: David Shepardson, "California Regulator Sees 2035 EV Mandate as 'Sweet Spot,'" Reuters, September 21, 2022, https://www.reuters.com/business/autos-transportation/california-regulator-sees-2035-ev-mandate-sweet-spot-2022-09-20/.

climate action framework: "CARB Approves Unprecedented Climate Action Plan to Shift World's 4th Largest Economy from Fossil Fuels to Clean and Renewable Energy," press release, California Air Resources Board, December 15, 2022, https://ww2.arb.ca.gov/news/carb-approves-unprecedented-climate-action-plan-shift-worlds-4th-largest-economy-fossil-fuels.

cosponsored Senate Bill 987: U.S. Congress, Senate, 100 by '50 Act, S. 987, 115th Congress, introduced in Senate April 27, 2017, https://www.congress.gov/bill/115th-congress/senate-bill/987.

A recent federal study: Paul Denholm et al., *Examining Supply-Side Options to Achieve 100% Clean Electricity by 2035*, NREL/TP-6A40-81644, National Renewable Energy Laboratory, Golden, CO, 2022, https://www.nrel.gov/docs/fy22osti/81644.pdf.

have now registered 360 companies: RE100, home page, The Climate Group and CDP, accessed January 27, 2023, https://www.there100.org/. RE100 is a global network of corporations committed to 100 percent renewable electricity. CDP, the former Carbon Disclosure Project, is a global nonprofit that tracks, analyzes, and discloses the environmental impacts of companies, cities, and governments to enhance ambition and increase transparency and accountability regarding climate and environmental goals. CDP, *Accelerating the Rate of Change: CDP Strategy 2021–2025*, CDP, n.d., https://cdn.cdp.net/cdp-production/comfy/cms/files/files/000/005/094/original/CDP_STRATEGY_2021-2025.pdf.

6. BEYOND FUMES: 100 PERCENT CLEAN TRANSPORTATION

single largest source: USEPA, "Fast Facts on Greenhouse Gas Emissions," Green Vehicle Guide, last updated July 14, 2022, https://www.epa.gov/greenvehicles/fast-facts-transportation-greenhouse-gas-emissions.

seventy to one hundred trillion dollars: "Sustainable Transport to Save the World US\$70 Trillion by 2050," Movin' On, accessed February 9, 2023, https://www.movinonconnect.com/news/sustainable-transport-save-world-us70-trillion-2050/; "Land Transport's Contribution to a 2°C Target," in *Bridging the Gap: Pathways for Transport in the Post-2020 Process*, November 2014, https://slocat.net/wp-content/uploads/legacy/u10/land_transports_contribution_to_a_2c_target_4pager_final.pdf; and update of Heather Allen et al., "Land Transport's Contribution to a 2°C Target: Key Messages on Mitigation Potential, Institutions and Financing of Low-Carbon Land Transport for Policy Makers on Transport and Climate Change," advance draft, Bridging the Gap Initiative and the Partnership on Sustainable, Low Carbon Transport, September 23, 2014. See also *Tracking Trends in a Time of Change: The Need for Radical Action Towards Sustainable Transport Decarbonisation*, SLOCAT Transport and Climate Change Global Status Report, 2nd ed., 2021, https://tcc-gsr.com.

a quarter of the world's greenhouse gases: Drew Kodjak, Francisco Posada Sanchez, and Laura Segafredo, *Policies That Work: How Vehicle Standards and Fuel Fees Can Cut CO2 Emissions and Boost the Economy*, Climate Works and the International Council on Clean Transportation, San Francisco and Washington, DC, 2013, https://issuu.com/icct/docs/policies_that_work_vehicles_and_fuels.

1.5 *billion* cars and light trucks: "How Many Cars Are There in the World in 2023?" Automotive Market Research, Hedges and Company, accessed on February 11, 2023, https://hedgescompany.com/blog/2021/06/how-many-cars-are-there-in-the-world/; and Brad Plumer et al., "Electric Cars Are Coming. How Long Until They Rule the Road?" *New York Times*, March 10, 2021, https://www.nytimes.com/interactive/2021/03/10/climate/electric-vehicle-fleet-turnover.html.

98 percent of them: "Projected Size of the Global Electric Vehicle Fleet Between 2021 and 2025," Statista, accessed February 14, 2023, https://www.statista.com/statistics/970958/worldwide-number-of-electric-vehicles/.

nearly 6 billion metric tons: "Annual Carbon Dioxide (CO2) Emissions Worldwide from 1940 to 2021 (in Billion Metric Tons)," Statista, accessed February 11, 2023, https://www.statista.com/statistics/276629/global-co2-emissions/; and Hannah Ritchie, "Cars, Planes, Trains: Where Do CO2 Emissions from Transport Come From?" Our World in Data, October 6, 2020, https://ourworldindata.org/co2-emissions-from-transport.

high-voltage charging stations: Jack Ewing, "World's Largest Long-Haul Truckmaker Sees Hydrogen-Fueled Future," *New York Times*, May 23, 2021, https://www.nytimes.com/2021/05/23/business/hydrogen-trucks-semis.html.

The shorter-range model: Jason Cannon, "What Does a Class 8 Truck Really Cost?" *Commercial Carrier Journal*, January 25, 2016, https://www.ccjdigital.com/business/article/14932561/what-does-a-class-8-truck-really-cost; Simon Alvarez, "Tesla Semi Price at Around $250,000: Report," Teslarati, April 14, 2023, https://www.teslarati.com/tesla-semi-production-price-revealed-pepsico/.

for between $80 and $100 per kWh: Norihiko Shirouzu and Paul Lienert, "Exclusive: Tesla's Secret Batteries Aim to Rework the Math for Electric Cars and the Grid," Reuters, May 14, 2020, https://www.reuters.com/article/us-autos-tesla-batteries-exclusive/exclusive-teslas-secret-batteries-aim-to-rework-the-math-for-electric-cars-and-the-grid-idUSKBN22Q1WC.

ban their sale in California: Rachel Cormack, "New York Is Banning the Sale of New Gas-Powered Cars and Trucks Starting in 2035," *Robb Report*, September 10, 2021, https://robbreport.com/motors/cars/new-york-bans-sale-new-gas-powered-cars-trucks-1234635358/.

to make the ban national: Peter Jones, "What States Are Banning Gas Cars? (Timeline and Facts)," *Motor and Wheels*, August 15, 2022, https://motorandwheels.com/what-states-banning-gas-cars/.

banning new gasoline and diesel vehicle sales: "Austria Seeking to Ban New-ICE Registrations by 2030," Autovista24, July 20, 2021, https://autovista24.autovistagroup.com/news/austria-seeking-to-ban-new-ice-registrations-by-2030/.

new EVs will reach price parity: Nic Lutsey and Michael Nicholas, "Update on Electric Vehicle Costs in the United States Through 2030," working paper, International Council on Clean Transportation, April 2, 2019, https://theicct.org/publication/update-on-electric-vehicle-costs-in-the-united-states-through-2030/; see also "California Moves to Accelerate to 100% New Zero-Emission Vehicle Sales by 2035," press release, California Air Resources Board, August 25, 2022, https://ww2.arb.ca.gov/news/california-moves-accelerate-100-new-zero-emission-vehicle-sales-2035.

seventy thousand charging stations: "Fact Sheet: The American Jobs Plan," Briefing Room, The White House, March 31, 2012, https://www.whitehouse.gov/briefing-room/statements-releases/2021/03/31/fact-sheet-the-american-jobs-plan/.

an estimated thirty million electric cars: "Projected Size of the Global Electric Vehicle Fleet Between 2021 and 2025."

60 percent of all new-vehicle car sales: Stephen Edelstein, "EVs Predicted to Gain Market Share, Despite 18% Sales Drop in 2020," Green Car Reports, May 21, 2020, https://www.greencarreports.com/news/1128234_evs-predicted-to-gain-market-share-despite-18-sales-drop-in-2020.

"sweep gasoline and diesel": Daniel Sperling, *Three Revolutions: Steering Automated, Shared, and Electric Vehicles to a Better Future* (Washington, DC: Island Press, 2018), 40.

program requires fuel providers: J. Bushnell et al., *Uncertainty, Innovation, and Infrastructure Credits: Outlook for the Low Carbon Fuel Standard Through 2030*, Office of the President, University of California Institute of Transportation Studies, Davis, CA, 2020, https://escholarship.org/uc/item/7sk9628s.

ten billion will live in urban areas: "Growing at a Slower Pace, World Population Is Expected to Reach 9.7 Billion in 2050 and Could Peak at Nearly 11 Billion Around 2100," News, United Nations Department of Economic and Social Affairs, June 17, 2019, https://www.un.org/development/desa/en/news/population/world-population-prospects-2019.html. The current world population is 8 billion. See "Current World Population," Worldometer, accessed February 28, 2023, https://www.worldometers.info/world-population/.

freight and passenger trains: Ralph Vartabedian, "Battery-Powered Trains Could Be a Climate Game Changer. Is Everyone All Aboard?" *Los Angeles Times*, July 5, 2021, https://www.latimes.com/california/story/2021-07-05/battery-powered-locomotives-zero-emission-train-future.

battery-electric or fuel cell trains: Vartabedian, "Battery-Powered Trains Could Be a Climate Game Changer."

electric bus's higher purchase price: According to a study by the Union of Concerned Scientists, even when powered by an electric grid that depends on coal and natural-gas power, the electric bus produces between an eighth and a third the pollution of the diesel, depending on how much fossil fuel is on the electric grid. Because it eliminates diesel fumes and tailpipe CO_2 emissions, an electric bus makes environmental as well as financial sense. See Jimmy O'Dea, "Electric vs. Diesel vs. Natural Gas: Which Bus Is Best for the Climate?" *The Equation* (blog), Union of Concerned Scientists, July 19, 2018, https://blog.ucsusa.org/jimmy-odea/electric-vs-diesel-vs-natural-gas-which-bus-is-best-for-the-climate.

as battery costs continue plunging: Shirouzu and Lienert, "Exclusive: Tesla's Secret Batteries Aim to Rework the Math."

43 percent annually: MarketsandMarkets, "Electric Bus Market by Propulsion (BEV, PHEV & FCEV) Application (Intercity & Intro-city) Consumer, Range, Bus Power Output, Battery Capacity, Component, Level of Autonomy, Battery type & Region – Global Forecast to 2027," Report AT 7483, *Electric Bus Market*, March 2022, https://www.marketsandmarkets.com/Market-Reports/electric-bus-market-38730372.html.

grew to 26 percent: "Electric bus market Europe 2022, all the figures. Guess the leaders!" Sustainable Bus, February 10, 2023, https://www.sustainable-bus.com/news/electric-bus-market-europe-2022/.

United States, sales growth: Kea Wilson, "US DOT Will Double the Nation's Electric Bus Fleet (But It Will Still Be Tiny)," Streetsblog USA, August 16, 2022, https://usa.streetsblog.

org/2022/08/16/us-dot-seeks-to-double-the-nations-electric-bus-fleet-which-is-currently-tiny-and-will-still-be/.

deficient cold-weather performance: Alon Levy, "The Verdict's Still Out on Battery-Electric Buses," *City Lab Transportation* (blog), Bloomberg, January 17, 2019, https://www.bloomberg.com/news/articles/2019-01-17/battery-electric-buses-yield-mixed-results-for-cities.

The number one electric bus: "Our Story," Proterra, accessed February 9, 2023, https://www.proterra.com/about/.

provided forty million dollars to transit agencies: "TIGGER Program," Federal Transit Administration, last updated May 14, 2018, https://www.transit.dot.gov/funding/grants/tigger-program.

Starting in 2029: "California Pushes for the Transition to Electric Buses: Only Zero Emission Purchases from 2029," Sustainable Bus, December 15, 2018, https://www.sustainable-bus.com/news/california-pushes-for-the-transition-only-zero-emission-purchases-from-2029/.

rock-bottom off-peak-power prices: The rate is under 3.5 cents a kilowatt-hour, according to Mr. Moynahan.

about three quarters of it still come: "Texas' Energy Profile: A Review of the State's Current Traditional and Renewable Energy Capabilities," Fiscal Notes, Comptroller.texas.gov, September 2022, https://comptroller.texas.gov/economy/fiscal-notes/2022/sep/energy.php.

50 percent more expensive: "U.S. Retail Diesel Prices," YCharts, accessed February 10, 2023, https://ycharts.com/indicators/us_retail_diesel_price.

280,000 conventional heavy trucks: Brian Wang, "Global Large and Semi Truck Market Is 4 Million Per Year," Next Big Future, December 20, 2022, https://www.nextbigfuture.com/2022/12/global-large-and-semi-truck-market-is-4-million-per-year.html. Globally, heavy trucks today are roughly a half-trillion-dollar a year market: 4 million are sold annually, and their price is around $165,000 in the United States (although less in China, where a great many heavy trucks are sold). See Tamara Warren, "Topsy-Turvy Times for Used Trucks," *FleetOwner*, September 21, 2022, https://www.fleetowner.com/equipment/article/21251002/why-buying-used-trucks-in-2022-has-been-like-no-other-year; and Wang, "Global Large and Semi Truck Market Is 4 Million Per Year."

truck industry will have a mixture: *Electric Trucks—Where They Make Sense*, guidance report, North American Council for Freight Efficiency, May 2018, https://nacfe.org/research/electric-trucks/#electric-trucks-where-they-make-sense.

EVs are "the inevitable future": Jim Stinson, "Electric Trucks Are the 'Inevitable Future,' Fleets Say," Transport Dive, May 19, 2020, https://www.transportdive.com/news/Electric-Class-8-trucks-CARB-2020-coronavirus/577687/.

in stop-and-go urban traffic: *Medium-Duty Electric Trucks: Cost of Ownership*, guidance report, North American Council for Freight Efficiency, October 2018, https://nacfe.org/research/electric-trucks/#medium-duty-electric-trucks.

a $593 million network: Rebecca Bellan, "Volvo, Daimler, Traton Invest $593 Million to Build Electric Truck Charging Network," TechCrunch, July 5, 2021, https://techcrunch.com/2021/07/05/volvo-daimler-traton-invest-593-million-to-build-electric-truck-charging-network/.

eliminate a truck's engine: The extraction and refining of lithium and cobalt for electric batteries, however, create significant environmental impacts. Lithium extraction uses a great deal of water, which is in scarce supply in the salt flats of Argentina and Chile and the regions of Australia from which most lithium comes. Lithium is also extracted in a non-sustainable, energy-intensive manner in China and Australia, nations that consume large amounts of coal. Most cobalt comes from the

Democratic Republic of the Congo, where environmental impacts are compounded by human rights violations and the use of child labor. See "Lithium-Ion Batteries Need to be Greener and More Ethical," editorial, *Nature*, June 29, 2021, https://www.nature.com/articles/d41586-021-01735-z.

Toyota Motor Company: Katie Fehrenbacher, "The Big Truck Makers Are Starting to Take Electric Trucks Seriously," Greenbiz, April 25, 2019, https://www.greenbiz.com/article/big-truck-makers-are-starting-take-electric-trucks-seriously.

a hydrogen fuel cell truck: Damien Cave, "Can a Carbon-Emitting Iron Ore Tycoon Save the Planet?" *New York Times*, October 16, 2021, https://www.nytimes.com/2021/10/16/business/energy-environment/green-energy-fortescue-andrew-forrest.html.

Charging a large e-truck fleet: Jack Roberts, "Q&A: Electric Vehicle Infrastructure with Paul Stith of Black and Veatch," *Heavy Duty Trucking*, September 28, 2018, https://www.truckinginfo.com/314493/hdt-interview-paul-stith-black-and-veatch.

These powerful, fast-charging systems: Kevin Walkowicz et al., *R&D Insights for Extreme Fast Charging of Medium- and Heavy-Duty Vehicles: Insights from the NREL Commercial Vehicles and Extreme Fast Charging Research Needs Workshop, August 27–28, 2019*, Technical Report NREL/TP-5400-75705, National Renewable Energy Laboratory, Golden, CO, March 2020, https://www.nrel.gov/docs/fy20osti/75705.pdf.

new utility substations: Roberts, "Q&A: Electric Vehicle Infrastructure with Paul Stith of Black and Veatch."

For large-scale electric truck deployment: Roberts, "Q&A: Electric Vehicle Infrastructure with Paul Stith of Black and Veatch."

major modernization of the utility system: Utilities will need to upgrade transformers and circuits and implement flexible pricing to encourage optimal use of their assets. Grid operators and energy market managers will need sophisticated, new distributed energy management software that will inform them in real time of power supply and demand and of prices throughout the utility system at a very granular level so that they will be able to manage power flow and schedule demand efficiently. See Robert MacDonald, "Charging Forward with the EV Revolution in the U.S.," *Renewable Energy World*, June 11, 2021, https://www.renewableenergyworld.com/storage/charging-forward-with-the-ev-revolution-in-the-us/.

vigorous organizing and lobbying: Hieu Le and Ramón Cruz, "It's Been One Year Since Two Historic Actions on Electrifying Trucks and Buses. Where Are We Now?" *Sierra Club Insider* (newsletter), Summer 2021, https://www.sierraclub.org/articles/2021/07/it-s-been-one-year-two-historic-actions-electrifying-trucks-and-buses-where-are-we.

standards designed to cut CO$_2$ emissions: Unfortunately, in April 2020, when most of the nation was focused on the novel coronavirus pandemic, President Trump abandoned the 2010 and 2012 national fuel efficiency standards that would have brought corporate average fuel efficiency to 54 miles per gallon, at a cumulative savings of $1.7 trillion to consumers—thereby avoiding 6 billion tons of CO$_2$ emissions. See Marianne Lavelle and Phil McKenna, "Trump's Fuel Efficiency Reduction Would Be Largest Anti-Climate Rollback Ever," Inside Climate News, March 31, 2020, https://insideclimatenews.org/news/31032020/fuel-efficiency-rollback-trump-administration-california-cafe-climate-change-tailpipe-emissions/; and David Roberts, "Gutting Fuel Economy Standards During a Pandemic Is Peak Trump," Vox, April 2, 2020, https://www.vox.com/energy-and-environment/2020/4/2/21202509/trump-climate-change-fuel-economy-standards-coronavirus-pandemic-peak.

cost about $0.57 a mile: Sperling, *Three Revolutions*, 14.

clean hydrogen fuel is falling: Jon LeSage, "How Long Until Hydrogen Is Competitive at the Pump?" OilPrice.com, May 28, 2020, https://oilprice.com/Alternative-Energy/Fuel-Cells/How-Long-Until-Hydrogen-Is-Competitive-At-The-Pump.html; Alan Mammoser, "Is Hydrogen the New LNG?," OilPrice.com, June 27, 2019, https://oilprice.com/Alternative-Energy/Fuel-Cells/Is-Hydrogen-The-New-LNG.html; "SGH2 building largest green hydrogen production facility in California; gasification of waste into H2," Green Car Congress, May 21, 2020, https://www.greencar-congress.com/2020/05/20200521-sgh2.html.

7. SOARING THE SKIES, SAILING THE WAVES—CLEANLY

Ubers and Lyfts of the air: Cade Metz and Erin Griffith, "What Is a Flying Car?" *New York Times*, last updated June 24, 2021, https://www.nytimes.com/2021/06/12/technology/flying-cars.html.

cruise at 23 knots: Naida Hakirević Prevljak, "World's First Electric Fast Ferry Is Here," Offshore Energy, July 14, 2022, https://www.offshore-energy.biz/worlds-first-electric-fast-ferry-is-here/; and Naida Hakirević Prevljak, "SINTEF Designs World's First Electric Speedboat," Offshore Energy, June 3, 2022, https://www.offshore-energy.biz/sintef-designs-worlds-first-electric-speedboat/.

more than forty billion tons: Hiroko Kabuchi, "'Worse than Anyone Expected': Air Travel Emissions Vastly Outpace Predictions," *New York Times*, September 19, 2019, https://www.nytimes.com/2019/09/19/climate/air-travel-emissions.html; Roz Pidcock and Sophie Yeo, "Analysis: Aviation Could Consume a Quarter of Remaining 1.5C Carbon Budget by 2050," Carbon Brief, August 8, 2016, https://www.carbonbrief.org/aviation-consume-quarter-carbon-budget/.

the seventh largest CO2 emitter: *Up in the Air: How Airplane Carbon Pollution Jeopardizes Global Climate Goals*, Center for Biological Diversity, December 2015, available at https://www.biological-diversity.org/programs/climate_law_institute/transportation_and_global_warming/airplane_emissions/.

"70,000 pounds of lead": Kevin Smith, "This LA Company Is Developing an All-Electric, Zero-Emission Airplane," *Los Angeles Daily News*, November 7, 2017, https://www.dailynews.com/2017/11/07/this-la-company-is-developing-all-electric-zero-emission-airplane; see also home page, Ampaire, accessed January 30, 2023, https://www.ampaire.com/.

EPA has yet to adopt: U.S. Environmental Protection Agency, "Final Rule for Finding that Green-house Gas Emissions from Aircraft Cause or Contribute to Air Pollution that May Reasonably Be Anticipated to Endanger Public Health and Welfare," Regulations for Emissions from Vehicles and Engines, EPA, last updated May 24, 2022, https://www.epa.gov/regulations-emissions-vehicles-and-engines/final-rule-finding-greenhouse-gas-emissions-aircraft.

aviation emissions were excluded: Alex Dichter et al., "How Airlines Can Chart a Path to Zero-Carbon Flying," McKinsey, May 13, 2020, https://www.mckinsey.com/industries/travel-logistics-and-infrastructure/our-insights/how-airlines-can-chart-a-path-to-zero-carbon-flying.

could reduce it by half again: Dichter et al., "How Airlines Can Chart a Path to Zero-Carbon Flying."

sustainable biofuels could cut emissions: Dichter et al., "How Airlines Can Chart a Path to Zero-Carbon Flying."

cost, supply, and environmental issues: Dichter et al., "How Airlines Can Chart a Path to Zero-Carbon Flying."

might turn out to be transitory: Carbon credits can be generated with the planting of certain crops and trees, or by restoring forests to store carbon in the soil. However, whereas carbon emissions from a jet have immediate impacts on the climate, trees take decades or centuries to reach their full carbon storage capacity and are subject to destruction by fires or illegal logging, while soils are subject to abuse that can result in the release of their stored carbon. Broadly speaking, carbon credits raise other accounting and monitoring issues as well as the question of "additionality"—namely, whether an action would have been taken anyway, even if carbon credits had not been sold based upon it. Other measures that can be used for offsetting jet emissions—generating power renewably instead of with fossil fuels or capturing methane from landfills or dairies—arguably should be required under sound climate policies rather than being made contingent upon voluntary payments from airline passengers.

the largest all-electric airplane: "First Successful Flight for World's Largest All-Electric Plane," Reuters, May 28, 2020, https://news.yahoo.com/first-successful-flight-worlds-largest-032555612.html.

Eviation, an Israeli company: William Johnson, "Eviation Orders Continue to Surge Past $2 Billion," Teslarati, January 17, 2023, https://www.teslarati.com/eviation-electric-plane-aerus-order/.

plans to have its four-propeller: See home page, Heart Aerospace, accessed February 28, 2023, https://heartaerospace.com/.

Major investors include: "Heart Aerospace, ES-30," FutureFlight, accessed February 16, 2023, https://www.futureflight.aero/aircraft-program/heart-electric-airliner.

or to a medium-range destination: Metz and Griffith, "What Is a Flying Car?"

a deal to be valued at $6.6 billion: At this writing, Joby is set to be acquired by Reinvent Technology Partners, a special purpose acquisition company (SPAC), thereby going public through the merger. See Amber Deter, "Joby Aviation Stock Is Almost Here: What We Need to Know About the SPAC IPO," Investment U, May 28, 2021, https://investmentu.com/joby-aviation-stock-ipo/.

like an airborne Uber: Other companies, like Archer, Kitty Hawk, Lilium, and Wisk Aero, have vehicles under development. Kitty Hawk, founded by Stanford engineering professor Sebastian Thrun, plans to compete in the urban air taxi business (as does Wisk). Thrun was the head of Google's driverless car program and founded Kitty Hawk with financial assistance from Google's Larry Page.

"a hundred times quieter": "Joby Completes Flight of More than 150 Miles with Electric Vertical-Takeoff Air Taxi," press release, Joby Aircraft, July 27, 2021, https://www.jobyaviation.com/news/joby-completes-flight-of-more-than-150-miles/.

a successful 150-mile test flight: "Joby Completes Flight of More than 150 Miles with Electric Vertical-Takeoff Air Taxi."

10 orders from UPS: Ben Ryder-Howe, "The Battery That Flies," *New York Times*, April 17, 2022, https://www.nytimes.com/2022/04/16/business/beta-electric-airplane.html.

nearly 13 *kilo*watt-hours: Umair Irfan, "Forget Cars. We Need Electric Airplanes," Vox, April 9, 2019, https://www.vox.com/2019/3/1/18241489/electric-batteries-aircraft-climate-change.

world's first nonstop transatlantic flight: Jason Paur, "June 15, 1919: First Nonstop Flight Crosses Atlantic," *Wired*, June 15, 2018, https://www.wired.com/2010/06/0615alcock-brown-fly-atlantic/.

Norway is so confident: Matthew Beedham, "Norway Pushes to Electrify All Local Flights by 2040," *Shift*, The Next Web, March 9, 2020, https://thenextweb.com/shift/2020/03/09/norway-pushes-to-electrify-all-domestic-flights-by-2040/.

improving the energy density: Westenhaus, "Sydney Based Researchers Announce Lithium Ion Battery Rival."

90 percent of the world's freight: Isabelle Gerretsen, "Shipping Is One of the Dirtiest Industries. Now It's Trying to Clean Up Its Act," CNN Business, October 3, 2019, https://www.cnn.com/2019/10/03/business/global-shipping-climate-crisis-intl/index.html.

"large floating shoeboxes": Professor James Corbett, University of Delaware, communication with author, May 6, 2020.

has allowed the shipping industry: Zheng Wan et al., "Decarbonizing the International Shipping Industry: Solutions and Policy Recommendations," *Marine Pollution Bulletin* 126 (January 2018): 428–35, https://doi.org/10.1016/j.marpolbul.2017.11.064.

an "aspirational" target: Gerretsen, "Shipping Is One of the Dirtiest Industries."

"watered down climate regulations": Matt Apuzzo and Sarah Hurtes, "Tasked to Fight Climate Change, a Secretive U.N. Agency Does the Opposite," *New York Times,* June 3, 2021, https://www.nytimes.com/2021/06/03/world/europe/climate-change-un-international-maritime-organization.html.

damaging emissions of black carbon: Sara Nilsson, "UN Shipping Agency Slammed for Fiddling While Arctic Melts," press release, Eco Age Limited, Gemini House, London, March 26, 2021, https://www.hellenicshippingnews.com/un-shipping-agency-slammed-for-fiddling-while-arctic-melts/.

carbon neutrality by 2050: Gerretsen, "Shipping Is One of the Dirtiest Industries."

the combustion of methanol: "Maersk Plans to Operate First Carbon-Neutral Container Ship by 2023," S&P Global, February 17, 2021, https://www.spglobal.com/commodityinsights/en/market-insights/latest-news/coal/021721-maersk-plans-to-operate-first-carbon-neutral-container-ship-by-2023.

a battery-hybrid-powered fleet: Gerretsen, "Shipping Is One of the Dirtiest Industries"; and Naida Hakirević Prevljak, "Hurtigruten Norway Launches Its First Ship Upgraded to Battery-Hybrid Op," Offshore Energy, September 23, 2022, https://www.offshore-energy.biz/hurtigruten-norway-launches-its-first-ship-upgraded-to-battery-hybrid-op/.

already launched the HydroTug 1: Ajsa Habibic, "Port of Antwerp-Bruges Readies for First Hydrogen-Powered Tugboat en Route to Becoming Climate-Neutral," Offshore Energy, May 20, 2022, https://www.offshore-energy.biz/port-of-antwerp-bruges-readies-for-first-hydrogen-powered-tugboat-en-route-to-becoming-climate-neutral. BeHydro is a joint venture of CMB and engine developer Anglo Belgian Corporation to research and develop hydrogen engines for maritime, railway, and power use.

fully operational in 2023: Habibic, "Port of Antwerp-Bruges Readies for First Hydrogen-Powered Tugboat"; and Gerretsen, "Shipping Is One of the Dirtiest Industries."

fairly easy to convert: "BeHydro Launches the First Hydrogen-Powered Dual-Fuel Engine," CMB Tech, September 17, 2020, https://cmb.tech/news/behydro-launch; and "Hydrotug," CMB Tech, accessed February 14, 2023, https://cmb.tech/hydrotug-project.

at close to $100 a ton in early 2023: "Live Carbon Prices Today: European Carbon Credit Market," CarbonCredits.com, accessed February 11, 2023, https://carboncredits.com/carbon-prices-today; and "Hydrotug," CMB Tech.

largest industrial source of CO_2: Royal Society, "Ammonia: Zero-Carbon Fertilizer, Fuel, and Energy Store," policy briefing, The Royal Society, February 2020, https://royalsociety.org/-/media/policy/projects/green-ammonia/green-ammonia-policy-briefing.pdf.

nearly 2 percent of all CO_2: Royal Society, "Ammonia: Zero-Carbon Fertilizer, Fuel, and Energy Store."

Research on how to combust ammonia: A. Alnasif et al., "Experimental and Numerical Analyses of Nitrogen Oxides Formation in a High Ammonia-Low Hydrogen Blend Using a Tangential Swirl Burner," *Carbon Neutrality* 1, no. 24 (June 23, 2022), https://doi.org/10.1007/s43979-022-00021-9.

could nonetheless be the basis: Royal Society, "Ammonia: Zero-Carbon Fertilizer, Fuel, and Energy Store."

food for half the earth's population: Royal Society, "Ammonia: Zero-Carbon Fertilizer, Fuel, and Energy Store."

its low flammability: Hideaki Kobayashi et al., "Science and Technology of Ammonia Combustion," *Proceedings of the Combustion Institute* 37, no. 1 (2019): 109–33, https://doi.org/10.1016/j.proci.2018.09.029.

by one estimate, $1.9 trillion: Marta Gallucci, "Why the Shipping Industry Is Betting Big on Ammonia," *IEEE Spectrum*, February 23, 2021.

an ammonia engine commercially available: "Unlocking Ammonia's Potential for Shipping," MAN Energy Solutions, accessed February 11, 2023, https://www.man-es.com/discover/two-stroke-ammonia-engine; and Gallucci, "Why the Shipping Industry Is Betting Big on Ammonia."

an ammonia-powered oil tanker: Gallucci, "Why the Shipping Industry Is Betting Big on Ammonia."

8. PUTTING CAPTURED CO2 TO WORK

plant materials instead of fossil fuels: Julie B. Zimmerman et al., "Designing for a Green Chemistry Future," *Science* 367, no. 6476 (January 24, 2020), 397–400.

brilliant overview of green chemistry: Zimmerman et al., "Designing for a Green Chemistry Future."

a trillion-dollar global business: David Roberts, "These Uses of CO2 Could Cut Emissions—and Make Trillions of Dollars," Vox, November 27, 2019, https://www.vox.com/energy-and-environment/2019/11/13/20839531/climate-change-industry-co2-carbon-capture-utilization-storage-ccu.

Polyvinyl chloride is a good example: Paul Wesley Brandt-Rauf et al., "Plastics and Carcinogenesis: The Example of Vinyl Chloride," *Journal of Carcinogenesis* 11, no. 5 (March 12, 2012), https://doi.org./10.4103/1477-3163.93700; Li Cohen, "Vinyl Chloride's Invisible Threat: Thousands of Pounds Are Released Every Year in the U.S. as Part of 'Poison Plastic' Manufacturing," CBS News, updated February 20, 2023, https://www.cbsnews.com/news/vinyl-chloride-ohio-train-derailment-thousands-pounds-released-annually-plastics/; and Malihe Razazan and Rose Aguilar, "One Planet: The Norfolk Southern Ohio Train Derailment Reveals the Dangers of Plastic Production," *Your Call*, KALW Public Media, February 20, 2023, https://www.kalw.org/show/your-call/2023-02-20/one-planet-the-norfolk-southern-ohio-train-derailment-reveals-the-dangers-of-plastic-production.

Newlight feeds bacteria: See "Technology," Newlight, accessed February 28, 2023, https://newlight.com/technology.

Carbon capture and reuse: "To be clear, CCU will never reduce enough CO2 to avoid the need for CCS (i.e., burying carbon). Not even close. The tonnage of CO2 [that] humanity emits simply dwarfs the tonnage of [the] carbon-based products it consumes." David Roberts, "Pulling CO2 Out of the Air and Using It Could Be a Trillion-Dollar Business," Vox, November 22, 2019, https://www.vox.com/energy-and-environment/2019/9/4/20829431/climate-change-carbon-capture-utilization-sequestration-ccu-ccs.

Gulf Oil storage tanks: Gregg Macey, "Negotiating with a Captive Audience in Kennedy Heights, TX: Settling Environmental Justice Ligation with a Special Master," in *Using Dispute Resolution Techniques to Address Environmental Justice Concerns: Case Studies: Four Cases* (Cambridge, MA: Consensus Building Institute, 2003), available at National Service Center for Environmental

Publications (NSCEP), United States Environmental Protection Agency, accessed February 6, 2023, https://nepis.epa.gov/Exe/ZyPURL.cgi?Dockey=100049NY.txt; Gregg P. Macey, "The Politics of Risk: Pre-Litigation Site Assessment in Houston, Texas," *Environmental Law* 37 (2007): 15–59, https://brooklynworks.brooklaw.edu/faculty/134/; Frances Nixon, "Houston v. Chevron: A Case of Environmental Justice," Colby College Wiki: Case Studies in Environmental Justice, https://wiki.colby.edu/display/es298b/Houston+Oil+Pits; and Sam Howe Verhovek, "Racial Rift Slows Suit for 'Environmental Justice,'" *New York Times*, September 7, 1997, https://www.nytimes.com/1997/09/07/us/racial-rift-slows-suit-for-environmental-justice.html. Preceding four sources are cited in Emil Holt, "Case Study: Houston v. Chevron," accessed February 6, 2023, https://silo.tips/download/case-study-houston-v-chevron.

a competitive federal fellowship: Home page, Cyclotron Road, Lawrence Berkeley National Laboratory, accessed February 6, 2023, https://www.cyclotronroad.org/.

"a five-to-one return": Sean Pool and Jennifer Erickson, "The High Return on Investment for Publicly Funded Research," Center for American Progress, December 10, 2012, https://www.americanprogress.org/article/the-high-return-on-investment-for-publicly-funded-research/.

over ten million gallons of jet fuel: Sheila Remes, vice president of strategy, Boeing Commercial Airplanes, quoted in "LanzaTech Moves Forward on Sustainable Aviation Scale Up in the USA and Japan," press release, LanzaJet, November 20, 2019, https://www.lanzajet.com/lanzatech-moves-forward-on-sustainable-aviation-scale-up-in-the-usa-and-japan/.

"the circular economy in action": LanzaTech, in the near future, is planning to convert some of its ethanol in China into polyester, which is used in clothing and packaging. It is also in negotiation with the U.S. Department of Energy for a $14 million investment in the company's proposed demonstration-scale integrated biorefinery in Soperton, Georgia.

another exciting claim: The company also has a process to turn "unrecyclable" and unwashed household waste into aviation fuel and has a contract to provide it to Nippon Airways. "Novo Holdings Invests $72 Million in Sustainable Products Leader LanzaTech," press release, Novo Holdings, August 6, 2019, https://www.novoholdings.dk/news/novo-holdings-invests-72-million-in-sustainable-products-leader-lanzatech/.

could be produced from CO2 and water: At some point, Twelve may make methane, specifically for upgrading biogas. "If you go into a dairy today," Flanders said, "very valuable methane is being captured from dairy manure, but about half of what comes out from the digester is CO2 mixed with the methane, and today we just throw away the CO2." Twelve believes their technology could turn it into more biomethane. Nicholas Flanders, CEO of Twelve, Inc., author interview, Berkeley, California, July 2, 2019.

9. A PLANT-BASED ECONOMY

Many common consumer products: For biobased insulation from hemp, see home page, Hemp in a Box, accessed February 6, 2023, https://hempinabox.be/en; for biobased and biodegradable plastics, see home page, B4Plastics, accessed February 6, 2023, https://www.b4plastics.com; and home page, BIOPLA, accessed February 6, 2023, https://biopla.be/?lang=en; for mycelium applications like mycelium lampshades or flowerpots, see home page, PermaFungi, accessed February 6, 2023, https://www.permafungi.be/en/permateria/; for human milk oligosaccharides, see home page, Inbiose, accessed February 6, 2023, https://www.inbiose.com/; for useful products formulated using sunflower oil, see Pierre Gielen, "Coatings and Cosmetics from French Sunflowers," *Agro and Chemistry*, June 26, 2018, https://www.agro-chemistry.com/articles/coatings-and-cosmetics-from-french-sunflowers. Finally, in the "what will they think of next" department, see "Komrads APL: A High-Quality Sneaker Made of Apples," IndieGogo campaign by Mark MJ Vandevelde, accessed February 6, 2023, https://www.indiegogo.com/projects/komrads-apl-a-high-quality-sneaker-made-of-apples#/ for sneakers made of apple leather.

€750 million had already been invested: Sofie Dobbelaere, PhD, managing director, Flanders Biobased Valley, email communication, February 3, 2020.

its capacity continues expanding: "Alco Bio Fuel, Ghent Celebrates 10 Years of Success," International Bulk Journal, June 8, 2018, https://www.ibj-online.com/alco-bio-fuel-ghent-celebrates-10-years-of-success/190.

not an ideal solution: Mark Z. Jacobson, *No Miracles Needed*, 177–81.

all sited close together: The North Sea Port is a merger of the Port of Ghent and Zeeland Seaports in Terneuzen, The Netherlands.

is not so clear-cut: Keisuke Hanaki and Joana Portugal-Pereira, "The Effect of Biofuel Production on Greenhouse Gas Emission Reductions," chapter 6 in Kazuhiko Takeuchi et al., eds., *Biofuels and Sustainability*, Science for Sustainable Societies series (Tokyo: Springer, 2018): 53–71, https://doi.org/10.1007/978-4-431-54895-9_6.

a careful life cycle analysis: Hanaki and Portugal-Pereira, "The Effect of Biofuel Production on Greenhouse Gas Emission Reductions."

the Bio Base Europe Pilot Plant: For information on the Europe Pilot Plant, see home page, Bio Base Europe Pilot Plant, accessed February 7, 2023, http://www.bbeu.org.

Surfactants are the active ingredients: Fakruddin Md, "Biosurfactant: Production and Application," *Journal of Petroleum and Environmental Technology* 3, no. 4, 2012, https://doi.org/10.4172/2157-7463.1000124.

potent broad-spectrum antibiotics: For further information, see home page, Centre for Industrial Biotechnology and Biocatalysis, accessed February 7, 2023, https://www.inbio.be.

XCarb Innovation Fund: "ArcelorMittal Inaugurates Flagship Carbon Capture and Utilisation Project at Its Steel Plant in Ghent, Belgium," press release, ArcelorMittal, December 8, 2022, https://corporate.arcelormittal.com/media/press-releases/arcelormittal-inaugurates-flagship-carbon-capture-and-utilisation-project-at-its-steel-plant-in-ghent-belgium.

a €35 million reactor: "ArcelorMittal inaugurates flagship carbon capture and utilisation project at its steel plant in Ghent, Belgium."

multitrillion-dollar bulk chemical market: "LanzaTech Announces Full Year 2022 Financial Results and Provides Full Year 2023 Financial Outlook," March 29, 2023, https://ir.lanzatech.com/news-releases/news-release-details/lanzatech-announces-full-year-2022-financial-results-and.

10. GREEN CEMENT AND STEEL

account for about 8 percent: Lucy Rodgers, "Climate Change: The Massive CO2 Emitter You May Not Know About," BBC News, December 17, 2018, https://www.bbc.com/news/science-environment-46455844.

total floor area of buildings: "Why the Built Environment?" Architecture 2030, accessed February 17, 2023, https://architecture2030.org/why-the-building-sector.

Solidia Concrete doesn't react: Nicholas DeCristofaro, author communication, email, January 9, 2023. "Precast concrete waste is reduced because ALL green concrete scrap (that is, Solidia concrete before it is introduced to CO2) can be recycled. This includes leftover concrete in the mixer, residue in the molds, and mis-shaped parts. The residue in the molds is probably the smallest of these three components. This is possible because, unlike Portland cement, Solidia Cement does not react with mixing water."

a trillion concrete blocks a year: Christine Wei-li Lee, "Team Led by UCLA Professor Wins $7.5 Million NRG Cosia Carbon XPRIZE," press release, UCLA Newsroom, April 19, 2021, https://newsroom.ucla.edu/releases/ucla-nrg-cosia-carbon-xprize-concrete.

the carbon footprint of cement: "Solidia Technologies, Inc. website, https://www.solidiatech.com/impact.html.

supplementary cementitious material: DeCristofaro, author communication, January 9, 2023.

"a carboxylate-cured concrete": DeCristofaro, author communication, January 9, 2023.

can, at best, offset 30 only percent: Solidia's current process eliminates about 30 percent of the thermal and process emissions incurred in making cement. "Solidia cement production can be accomplished with CO_2 emissions up to 30% less than that for Portland cement," according to Nick DeCristofaro and Sada Sahu, "CO_2 Reducing Cement, Part One: Solidia Cement Composition and Synthesis," *World Cement*, January 9, 2014.

less-corrosion-resistant electrodes: Bella Peacock, "Melbourne Researchers Use Sound Waves to Boost Green Hydrogen Production 14-fold," *PV Magazine*, December 13, 2022, https://www.pv-magazine-australia.com/2022/12/13/melbourne-researchers-use-sound-waves-to-boost-green-hydrogen-production-14-fold/.

to grow cement bricks and floor tiles: Amy Feldman, "Startup Biomason Makes Biocement Tiles. Retailer H&M Group Plans to Outfit Its Stores' Floors with Them," *Forbes*, June 14, 2021, https://www.forbes.com/sites/amyfeldman/2021/06/14/startup-biomason-makes-bio-cement-tiles-retailer-hm-group-plans-to-outfit-its-stores-floors-with-them/?sh=482dde9457c9.

Nearly twenty other companies: "Top Players in the Global Green Cement Industry," IMARC Group, June 17, 2021, https://www.imarcgroup.com/green-cement-companies.

without the need for a kiln: Nadine M. Post et al., "New Technologies for Reduced-Carbon Concrete Are on the Horizon," *Engineering News-Record*, September 15, 2022, https://www.enr.com/articles/54809-new-technologies-for-reduced-carbon-concrete-are-on-the-horizon.

the $2.5 trillion global steel industry: "Worldsteel Recently Concluded a Global Economic Modelling Exercise with Oxford Economics That Found That in 2017 the Steel Industry Sold US$2.5 Trillion Worth of Products and Created US$500 Billion Value Added." See *World Steel in Figures 2019*, report, World Steel Association, https://worldsteel.org/wp-content/uploads/2019-World-Steel-in-Figures.pdf.

at least 1.8 metric tons: In a video produced by Bill Gates, Dr. Sadoway states, "For every ton of steel you make [today], you make several tons of CO_2." Bill Gates, "Better Metal," YouTube, August 28, 2019, video, 2:08 (quotation at 1:01), https://www.youtube.com/watch?v=-pYNzxorJso&t=3s.

3.6 billion metric tons of steel: "Global Crude Steel Output Increases by 3.4% in 2019," World Steel Association, January 27, 2020, https://www.worldsteel.org/media-centre/press-releases/2020/Global-crude-steel-output-increases-by-3.4--in-2019.html.

slashed by at least two thirds: Yongqi Sun et al., "Decarbonising the Iron and Steel Sector for a 2 °C Target Using Inherent Waste Streams," *Nature Communications* 3, no. 297 (2022), https://doi.org/10.1038/s41467-021-27770-y.

new technology and new investment: M. Kundak, L. Lazić, and J. Črinko, "CO_2 Emissions in the Steel Industry," *Metalurgija* 48, no. 3 (2009): 193–97.

hydrogen will be produced in situ: "ArcelorMittal Sestao to Become the World's First Full-Scale Zero Carbon-Emission Steel Plant," press release, ArcelorMittal, July 13, 2021, https://corporate. arcelormittal.com/media/press-releases/arcelormittal-sestao-to-become-the-world-s-first-full-scale-zero-carbon-emissions-steel-plant.

fuel Fortescue's own operations: Cave, "Can a Carbon-Emitting Iron Ore Tycoon Save the Planet?"

energy required to produce green steel: Artem Baroyan et al., *The Resilience of Steel: Navigating the Crossroads*, McKinsey, April 18, 2023. Producing steel using direct-reduced iron plus an electric arc furnace requires more than 3 megawatt-hours per metric ton versus 0.1 MWh for conventional steel.

Colorado Fuel and Iron: Gregory Howell, "Colorado Fuel & Iron: The Largest Consolidated Steel Mill West of the Mississippi," accessed March 3, 2022, http://www.gregoryhowell.com/colorado-fuel-iron-company.

the U.S. labor movement: Howell, "Colorado Fuel & Iron," accessed March 3, 2022.

fluctuating coal demand: Simone Liedtke, "Coal Prices to Remain Volatile Heading into 2022," *Mining Weekly*, November 12, 2021, https://www.miningweekly.com/article/coal-prices-to-remain-volatile-heading-into-2022-2021-11-12.

cheaper than the coal power: Derek Brower, "Solar-Powered Steel Mill Blazes Trail for Green Energy Transition," *Financial Times*, October 11, 2021, https://www.ft.com/content/f6693948-2c3d-4508-96cf-c374efofa6ad.

the length of a football field: Allen Best, "Making Steel with Solar Energy," Big Pivots, August 7, 2020, https://bigpivots.com/evraz-and-solar-project/.

There's simply not enough scrap steel: "Dealing with Environmental Pollution in the Iron and Steel Industry: The China Case Study," Southeast Asia Iron and Steel Institute, September 29, 2008. Electric arc furnace technology works only for recycled steel, so it cannot be used to replace Boston Metal's technology.

in Sadoway's MIT laboratory: Antoine Allanore, Lan Yin, and Donald R. Sadoway, "A New Anode Material for Oxygen Evolution in Molten Oxide Electrolysis," *Nature* 497 (2013): 353–56, https://doi.org/10.1038/nature12134.

competitive . . . even without a carbon tax: The company's economic projections are also conservative in ignoring the potential revenue from the large quantities of oxygen produced as a by-product of its process. Valentin Vogl et al., "Assessment of Hydrogen Direct Reduction for Fossil-Free Steelmaking," *Journal of Cleaner Production* 203 (August 29, 2018), https://doi.org/10.1016/j.jclepro.2018.08.279.

steel would be 35 percent cheaper: The aluminum industry today pays 1.5 to 3.5 cents per kilowatt-hour. "The energy is there," said Adam Rauwerdink, Boston Metal's VP of business development. "They're signing solar at 1.5 cents in Portugal last week." Large wind plants, too, can now be very cost-effective; the 850-foot-high GE Haliade-X 12 MW offshore wind turbine is expected to produce enough power to supply sixteen thousand homes. "Haliade-X 12 MW Offshore Wind Turbine Platform," GE Renewable Energy, accessed February 28, 2023, https://www.ge.com/renewableenergy/wind-energy/offshore-wind/haliade-x-offshore-turbine.

the cost of that clean hydrogen: Kundak, Lazić, and Črinko, "CO2 Emissions in the Steel Industry."

a road map for decarbonizing: Eurofer, *Low Carbon Roadmap: Pathways to a CO2-Neutral European Steel Industry*, Eurofer, Brussels, Belgium, November 2019.

a climate-related trade deal: Ana Swanson, "U.S. Proposes Green Steel Club That Would Levy Tariffs on Outliers," *New York Times*, December 7, 2022, https://www.nytimes.com/2022/12/07/business/economy/steel-tariffs-climate-change.html.

has to transport the pig iron: Terence Bell, "The Modern Steel Manufacturing Process," Thought-Co., last updated August 21, 2020, https://www.thoughtco.com/steel-production-2340173.

oxygen is blown into the melt: "Basic Oxygen Steelmaking," Wikimedia Foundation, last edited January 2, 2023, 22:59, https://en.wikipedia.org/wiki/Basic_oxygen_steelmaking.

11. VISIONARY MAYORS, GREEN CITIES

rise as much as five feet: Daphne Wysham, "U.S. Climate Envoy Jonathan Pershing: Five Feet of Sea Level Rise by 2050 Possible," *HuffPost*, last updated December 6, 2017. In more recent personal communication with the author (October 17, 2019), Pershing qualified and underscored his concerns, noting, "The volume of ice in the ice sheets would be enough to raise sea level by substantially more than five feet, and we'd be committed to that once we reached a tipping point—which could well be before 2050."

half the world's population: United States Census Bureau, "U.S. Cities Are Home to 62.7 Percent of the U.S. Population, but Comprise Just 3.5 Percent of Land Area," Release Number CB15-33, U.S. Census Bureau, March 4, 2015.

produce about three quarters: United States Census Bureau, "U.S. Cities Are Home to 62.7 Percent of the U.S. Population, but Comprise Just 3.5 Percent of Land Area."

the world is urbanizing so fast: Andy Gouldson et al., "Accelerating Low-Carbon Development in the World's Cities," contributing paper for *Seizing the Global Opportunity: Partnerships for Better Growth and a Better Climate* (London and Washington, DC: New Climate Economy, 2015).

a coalition of leading cities: "Mayors Announce Support for Global Green New Deal; Recognize Global Climate Emergency," statement issued by C40 Cities, C40 World Mayors Summit, Copenhagen, Denmark, October 9, 2019.

75 billion metric tons: *Further and Faster Together: The 2021 Global Covenant of Mayors' Impact Report*, Global Covenant of Mayors, September 11, 2021, https://www.globalcovenantofmayors.org/impact2021/.

challenges ahead are so vast: Michael Bloomberg and Carl Pope, *Climate Hope: How Cities, Businesses, and Citizens Can Save the Planet* (New York: St. Martin's Press, 2017), 196–98.

57 percent since 1990: *Strategy for a Fossil Fuel Free Stockholm by 2040*, Report 134-175/2015, City of Stockholm Executive Office, December, 2016; *Stockholm Action Plan for Climate and Energy 2012–2015 with an Outlook to 2030*, City of Stockholm Environment and Health Administration, n.d.

despite robust economic growth: Jørgen Abildgaard, climate program manager, City of Copenhagen, author communication, email, November 3, 2019; Derek Robertson, "Inside Copenhagen's Race to Be the First Carbon-Neutral City," *Guardian* (London), October 11, 2019, https://www.theguardian.com/cities/2019/oct/11/inside-copenhagens-race-to-be-the-first-carbon-neutral-city.

gave up on its 2025 goal: Kirstine Lund Christiansen and Inge-Merete Hougaard, "Copenhagen's Failure to Meet 2025 Net Zero Target Casts Doubt on Other Pledges," *Climate Home News*, September 16, 2022, https://www.climatechangenews.com/2022/09/16/copenhagens-failure-to-meet-2025-net-zero-target-casts-doubt-on-other-city-pledges/.

between 2026 and 2028: Christiansen and Hougaard, "Copenhagen's Failure to Meet 2025 Net Zero Target."

a European thermal grid: Brian Vad Mathiesen, Ida Auken, and Jens Martin Skibsted, "This Is How Copenhagen Plans to Go Carbon-Neutral by 2025," *Energy Transition* (blog), World Economic Forum, May 21, 2019, https://www.weforum.org/agenda/2019/05/the-copenhagen-effect-how-europe-can-become-heat-efficient/.

who daily cycle to work or school: Author interview with former Copenhagen mayor Klaus Bondam, CEO, Danish Cyclists' Federation, via Skype, September 16, 2019. For 2017 data, see "Copenhagen City of Cyclists: Facts and Figures 2017," Cycling Embassy of Denmark, July 4, 2017, https://kk.sites.itera.dk/apps/kk_pub2/pdf/2268_9bc34ada85c8.pdf/.

Amager Resource Center: John J. Berger, "Heat, Power, Water, Resources, Plus Recreational Opportunities—What Can't This Iconic New Waste-to-Energy Plant Do?" *Huffington Post,* last updated March 23, 2017, https://www.huffpost.com/entry/heat-power-water-resources-plus-recreational-opportunities_b_5898df73e4b02bbb1816bd7d.

"an extremely homogeneous society": Author interview with Klaus Bondam, September 16, 2019.

"the environmental capital of Europe": Author interview with former Copenhagen mayor Bo Asmus Kjeldgaard, via Skype, September 19, 2019.

a landmark global climate agreement: The Kyoto Protocol set emission limits for greenhouse gases and bound the parties to a suite of progressive climate policies to reduce those gases by at least 5 percent below 1990 levels from 2008 to 2012. See *Kyoto Protocol to the United Nations Framework Convention on Climate Change*, United Nations, 1998, https://unfccc.int/resource/docs/convkp/kpeng.pdf.

"Copenhagen will demonstrate": *Copenhagen as the World's Eco-Metropolis: Copenhagen's Green Accounts 2007*, PDF, Technical and Environmental Administration, City of Copenhagen, n.d., http://kk.sites.itera.dk/apps/kk_pub2/pdf/659_82BJRAzhBF.pdf.

another landmark vision statement: *A Metropolis for People: Vision and Goals for Urban Life in Copenhagen 2015*, PDF, Technical and Environmental Administration, City of Copenhagen, 2009, https://kk.sites.itera.dk/apps/kk_pub2/pdf/646_mIrodQ6Wdu.pdf.

an average of 4.69 metric tons: "Denmark—CO_2 Emissions per Capita (2021)," Denmark, Environment, World Data Atlas, accessed February 19, 2023, https://knoema.com/atlas/Denmark/CO2-emissions-per-capita.

15 metric tons of CO_2 per person: Table 1.7: "Primary Consumption, Energy Expenditures, and Carbon Dioxide Emissions Indicators," *April 2019 Monthly Energy Review*, U.S. Energy Information Administration, April 25, 2019.

within half a mile of a subway: *CPH 2025 Climate Plan*, PDF, Technical and Environmental Administration, City of Copenhagen, September 2012, https://kk.sites.itera.dk/apps/kk_pub2/pdf/983_jk-PoekKMyD.pdf.

an organic extension of its goals: John J. Berger, "Copenhagen, Striving to Be Carbon-Neutral," parts 1–6, *Huffington Post*, March 13, 2017, complete series available at https://www.huffpost.com/author/johnjberger1-422.

electric vehicles are already tax-free: Jesper Berggren, "The New Danish Climate Plan—Together for a Greener Future," CleanTechnica, October 12, 2018, https://cleantechnica.com/2018/10/12/the-new-danish-climate-plan-together-for-a-greener-future/. Denmark as a whole will be done selling gasoline and diesel cars by 2030, and sales of hybrids will be banned by 2035.

a million fewer tons: "Bio4: Towards Carbon Neutrality in Copenhagen," Ramboll, accessed February 7, 2023, https://ramboll.com/projects/re/bio4-towards-carbon-neutrality-in-copenhagen.

"Less than one percent": Author interview with former Copenhagen mayor Morten Kabell, Copenhagen, Denmark, November 23, 2015.

largely escaped the global recession: Author interview with Morten Kabell, November 23, 2015.

a net return of close to a billion: Author interview with Jørgen Abildgaard, executive director of Copenhagen's Climate Project, Copenhagen, Denmark, November 23, 2015.

a "very conservative Republican": Author interview with Dale Ross, Las Vegas, Nevada, October 15, 2017.

about $2.75 million a year: The city's Energy Fund is created by revenue from the city's citizen-owned electric utility, which, since 2008, has generated fifty million dollars in revenue, allowing the city to reduce citizens' tax burdens.

driving force behind the plan: City and County of San Francisco, *San Francisco's Climate Action Plan 2021*, https://sfenvironment.org/sites/default/files/events/cap_fulldocument_wappendix_web_220124.pdf.

it also proposes to remove carbon: City and County of San Francisco, *San Francisco's 2021 Climate Action Plan*; "San Francisco's Climate Storyboard," San Francisco Department of the Environment, accessed February 7, 2023, https://sfenvironment.org/sf-climate-dashboard.

city is reportedly "well on its way": City and County of San Francisco, *San Francisco's 2021 Climate Action Plan*.

less than 3 percent of the city's emissions: "Municipal GHG emissions for the 2016 calendar year totaled 117,394 metric tons of CO_2e, representing less than 3% of total emissions in San Francisco." *San Francisco Municipal Progress Report: Climate and Sustainability*, San Francisco Department of the Environment, November 2018, 5, https://sfenvironment.org/sites/default/files/fliers/files/sfe_municipal_progress_report_19.pdf.

report their energy and water consumption publicly: "Minneapolis Works to Beat the Heat," We Are Still In, accessed November 11, 2019, https://www.wearestillin.com/success/minneapolis-works-beat-heat.

12. REINVENTING AGRICULTURE

"My wife and I had a decision": Author interview with Gabe Brown, via telephone, September 4, 2018.

the nonprofit Soil Health Academy: "Who We Are," Soil Health Consultants, accessed November 12, 2018, https://soilhealthconsultants.com/who-we-are.

"If you have healthy soil": Gabe Brown, "Five Keys to Building a Healthy Soil," speech, Idaho Center for Sustainable Agriculture, November 18, 2014, Boise, ID, YouTube, March 3, 2015, 36:42, https://www.youtube.com/watch?v=K5IDaRKET20.

Creque's advice to Wick: Author interview with John Wick, El Cerrito, CA, October 23, 2018.

Wick had also recently heard: Author interview with John Wick, via telephone, November 21, 2018.

to triple the carbon content: U.S. Department of Agriculture Natural Resource Conservation Service website, http://comet-planner.com, documents many other agricultural management practices that can also increase soil carbon.

taking six million cars off the road: Marcia S. DeLonge, Rebecca Ryals, and Whendee L. Silver, "A Lifecycle Model to Evaluate Carbon Sequestration Potential and Greenhouse Gas Dynamics of

Managed Grasslands," *Ecosystems* 16 (September 2013): 962–79, https://doi.org/10.1007/s10021-013-9660-5.

a five-pronged statewide Healthy Soils Initiative: "California's Healthy Soils Initiative," California Department of Food and Agriculture, accessed February 12, 2023, https://www.cdfa.ca.gov/healthysoils/.

"I am trying to imitate": Author interview with David C. Johnson, via telephone, February 18, 2019.

the entire world's annual human-caused carbon emissions: Quoted in John J. Berger, "Can Soil Microbes Slow Climate Change?" *Scientific American*, March 26, 2019, https://www.scientificameri-can.com/article/can-soil-microbes-slow-climate-change/?print=true.

to validate Johnson's work: Quoted in Berger, "Can Soil Microbes Slow Climate Change?"

he has not seen strong evidence: Quoted in Berger, "Can Soil Microbes Slow Climate Change?"

the rate of carbon gain will depend on: Quoted in Berger, "Can Soil Microbes Slow Climate Change?"

the soil seemed to be capturing about six metric tons of CO_2: Dr. David C. Johnson, author communication (email), March 2, 2023.

$360 to $445 additional profit: Dr. David C. Johnson, author communication (email), March 2, 2023.

the cost of incentivizing the capture: The data on global industrial CO_2 emissions is from the 2022 Global Carbon Budget by the Global Carbon Project. See Pierre Friedlingstein et al., "Global Carbon Budget," *Earth System Science Data* 14, no. 11 (2022): 4811–900, https://doi.org/10.5194/essd-14-4811-2022); and also Zeke Hausfather and Pierre Friedlingstein, "Analysis: Global CO_2 Emissions from Fossil Fuels Hits Record High in 2022," World Economic Forum, November 11, 2020, https://www.weforum.org/agenda/2022/11/global-co2-emissions-fossil-fuels-hit-record-2022.

13. FOREST SAVERS AND FIXERS

up to seven hundred metric tons of carbon dioxide: Andrea Jane Leys, "How Is Carbon Stored in Trees and Wood Products?" report, Forest Learning, Forest and Wood Products Australia, accessed November 13, 2018, https://forestlearning.edu.au/images/resources/How%20carbon%20is%20stored%20in%20trees%20and%20wood%20products.pdf.

1.5 million acres of forest are converted: Laurie A. Wayburn, "U.S. Forest Carbon Policy: The Role of State and Federal Governments," speech, Carbon Finance Speaker Series, Burke Auditorium, Yale University, New Haven, CT, March 30, 2009.

logging rates were four times that of . . . rain forests: M. C. Hansen et al., "High-Resolution Global Maps of 21st-Century Forest Cover Change," *Science* 342, (2013): 850–53, https://www.science.org/doi/10.1126/science.1244693.

chip mills . . . were popping up: Wood removals from the Atlantic region of the Southeast in 2014 were up relative to 2000. See Pete Stewart, *Wood Supply Market Trends in the US South, 1995–2015*, U.S. Industrial Pellet Association, November 19, 2015, https://www.forest2market.com/hubfs/2016_Website/Documents/20151119_Forest2Market_USSouthWoodSupplyTrends.pdf.

50 *percent more* carbon than coal: Sam L. Davis, "Seeing the Forest: Nature's Solution to Climate Change," report, Dogwood Alliance, PDF, 2018, https://stand4forests.org/wp-content/uploads/2018/09/The-Climate-Plan.pdf.

The Great American Stand: Bill Moomaw and Danna Smith, *The Great American Stand: U.S. Forests and the Climate Emergency* (Asheville, NC: Dogwood Alliance, 2017), PDF, https://www.dogwoodalliance.org/wp-content/uploads/2017/03/The-Great-American-Stand-Report.pdf.

"A number of years ago": Author interview with William Moomaw, via telephone, June 4, 2018.

we are emitting about 9.4 billion metric tons: Corinne Le Quéré et al., "Global Carbon Budget 2017," *Earth System Science Data* 10 (2018): 405–48, https://doi.org/10.5194/essd-10-405-2018.

4.4 billion metric tons of carbon: Bronson W. Griscom et al., "Natural Climate Solutions," *Proceedings of the National Academy of Sciences* 114, no. 44 (October 2017): 11645–50, https://doi.org/10.1073/pnas.1710465114.

U.S. forests hold only 10 to 50 percent: Hansen et al., "High-Resolution Global Maps of 21st-Century Forest Cover Change."

"only some forests get paid": Quotations attributed to Laurie Wayburn are taken from author interviews via telephone or Skype on March 13, 2018, and March 20, 2018, as well as dozens of email exchanges between 2018 and 2022.

14. NEW LAWS AND POLICIES

political donations were a form of speech: Tim Lau, "Citizens United Explained," Brennan Center for Justice, December 12, 2019, https://www.brennancenter.org/our-work/research-reports/citizens-united-explained.

more than two billion dollars in 2020: Kenneth P. Vogel and Shane Goldmacher, "Democrats Decried Dark Money. Then They Won with It in 2020," *New York Times*, August 21, 2022, https://www.nytimes.com/2022/01/29/us/politics/democrats-dark-money-donors.html.

at the federal, state, and local levels: Kimberly Adams, David Brancaccio, and Rose Conlon, "Inside the ever-growing power of dark money in U.S. politics," *Marketplace*, March 4, 2022, https://www.marketplace.org/2022/03/04/inside-the-ever-growing-power-of-dark-money-in-u-s-politics/.

Apart from super PACs: Chisun Lee, Douglas Keith, and Ava Mehta, "Elected Officials, Secret Cash," Brennan Center for Justice, March 15, 2018, https://www.brennancenter.org/our-work/research-reports/elected-officials-secret-cash.

donated $1.6 billion: David Sirota and Joel Warner, "Billions in 'Dark Money' Is Influencing US Politics. We Need Disclosure Laws," *Guardian* (London), August 29, 2022, https://www.theguardian.com/commentisfree/2022/aug/29/billions-in-dark-money-is-influencing-us-politics-we-need-disclosure-laws.

as much wealth as the bottom 80 percent: Sawhill and Pulliam, "Six Facts About Wealth in the U.S."

astonishing speed for a demilitarized nation: Jacobson and DeLucchi, "A Path to Sustainable Energy by 2030."

H.Res. 109, the Green New Deal: House Resolution 109, "Recognizing the Duty of the Federal Government to Create a Green New Deal," introduced February 7, 2019, in the House by Rep. Alexandria Ocasio-Cortez with 66 cosponsors (https://www.congress.gov/bill/116th-congress/house-resolution/109/text), and in the Senate by Sen. Ed Markey with 11 cosponsors, https://www.markey.senate.gov/imo/media/doc/Green%20New%20Deal%20Resolution%20SIGNED.pdf.

previous authoritative studies: Committee on America's Energy Future, National Academy of Sciences, National Academy of Engineering, and National Research Council of the National Academy of Sciences, *America's Energy Future: Technology, Risks, Opportunities, and Tradeoffs* (Washington, DC: National Academies Press, 2009).

The climate plan that President Biden outlined: "The Biden Plan for a Clean Energy Revolution and Environmental Justice," JoeBiden.com, accessed March 3, 2023, https://joebiden.com/climate-plan/#.

climate challenges vary by region: Washington State governor Jay Inslee's plan, by contrast, would have established interagency "climate mobilization councils" in every state and territory to harmonize federal, state, and local efforts. See Sam Ricketts, Bracken Hendricks, and Maggie Thomas, "Evergreen Action Plan: A National Mobilization to Defeat the Climate Crisis and Build a Just and Thriving Clean Energy Economy" (adapted from Governor Inslee's 218-page Climate Mission plan), Medium, April 15, 2020, https://medium.com/@sam.t.ricketts/evergreen-action-plan-3f705ecb500a.

cabinet-level Climate Protection Department: Ricketts, Hendricks, and Thomas, "Evergreen Action Plan."

Clean Infrastructure Agency: This office should also coordinate with the Department of Energy, the Federal Energy Regulatory Commission, the Federal Power Commission, the National Renewable Energy Laboratory and other national laboratories, and the Department of Defense through DARPA and ARPA-E. See William B. Bonvillian, "DARPA and Its ARPA-E and IARPA Clones: A Unique Innovation Organization Model," *Industrial and Corporate Change* 27, no. 5 (October 2018): 897–914, https://doi.org/10.1093/icc/dty026.

a smart electrical grid: President Biden's climate action plan proposes that a new research agency, ARPA-C, should be formed to support the development of a smart grid. See "The Biden Plan for a Clean Energy Revolution and Environmental Justice," JoeBiden.com, accessed February 14, 2023, https://joebiden.com/climate-plan/#.

Clean Energy Bank: Governor Inslee's climate mission statement for Congress proposes a new Clean Energy Deployment Administration or Green Infrastructure Bank; see Ricketts, Hendricks, and Thomas, "Evergreen Action Plan."

Civilian Climate Corps: Governor Inslee's climate plan also proposes a Climate Corps, but with a focus on youth, skills development, global service (like the Peace Corps), and projects focused on community sustainability rather than on natural resources. See Ricketts, Hendricks, and Thomas, "Evergreen Action Plan."

Freeman and I collaborated: S. David Freeman and John J. Berger, "It's Up to Next President to Protect the Climate," *Detroit Free Press*, January 17, 2020.

by 8.5 percent of 2020 emissions: The Green New Deal even more ambitiously proposes that U.S. power come from clean renewable sources by 2030. See House Resolution 109, "Recognizing the Duty of the Federal Government to Create a Green New Deal."

must be equipped with solar: "Energy Commission Adopts Standards Requiring Solar Systems for New Homes, First in Nation," press release, California Energy Commission, May 9, 2018, https://www.energy.ca.gov/news/2018-05/energy-commission-adopts-standards-requiring-solar-systems-new-homes-first.

saves homeowners more than $25,000: "Zero Energy Homes Cost Less to Own," Zero Energy Project, accessed March 4, 2023, https://zeroenergyproject.com/buy/cost-less-to-own/.

eighty dollars in monthly savings: Noah Higgins-Dunn, "Net-Zero Energy Homes Have Arrived—and Are Shaking Up the US Housing Market," CNBC, February 14, 2019, https://www.cnbc.

com/2019/02/14/homes-that-produce-their-own-energy-might-be-the-future-and-california-is-inching-closer.html.

leased and financed with utility bill savings: Higgins-Dunn, "Net-Zero Energy Homes Have Arrived."

cleaning up building sector emissions: Gianna Melillo, "California Moves to Be First State to Ban Natural Gas Heaters and Furnaces," The Hill, updated September 27, 2022, https://thehill.com/changing-america/sustainability/energy/3658284-california-first-state-to-ban-natural-gas-heaters-and-furnaces/.

zero-emission vehicles (ZEVs) or plug-in hybrids by 2035: "Governor Newsom Announces California Will Phase Out Gasoline-Powered Cars & Drastically Reduce Demand for Fossil Fuel in California's Fight Against Climate Change," press release, Office of Governor Gavin Newsom, September 23, 2020, https://www.gov.ca.gov/2020/09/23/governor-newsom-announces-california-will-phase-out-gasoline-powered-cars-drastically-reduce-demand-for-fossil-fuel-in-california-as-fight-against-climate-change.

will stimulate consumer demand: Freeman proposed that, for the five years after the clean vehicle mandate went into effect, a ten-thousand-dollar federal tax credit or prorated subsidy be provided for EVs of forty thousand dollars or less. A yearly increasing federal surcharge, starting at a thousand dollars and increasing to a maximum of ten thousand, would be added to the price of new fossil fuel vehicles.

some seventeen other states: Comments of Margo Oge on "Baby, You Can Drive My Electric Car," hosted by Mina Kim, *Forum*, KQED, August 31, 2022, https://www.kqed.org/forum/2010101890398/baby-you-can-drive-my-electric-car.

carbon-tax bill was introduced in the U.S. Senate: "US Senators Propose Long-Shot Carbon Tax Bill for Big Polluters," Reuters, February 14, 2013, https://www.reuters.com/article/usa-climate-legislation-idUSL1N0BE7ZP20130214.

"I'm an emergency physician": Author interviews with Peter Joseph, via telephone, June 11, 2020, and June 19, 2020.

federal carbon-tax bill, H.R. 763: H.R. 763—Energy Innovation and Carbon Dividend Act, 116th Congress (2019–2020), 1st Sess., introduced January 24, 2019, by Representative Theodore Deutch, https://www.congress.gov/bill/116th-congress/house-bill/763.

"an almost insuperable political challenge": David Roberts, "At Last, a Climate Policy That Can Unite the Left," Vox, July 9, 2020, https://www.vox.com/energy-and-environment/21252892/climate-change-democrats-joe-biden-renewable-energy-unions-environmental-justice.

a $15-per-ton tax: A $15-per-ton tax would raise gasoline prices at the pump by only $0.13 per gallon, a small fraction of the price of gas and only 17 percent of gas taxes—about $0.74 in California. Erin Baldassari, "Gas Tax: What Californians Actually Pay on Each Gallon of Gas," *San Jose Mercury News*, September 4, 2018, https://www.mercurynews.com/2018/09/04/gas-tax-what-you-actually-pay-on-each-gallon-of-gas/.

report by the Rhodium Group: John Larsen et al., *Energy and Environmental Implications of a Carbon Tax in the United States*, report, SIPA Center on Global Energy Policy, Columbia University, July 2018, https://www.energypolicy.columbia.edu/publications/energy-and-environmental-implications-carbon-tax-united-states.

a ban on new fossil fuel extraction: "Vision for Equitable Climate Action," US Climate Action Network, accessed March 10, 2023, https://equitableclimateaction.org/.

the bill's carbon-reduction target: Despite the climate emergency, the bill has no emissions-reduction target at all for the years 2020 to 2024. From 2025 to 2034, the bill sets a carbon-reduction target of 5 percent per year, which is then cut in half from 2035 to 2050, to 2.5 percent annually. If the economy can tolerate a 5 percent-per-annum reduction, one might ask, why slash the target in half?

"fee-and-dividend" program: A dividend is a reward or incentive—usually cash—that a company pays its investors, generally reflecting its profits. Money charged as a percentage of a purchase is an excise tax, not a fee; therefore, the return of payment in question is a tax rebate.

Others critical of the dividends: David Roberts, "Energy Lobbyists Have a New PAC to Push for a Carbon Tax. Wait, What?" Vox, June 23, 2018, https://www.vox.com/energy-and-environment/2018/6/22/17487488/carbon-tax-dividend-trent-lott-john-breaux.

raise the price of fossil fuels internationally: William Nordhaus, "The Climate Club: How to Fix a Failing Global Effort," *Foreign Affairs*, May/June 2020, https://www.foreignaffairs.com/articles/united-states/2020-04-10/climate-club.

well-designed cross-border carbon adjustment: "The Baker Shultz Carbon Dividends Plan: A Bipartisan Climate Roadmap," PDF, Climate Leadership Council, Washington, DC, February 2021, https://clcouncil.org/Bipartisan-Climate-Roadmap.pdf.

CLC is an important climate policy lobbying group: "Economists' Statement on Carbon Dividends, Organized by the Climate Leadership Council," EconStatement.org, accessed February 14, 2023, https://www.econstatement.org/.

Revenue neutrality suits . . . oil companies: Rex Tillerson, "Exxon Mobil's CEO Rex Tillerson Looks at America's Energy Industry Future," speech to the Economic Club of Washington, DC, October 1, 2009, available at https://www.economicclub.org/sites/default/files/transcripts/Tillerson%20Transcript%202009.pdf.

"first and foremost a bid": David Roberts, "Energy Lobbyists Have a New PAC to Push for a Carbon Tax."

15. CLIMATE DIPLOMACY: GETTING TO YES

"These nations consider delay acceptable": Nigel Purvis and Andrew Stevenson, "Rethinking Climate Diplomacy: New Ideas for Transatlantic Cooperation Post-Copenhagen," Brussels Forum Paper Series, German Marshall Fund, Washington, DC, 2010.

they are 60 percent *higher*: Pierre Friedlingstein et al., "Global Carbon Budget 2022," *Earth System Science Data* 14, no. 11 (November 11, 2022): 4811–900.

Hansen courageously testified: U.S. Senate, 100th Congress, 1st Sess., 461, Part 2, Hearing Before the Committee on Energy and Natural Resources, "Greenhouse Effect and Global Climate Change," Testimony of Dr. James Hansen, director, NASA Goddard Institute for Space Studies, June 23, 1988. Citing overwhelming scientific evidence, Dr. Hansen told the Senate committee that due to the greenhouse effect, the earth was hotter in 1988 than at any previous time in modern temperature history; the cause, Hansen said, was humanity's use of fossil fuels. Hansen's 1988 testimony sent shockwaves around the world, brought climate change to widespread public attention for the first time, and led to calls for action. Philip Shabecoff, "Global Warming Has Begun, Expert Tells Senate," *New York Times*, June 24, 1988, https://www.nytimes.com/1988/06/24/us/global-warming-has-begun-expert-tells-senate.html.

atmospheric GHG concentrations: "The NOAA Annual Greenhouse Gas Index (AGGI)," NOAA Global Monitoring Laboratory, updated Spring 2022, https://gml.noaa.gov/aggi/aggi.html; Carl Edward Rasmussen, "Atmospheric Carbon Dioxide Growth Rate," University of Cambridge, February 7,

2023, https://mlg.eng.cam.ac.uk/carl/words/carbon.html; "Daily CO2," CO2.earth, accessed March 9, 2023, https://www.co2.earth/daily-co2.

international banks invested $2.7 trillion: Rainforest Action Network et al., *Banking on Climate Change: Fossil Fuel Finance Report 2020*, March 18, 2020, https://www.bankingonclimatechaos.org/bankingonclimatechange2020/. For a lower estimate, see Cameron Russell, "Dear Fossil Fuel Executives," in Ayana Elizabeth Johnson and Katharine K. Wilkinson, *All We Can Save: Truth, Courage, and Solutions for the Climate Crisis* (New York: One World, 2020), 205–212.

multinational banks supporting fossil fuel expansion: Rainforest Action Network et al., *Banking on Climate Change.*

multilateral funding bodies: Gaia Larsen et al., "4 Ways Development Banks Can Better Support the Paris Agreement," World Resources Institute, December 4, 2018, https://www.wri.org/insights/4-ways-development-banks-can-better-support-paris-agreement; Joe Thwaites, "4 Climate Finance Priorities for the Biden Administration," World Resources Institute, January 28, 2021, https://www.wri.org/insights/4-climate-finance-priorities-biden-administration.

Biden administration is committed to opposing: Jake Schmidt, "US Will Oppose Fossil Fuel Projects at Development Banks," Natural Resources Defense Council, August 20, 2021, https://www.nrdc.org/experts/jake-schmidt/us-will-oppose-fossil-fuel-projects-development-banks.

President Biden's 2021 executive order: President Joseph R. Biden, "Executive Order on Tackling the Climate Crisis at Home and Abroad," White House Briefing Room, January 27, 2021, https://www.whitehouse.gov/briefing-room/presidential-actions/2021/01/27/executive-order-on-tackling-the-climate-crisis-at-home-and-abroad/.

Kerry and the State Department: Alan Yu, "How U.S. Diplomacy and Diplomats Can Help Get International Climate Action Back on Track," Center for American Progress, December 8, 2020, https://www.americanprogress.org/article/u-s-diplomacy-diplomats-can-help-get-international-climate-action-back-track/.

Kerry further defined his mission: Special Envoy for Climate Change John Kerry in conversation with Bill Maher, *Real Time with Bill Maher*, aired January 29, 2021, on HBO.

"The conditions for access to GSP+": Pascal Lamy, Geneviève Pons, and Pierre Leturcq, "Greening EU Trade 4: How to 'Green' Trade Agreements?" policy paper, Jacques Delors Institute, November 2020, https://institutdelors.eu/en/publications/greening-eu-trade-4-how-to-green-trade-agreements/.

"setting the stage": George Marshall, *Don't Even Think About It: Why Our Brains Are Wired to Ignore Climate Change* (New York: Bloomsbury Publishing, 2014), 161.

the world will still experience: Lindsay Maizland, "Global Climate Agreements: Successes and Failures," Council on Foreign Relations, November 4, 2022; "COP27 Reaches Breakthrough Agreement on New 'Loss and Damage' Fund for Vulnerable Countries," press release, UN Climate Change, November 20, 2022, https://unfccc.int/news/cop27-reaches-breakthrough-agreement-on-new-loss-and-damage-fund-for-vulnerable-countries. The projected global failure to hold temperature increase to no more than 2°C and, aspirationally, to 1.5°C is a result both of the overall inadequacy of INDC pledges and the failure of some nations to make the requisite progress to fulfill their pledges.

once again, "the stage was set": UN Climate Change Conference (COP26), Glasgow, Scotland, November 1–12, 2021.

the Biden administration subsequently reinvigorated: UN Climate Change Conference (COP26).

reduce those emissions 30 percent: "FACT SHEET: Renewed U.S. Leadership in Glasgow Raises Ambition to Tackle Climate Crisis," White House Briefing Room, November 13, 2021, https://www.whitehouse.gov/briefing-room/statements-releases/2021/11/13/fact-sheet-renewed-u-s-leadership-in-glasgow-raises-ambition-to-tackle-climate-crisis/.

hasn't fully materialized: Leia Achampong, "Where Do Things Stand on the Global US$100 Billion Climate Finance Goal?" European Network on Debt and Development, September 7, 2022, https://www.eurodad.org/where_do_things_stand_on_the_global_100_billion_climate_finance_goal.

twenty-five-member "buyers' club": "FACT SHEET: Renewed U.S. Leadership in Glasgow Raises Ambition to Tackle Climate Crisis."

2022 not surprisingly being one of the hottest years: Hannah Ritchie and Max Roser, "Greenhouse Gas Emissions," Our World in Data, accessed February 8, 2021, https://ourworldindata.org/greenhouse-gas-emissions.

"Paris enjoys unique legitimacy": Charles F. Sabel and David G. Victor, *Fixing the Climate: Strategies for an Uncertain World* (Princeton, NJ: Princeton University Press, 2022), 166.

United States spends only $2.5 billion: Thwaites, "4 Climate Finance Priorities for the Biden Administration."

heads of important domestic agencies: These agencies will include the U.S. Agency for International Development, the U.S. International Development Finance Corporation, the Millennium Challenge Corporation, the United States Trade and Development Agency, the Office of Management and Budget, and other agencies providing foreign assistance and development financing.

the continued failure of the United States: Purvis and Stevenson, "Rethinking Climate Diplomacy."

The United Nations' Green Climate Fund: Sophie Yeo, "Green Climate Fund Attracts Record US$9.8 Billion for Developing Nations," *Nature*, November 1, 2019, https://www.nature.com/articles/d41586-019-03330-9.

many of them need replenishment: Thwaites, "4 Climate Finance Priorities for the Biden Administration."

as little as $5 per metric ton: Carol Browner and Nigel Purvis, "On Climate, Love Thy Neighbor," *Huffington Post*, September 29, 2016, https://www.huffpost.com/entry/on-climate-love-thy-neigh_b_8213430.

This is a paltry sum: "COP27 Reaches Breakthrough Agreement on New 'Loss and Damage' Fund for Vulnerable Countries."

long history of financing oil and gas: Larsen et al., "4 Ways Development Banks Can Better Support the Paris Agreement."

in urgent need of radical reform: Larsen et al., "4 Ways Development Banks Can Better Support the Paris Agreement."

its Belt and Road Initiative: These banks include the Asian Infrastructure Investment Bank, the China Development Bank, and the Export-Import Bank of China. See Joshua Busby and Nigel Purvis, "Climate Leadership in Uncertain Times," Atlantic Council Global Energy Center, Washington, DC, September 2018, https://www.atlanticcouncil.org/in-depth-research-reports/report/climate-leadership-in-uncertain-times/.

first time China had ever agreed: Taylor Dimsdale, "U.S. Climate Diplomacy Leadership: The Evidence," briefing paper, E3G, November 1, 2016, https://www.jstor.org/stable/resrep17902.

China's emissions are growing rapidly: Lindsay Maizland, "China's Fight Against Climate Change and Environmental Degradation," Council on Foreign Relations, updated May 19, 2021, https://www.cfr.org/backgrounder/china-climate-change-policies-environmental-deg-radation.

12 billion metric tons: Lauri Myllyvirta, "China's Carbon Emissions Grow at Fastest Rate in More than a Decade," *Carbon Brief*, May 20, 2021, https://www.carbonbrief.org/analysis-chinas-carbon-emissions-grow-at-fastest-rate-for-more-than-a-decade/.

massive GHG emissions leaks: Naureen S. Malik, "New Climate Satellite Spotted Giant Methane Leak as It Happened," Bloomberg Green, February 12, 2021, https://www.bloomberg.com/news/arti-cles/2021-02-12/new-climate-satellite-spotted-giant-methane-leak-as-it-happened.

inaccurate emissions reporting: John Schwartz, "U.S. Cities Are Vastly Undercounting Emissions, Researchers Find," *New York Times*, February 2, 2021, https://www.nytimes.com/2021/02/02/climate/cities-greenhouse-gas-emissions.html; Kevin Robert Gurney et al., "Under-reporting of Greenhouse Gas Emissions in U.S. Cities," *Nature Communications* 12, no. 553 (2021), https://doi.org/10.1038/s41467-020-20871-0.

China feels justified: "Chinese diplomats argued that China shouldn't have to sacrifice its economic development for environmental protection and that developed countries, such as the United States, should carry more of the burden because they were able to grow their economies without limitations." Maizland, "China's Fight Against Climate Change."

two thirds of China's electricity: Lucille Liu et al., "The Secret Origins of China's 40-Year Plan to End Carbon Emissions," Bloomberg News, November 22, 2020, https://www.bloomberg.com/news/features/2020-11-22/china-s-2060-climate-pledge-inside-xi-jinping-s-secret-plan-to-end-emissions.

China's coal production: Xiaoying You, "Analysis: What Does China's Coal Push Mean for Its Climate Goals?" *Carbon Brief*, March 29, 2022, https://www.carbonbrief.org/analysis-what-does-chi-nas-coal-push-mean-for-its-climate-goals/.

continuing to build . . . coal power plants: Jessie Yeung, "China Approved Equivalent of Two New Coal Plants a Week in 2022, Report Finds," CNN Business, February 27, 2023, https://www.cnn.com/2023/02/27/energy/china-new-coal-plants-climate-report-intl-hnk/index.html; see also "Electricity Sector in China," last edited March 7, 2023, 18:58, https://en.wikipedia.org/wiki/Electric-ity_sector_in_China.

countries participating in its Belt and Road Initiative: Lydia Powell, "Climate Superpowers," in Alexander Carius, Noah J. Gordon, and Lauren Herzer Risi, eds., *21st Century Diplomacy: Foreign Policy Is Climate Policy*, Adelphi, The Wilson Center, 2020.

cause global warming to exceed 2.7°C: Maizland, "China's Fight Against Climate Change."

two heads of state promised: "FACT SHEET: The United States and China Issue Joint Presidential Statement on Climate Change with New Domestic Policy Commitments and a Common Vision for an Ambitious Global Climate Agreement in Paris," White House Briefing Room, September 25, 2015, https://obamawhitehouse.archives.gov/the-press-office/2015/09/25/fact-sheet-united-states-and-china-issue-joint-presidential-statement.

discussions between the United States and China: Former California governor Jerry Brown, interview with Michael Krasny, *Forum*, KQED, February 3, 2021, https://www.kqed.org/fo-rum/2010101881879/former-governor-jerry-brown-talks-pandemic-climate-change-and-our-chang-ing-state.

"avert an outcome that destroys everyone": Huiyao Wang, "'Climate Superpowers?' Why the Cold War Is the Wrong Analogy for Our Heating Planet," in Carius, Gordon, and Risi, eds., *21st Century Diplomacy.*

"climate change is a more complex": Wang, "'Climate Superpowers?'"

United States might need to make concessions: Liu et al., "The Secret Origins of China's 40-Year Plan to End Carbon Emissions."

45 percent of its revenue: International Energy Agency, "Energy Fact Sheet: Why Does Russian Oil and Gas Matter?" March 21, 2022, https://www.iea.org/articles/energy-fact-sheet-why-does-russian-oil-and-gas-matter.

climate damage from Russia's fuel exports: At the start of the war, Russia was close to completing the Nord Stream 2 gas pipeline under the Baltic Sea to Germany and other nations, and both Germany and France were eager to secure the fuel. The Russian invasion of Ukraine led Germany to cancel the pipeline, and it was subsequently sabotaged under the North Sea, so it is inoperable. "Germany's Merkel Stands by Backing for Russia Gas Pipeline," Associated Press, February 5, 2021, https://apnews.com/article/europe-baltic-sea-angela-merkel-germany-russia-cb7a-3538c0aff20e28ii61db21bf2842.

European Union raised its emissions-reduction targets: "EU Agrees on Tougher Climate Goals for 2030," DW, November 12, 2020, https://www.dw.com/en/eu-agrees-on-tougher-climate-goals-for-2030/a-55901612. The European Union's previous goal was to reduce its emissions 40 percent below 1990 levels by 2030. By 2021, industrial emissions were already about 40 percent below 1990 emissions.

These troubled nations: Phillip H. Gordon, "How Curbing Reliance on Fossil Fuels Will Change the World," in Carius, Gordon, and Risi, eds., *21st Century Diplomacy.*

diplomacy will become increasingly vital: Dennis Tänzler and Noah Gordon, "The New Geopolitics of a Decarbonizing World," in Carius, Gordon, and Risi, eds., *21st Century Diplomacy.*

"Petrostates" could lose $9 trillion: Mike Coffin and Andrew Grant, "Petrostates Need to Bridge $9 Trillion Income Gap in Energy Transition," Carbon Tracker, February 11, 2021, https://carbontracker.org/petrostates-13-trillion-gap-energy-transition-pr/.

according to the International Energy Agency: If all oil and gas revenues are included, not just those of petrostates, total oil and gas revenues could decline by $13 trillion (Coffin and Grant, "Petrostates Need to Bridge $9 Trillion Income Gap"). The International Energy Agency has estimated the loss at $7 billion. See Helen Robertson, "Petro-States Have Got a Problem with Their Economies," Bloomberg News, October 24, 2018, https://www.bloomberg.com/news/articles/2018-10-24/petro-states-have-got-a-problem-with-their-economies.

incorporated in future EU trade agreements: "Non-paper from the Netherlands and France on Trade, Social Economic Effects and Sustainable Development," Trésor-Info, Ministry of the Economy, Finance and Industrial and Digital Sovereignty of France, accessed February 15, 2023, https://www.tresor.economie.gouv.fr/Articles/73ce0c5c-11ab-402d-95b1-5dbb8759d699/files/6b6ff3bf-e8fb-4de2-94f8-922eddd81d08.

eight hundred civil-society representatives: John Vidal and Fiona Harvey, "Green Groups Walk out of UN Climate Talks," *Guardian* (London), November 21, 2013, https://www.theguardian.com/environment/2013/nov/21/mass-walk-out-un-climate-talks-warsaw.

the Lima-Paris Action Agenda: Partners included the French COP21 president, Laurent Fabius, the secretary-general of the United Nations, Ban Ki-moon, and the UNFCC.

a global transformation to a low-carbon economy: "The Lima-Paris Action Agenda: Promoting Transformational Climate Action," United Nations Climate Change, October 21, 2015, https://unfccc.int/news/the-lima-paris-action-agenda-promoting-transformational-climate-action.

"we not only need universal awareness": "Lima-Paris Action Agenda Primer," United Nations Framework Convention on Climate Change, 2015, accessed February 15, 2023, https://unfccc.int/media/509508/lpaa-primer.pdf.

a testament to Pulgar-Vidal's vision: "Actor tracking," Global Climate Action, NAZCA, accessed February 15, 2023, https://climateaction.unfccc.int/Actors. The roster is described in the "Lima-Paris Action Agenda Primer" and is referred to as Non-State Actor Zone for Climate Action (NAZCA).

pay a lot more attention: Marshall, *Don't Even Think About It*, 168–74.

a "Fossil Fuel Non-Proliferation Treaty Initiative": "Climate Risks from Oil, Gas and Coal Production Must be Added Up to Avoid Locking in the Climate Emergency," press release, Carbon Tracker, February 5, 2021, https://carbontracker.org/climate-risks-from-oil-gas-and-coal-production-must-be-added-up-to-avoid-locking-in-the-climate-emergency/.

on a "pay-for-performance" basis: Purvis and Stevenson, "Rethinking Climate Diplomacy."

cumulative total of greenhouse gases: National Oceanic and Atmospheric Administration, U.S. Department of Commerce, "Carbon dioxide now more than 50% higher than pre-industrial levels," June 3, 2022, https://www.noaa.gov/news-release/carbon-dioxide-now-more-than-50-higher-than-pre-industrial-levels.

16. GEOENGINEERING: FRIEND OR FOE?

various forms of geoengineering: Author interview with Dr. Wil C. G. Burns, October 16, 2020.

cloud condensation nuclei: Increased cloud nucleation is caused by a phenomenon known as the Twomey effect.

changes to the world's weather systems: Fahad Saeed et al., "Why Geoengineering Is Not a Solution to the Climate Problem," *Climate Analytics*, February 12, 2018, https://climateanalytics.org/publications/2018/why-geoengineering-is-not-a-solution-to-the-climate-problem/.

the destruction of stratospheric ozone: Zaria Gorvett, "How a Giant Space Umbrella Could Stop Global Warming," BBC Future, April 26, 2016, https://www.bbc.com/future/article/20160425-how-a-giant-space-umbrella-could-stop-global-warming.

cheap, by geoengineering standards: Tingzhen Ming et al., "Fighting Global Warming by Climate Engineering. Is the Earth Radiation Management and the Solar Radiation Management Any Option for Fighting Climate Change?" *Renewable and Sustainable Energy Reviews* 31 (2014): 792–834, https://doi.org/10.1016/j.rser.2013.12.032.

as many as 26,000 additional deaths: Author interview with Dr. Wil C. G. Burns, October 16, 2020.

without acidifying the rain: David W. Keith et al., "Stratospheric Solar Geoengineering Without Ozone Loss," *Proceedings of the National Academy of Sciences* 113, no. 52 (December 27, 2016): 14910–14, https://doi.org/10.1073/pnas.1615572113.

"potential of unintended consequences": Ming et al., "Fighting Global Warming by Climate Engineering."

A recent MIT study found: Jennifer Chu, "Reflecting Sunlight to Cool the Planet Will Cause Other Global Changes," *MIT Science Daily*, June 2, 2020, https://news.mit.edu/2020/reflecting-sunlight-cool-planet-storm-0602.

"Solar geoengineering is not reversing climate change": Charles G. Gertler et al., "Weakening of the Extratropical Storm Tracks in Solar Geoengineering Scenarios," *Geophysical Research Letters* 47, no. 11 (2020), https://doi.org/10.1029/2020GL087348.

detrimental to the ozone layer: Hannah Osborne, "Climate Change and Geoengineering: Artificially Cooling Planet Earth by Thinning Cirrus Clouds," *Newsweek*, July 21, 2017, https://www.newsweek.com/climate-change-geoengineering-artificially-cool-planet-640124.

could also potentially cause warming: Ulrike Lohmann and Blaž Gasparini, "A Cirrus Cloud Climate Dial?" *Science* 357, no. 6348 (July 21, 2017), https://www.science.org/doi/10.1126/science.aan3325.

preclude these particular technologies: Michael Barnard, five-part CleanTechnica series: "Chevron's Fig Leaf Part 1: Carbon Engineering Burns Natural Gas to Capture Carbon from the Air," April 12, 2019, https://cleantechnica.com/2019/04/12chevrons-fig-leaf-part-1-carbon-engineering-burns-natural-gas-to-capture-carbon-from-the-air/; "Chevron's Fig Leaf Part 2: Carbon Engineering Burns Gas for 0.5 Tons of CO2 for Each Ton Captured," April 13, 2019, https://cleantechnica.com/2019/04/12chevrons-fig-leaf-part-2-carbon-engineering-burns-gas-for-5-tons-of-co2-for-each-ton-captured/; "Chevron's Fig Leaf Part 3: Carbon Engineering's Scale & Power Problems," April 14, 2019, https://cleantechnica.com/2019/04/12chevrons-fig-leaf-part-3-carbon-engineerings-scale-power-problems/; "Chevron's Fig Leaf Part 4: Carbon Engineering's Only Market Is Pumping More Oil," April 19, 2019, https://cleantechnica.com/2019/04/12chevrons-fig-leaf-part-4-carbon-engineerings-only-market-is-pumping-more-oil/; "Chevron's Fig Leaf Part 5: Who Is Behind Carbon Engineering and What Do Experts Say?" April 20, 2019, https://cleantechnica.com/2019/04/12chevrons-fig-leaf-part-5- Part-5-who- is- behind carbon-engineering-what-do-experts-say/.

an "artificial tree": Gaia Vince, "Scientists are looking at ways to lower the global temperature by removing greenhouse gases from the air. Could super-absorbent fake leaves be the answer?" BBC Future, October 3, 2012, https://www.bbc.com/future/article/20121004-fake-trees-to-clean-the-skies.

The newer, disc-shaped leaves: Zayna Sy, "Can a mechanical 'tree' help slow climate change? An ASU researcher built one to find out," *Arizona Republic*, April 22, 2022, https://www.azcentral.com/story/news/local/arizona-environment/2022/04/22/asu-researcher-builds-mechanical-tree-capture-carbon-dioxide/7398671001/.

one hundred million of these devices: R. P. Siegel, "Building the Ultimate Carbon Capture Tree," American Society of Mechanical Engineers, May 17, 2019, https://www.asme.org/topics-resources/content/building-ultimate-carbon-capture-tree.

about $100 per ton: "Lackner's carbon-capture technology moves to commercialization: Powerful 'mechanical trees' can remove CO2 to combat global warming at scale," *ASU News*, April 29, 2019, https://news.asu.edu/20190429-solutions-lackner-carbon-capture-technology-moves-commercialization.

"energetically and financially costly distraction": Sudipta Chatterjee and Kuo-wei Huang, "Unrealistic energy and materials requirements for direct air capture in deep mitigation paths," Matters Arising, *Nature Communications* 11, no. 3287 (2020), https://doi.org/10.1038/s41467-020-17203-7.

"It's too little, too late": Sy, "Can a mechanical 'tree' help slow climate change?"

they have begun a joint venture: "Lackner's carbon-capture technology moves to commercialization."

chemical reactions with the basalt: Ragnhildur Sigurdardottir and Akshat Rathi, "The Icelandic Startup Bill Gates Uses to Turn CO2 into Stone," Bloomberg Law, March 4, 2021, https://news.bloomberglaw.com/environment-and-energy/the-icelandic-startup-bill-gates-uses-to-turn-co%E2%82%82-into-stone.

geothermally active sites: Holmfridur Sigurdardottir et al., "The CO_2 Fixation into Basalt at Hellisheidi Geothermal Power Plant, Iceland," Proceedings of the World Geothermal Congress 2010, Bali, Indonesia, April 25–29, 2010, https://georg.cluster.is/wp-content/uploads/2017/02/0237.pdf.

cost of removing a metric ton: David Keith et al., "A Process for Capturing CO_2 from the Atmosphere," *Joule*, June 7, 2018, https://doi.org/10.1016/j.joule.2018.05.006.

This remains to be seen: Barnard, "Chevron's Fig Leaf," five-part CleanTechnica series.

"a nice big marketing win": Barnard, "Chevron's Fig Leaf Part 4."

the negative externalities of fossil fuels: Barnard, "Chevron's Fig Leaf Part 4."

the acidic CO_2 mixes with the oil: *Carbon Dioxide Enhanced Oil Recovery*, report, U.S. Department of Energy, National Energy Technology Laboratory, March 2010, https://www.netl.doe.gov/sites/default/files/netl-file/co2_eor_primer.pdf.

at two to three times: Michael Barnard, "The Billions Spent on Carbon Capture over 50 Years Would Have Been Better Spent on Wind and Solar," CleanTechnica, April 21, 2019, https://cleantechnica.com/2019/04/21/carbon-captures-global-investment-would-have-been-better-spent-on-wind-solar/.

more than a third of the CO_2: Mark Z. Jacobson, "Why Not Synthetic Direct Air Carbon Capture and Storage (SDACCS) as Part of a 100% Wind-Water-Solar (WWS) and Storage Solution to Global Warming, Air Pollution, and Energy Security," in Jacobson, *100% Clean, Renewable Energy and Storage for Everything* (New York: Cambridge University Press, 2020); Jacobson, *No Miracles Needed*, 185–87, 328.

nineteen larger carbon capture facilities: Barnard, "The Billions Spent on Carbon Capture over 50 Years."

direct air capture is unlikely: Carbon Engineering, "Our Story," Carbon Engineering, accessed February 15, 2023, https://carbonengineering.com/our-story/.

"critical new Direct Air Capture industry": Carbon Engineering, "Our Story."

a natural-gas-burning power plant: Barnard, "Chevron's Fig Leaf Part 1."

"not a rational choice for humanity": Michael Barnard, "Carbon Capture Is Expensive Because Physics," CleanTechnica, January 19, 2019, https://cleantechnica.com/2016/01/19/carbon-capture-expensive-physics/.

"drawdown [sic] more than 100%": Jeff Moyer et al., "Regenerative Agriculture and the Soil Carbon Solution," PDF, White Paper, Rodale Institute, Kutztown, PA, 2020, https://rodaleinstitute.org/wp-content/uploads/Rodale-Soil-Carbon-White-Paper_v11-compressed.pdf.

must be part of a suite: John J. Berger, "Can Soil Microbes Slow Climate Change?" *Scientific American*, March 26, 2019, https://www.scientificamerican.com/article/can-soil-microbes-slow-climate-change/.

based on a 500 percent overestimate: "In Depth: One Trillion Trees," *Forest News*, Spring 2020, 4–6.

can even be counterproductive: "In Depth: One Trillion Trees."

monocultures of single-age trees: John J. Berger, *Forests Forever: Their Ecology, Restoration, and Protection*, Columbia College, Chicago and San Francisco, CA: Center for American Places and Forests Forever Foundation, 2008, 122–25.

a trillion healthy, living trees: Michael Barnard, "What Would Planting 100 Million Trees Per Week Do in 5, 50, and 500 Years?" CleanTechnica, February 16, 2021, https://cleantechnica.com/2021/02/16/what-would-planting-100-million-trees-per-week-do-in-5-50-500-years/.

four hundred years **for a trillion trees:** Barnard, "What Would Planting 100 Million Trees Per Week Do?"

17. A JUST, CLEAN-ENERGY TRANSITION FOR ALL

Most of the displaced: Lauren Lluveras, "One Year Later, Puerto Rico Still Hasn't Recovered from Hurricane Maria," Global Citizen, September 20, 2018, https://www.globalcitizen.org/es/content/puerto-rico-slow-recovery-hurricane-maria/.

neighborhoods with little shade: "What Would Happen If It Were 116°F in San Francisco?" *Forum, KQED,* https://www.kqed.org/forum/2010101884331/what-would-happen-if-it-was-116-degrees-in-san-francisco.

create ground-level ozone: M. De Sario, K. Katsouyanni, and P. Michelozzi, "Climate Change, Extreme Weather Events, Air Pollution and Respiratory Health in Europe," *European Respiratory Journal* 42, no. 3 (2013): 826–43.

often gets *less* **government help:** Joe McCarthy, "Why Climate Change and Poverty Are Inextricably Linked," Global Citizen, February 19, 2020, https://www.globalcitizen.org/en/content/climate-change-is-connected-to-poverty/.

"a bridge to address other issues": Quotations attributed to Vien Truong are taken from an author interview with Vien Truong, Greenlining Institute, Oakland, CA, February 17, 2015; as well as subsequent exchanges over email or by telephone.

CONCLUSION: WHAT WE CAN DO TO RESTABILIZE THE CLIMATE

for a thousand years or more: Jennifer Chu, "Short-lived greenhouse gases cause centuries of sea-level rise," *Global Climate change: Vital Signs of the Planet,* NASA, January 12, 2017.

temperatures rose four times as fast: Chelsea Harvey, "The Arctic Is Warming Four Times Faster than the Rest of the Planet," *Scientific American,* August 12, 2022, https://www.scientificamerican.com/article/the-arctic-is-warming-four-times-faster-than-the-rest-of-the-planet.

reaching between 2.8°C (5°F) and 4.6°c: "Anticipating Future Sea Levels," Earth Observatory, National Oceanic and Atmospheric Administration, accessed March 9, 2023, https://earthobservatory.nasa.gov/images/148494/anticipating-future-sea-levels; Rebecca Lindsey, "Climate Change: Global Sea Level," National Oceanic and Atmospheric Administration, August 14, 2020, https://www.climate.gov/news-features/understanding-climate/climate-change-global-sea-level; Adam Vaughan, "Earth Will Hit 1.5°C Climate Limit Within 20 Years, Says IPCC Report," *New Scientist,* August 21, 2021, https://www.newscientist.com/article/2286454-earth-will-hit-1-5c-climate-limit-within-20-years-says-ipcc-report/.

unstoppable releases of greenhouse gases: Merritt R. Turetsky et al., "Permafrost Collapse Is Accelerating Carbon Release," *Nature,* April 30, 2019, https://www.nature.com/articles/d41586-019-01313-4.

declining by almost 70 percent: R. E. A. Almond et al., eds., *Living Planet Report 2022: Building a Nature-Positive Society,* World Wildlife Fund, Gland, Switzerland, 2022, available at https://livingplanet.panda.org/en-US/.

their highest peak in history: "Global CO2 Emissions Rebounded to Their Highest Level in History in 2021," press release, International Energy Agency, March 8, 2022, https://www.iea.org/news/global-co2-emissions-rebounded-to-their-highest-level-in-history-in-2021.

almost double the amount spent: "Support for Fossil Fuels Almost Doubled in 2021, Slowing Progress Toward International Climate Goals, According to New Analysis from OECD and IEA," Organization for Economic Co-operation and Development, August 29, 2022, https://www.oecd.org/newsroom/support-for-fossil-fuels-almost-doubled-in-2021-slowing-progress-toward-international-climate-goals-according-to-new-analysis-from-oecd-and-iea.htm#. When a different methodology was used—one that took account of environmental and climate damages—fossil fuel subsidies were projected to be $5.3 trillion in 2015 (6.5 percent of global GDP). See David Coady et al., "How Large Are Global Energy Subsidies?" working paper, International Monetary Fund, May 18, 2015, https://www.imf.org/en/Publications/WP/Issues/2016/12/31/How-Large-Are-Global-Energy-Subsidies-42940.

Ordinary citizens can join: The Climate Action Network (CAN) is a worldwide network of more than 430 nongovernmental organizations (NGOs) working to promote government and individual action to limit human-induced climate change to ecologically sustainable levels. See also home page, Climate Mobilization Network, accessed March 9, 2023, https://www.theclimatemobilization.org/climate-mobilization-network/.

We have the technology required: Denholm et al., *Examining Supply-Side Options to Achieve 100% Clean Electricity by 2035.*

About 60 percent of these: Damian Carrington and Matthew Taylor, "Revealed: The 'Carbon Bombs' Set to Trigger Catastrophic Climate Breakdown," *Guardian* (London), May 11, 2022, https://www.theguardian.com/environment/ng-interactive/2022/may/11/fossil-fuel-carbon-bombs-climate-breakdown-oil-gas.

leave current profit taking undisturbed: Rainforest Action Network et al., *Banking on Climate Chaos: Fossil Fuel Finance Report 2021,* March 24, 2021, https://www.ran.org/wp-content/uploads/2021/03/Banking-on-Climate-Chaos-2021.pdf.

more carbon than can be burned: Fossil Fuel Non-Proliferation Treaty Initiative, accessed March 9, 2023, https://fossilfueltreaty.org/.

a steadily rising national fee on carbon: James Hansen et al., "July 2022 Temperature Update & a Turning Point," via email, August 25, 2022.

40 percent of its electricity: Valerie Volcovici, "Solar Energy Can Account for 40% of U.S. Electricity by 2035—DOE," Reuters, September 8, 2021, https://www.reuters.com/business/energy/biden-administration-set-goal-45-solar-energy-by-2050-nyt-2021-09-08/.

applying "deep energy retrofits": Amory B. Lovins, "How Big Is the Energy Efficiency Resource?"

Thousands of miles of new high-capacity: Denholm et al., *Examining Supply-Side Options to Achieve 100% Clean Electricity by 2035.* This supply-side study does not incorporate integrative energy efficiency measures in its modeling and, therefore, overestimates the necessary cost and difficulty of the energy transition by overestimating the total amount of power required.

far more than we have currently budgeted: Haley Zaremba, "The U.S. Power Grid Can't Support Its Climate Pledges," OilPrice.com, September 1, 2022, https://www.yahoo.com/news/u-power-grid-t-support-170000825.html.

a major expansion of the hydrogen industry: Radowitz, "'Warp Speed Away from Russian Energy'"; Greg Avery, "Toyota, NREL Team to Make Green Hydrogen on a Large Scale in the

Denver Metro Area," *Denver Business Journal*, August 24, 2022, https://www.bizjournals.com/denver/news/2022/08/24/toyota-nrel-make-hydrogen-energy-denver.html.

57,000 confirmed abandoned: According to Mitch Bernard, chief counsel for the Natural Resources Defense Council, the true number could be ten times higher (undated letter). After more than a hundred years of oil and gas drilling in the United States, there may be as many as three million abandoned wells, according to one report, two thirds of which are uncapped.

consumers would save $2.7 trillion: Amol Phadke et al., "2035 Report 2.0, Plummeting Costs and Dramatic Improvements in Batteries Can Accelerate Our Clean Transportation Future," Goldman School of Public Policy, Grid Lab, and Energy Innovation, University of California, Berkeley, April 2021.

$1.3 trillion in environmental and health: Phadke et al., "2035 Report 2.0."

$54.2 billion in subsidies: The CHIPS and Science Act, Public Law 117-167, United States Congress, August 9, 2022, https://congress.gov/117/plaws/pub167/PLAW-117pub167.pdf.

paid at single-digit tax rates: Matthew Gardner, "Amazon Has Record-Breaking Profits in 2020, Avoids $2.3 Billion in Federal Income Taxes," Institute on Taxation and Economic Policy, February 3, 2021, https://itep.org/amazon-has-record-breaking-profits-in-2020-avoids-2-3-billion-in-federal-income-taxes/.

According to the Roosevelt Institute: Tim Mullaney, "The market's biggest winners and losers in the Inflation Reduction Act," The Bottom Line, CNBC, August 8, 2022, https://www.cnbc.com/2022/08/08/the-market-winners-and-losers-in-the-climate-health-and-tax-bill.html.

powerful, broad-based, national coalition: Readers frustrated with the slow pace of progress on climate issues and having the impulse to help create a powerful new climate coalition may want to look for guidance beyond this book, to writings by those who have devoted their lives to political organizing and analysis and who can thus provide wise guidance. Such materials might include Jane F. McAlevey's *No Shortcuts: Organizing for Power in the New Gilded Age* (New York: Oxford University Press, 2016), Becky Bond and Zack Exley's *Rules for Revolutionaries: How Big Organizing Can Change Everything* (White River Junction, VT: Chelsea Green, 2016), George Lakey's *How We Win: A Guide to Nonviolent Direct Action Campaigning* (Brooklyn, NY: Melville House Publishing, 2018), or Sam Harris's *Reclaiming our Democracy: Healing the Break Between People and Government* (Philadelphia: Camino Books, Inc., 1994), or reading the works of other master organizers, or studying the subject through training programs.

forged into a single, massive coalition: Coalition builders may be tempted to create their coalition from similar groups, with activists seeking activist allies (see https://www.theclimatemobilization.org/climate-mobilization-network/) or with those in scientific, academic, or industry groups seeking like-minded organizations in their sector of the economy or political landscape. The most powerful coalitions, however, are likely to be those that extend beyond the founder's silo to many different types of stakeholders.

about a quarter of Americans: Scott Newman, "1 in 4 Americans Think the Sun Revolves Around the Earth, Survey Says," NPR, February 14, 2014.

population stabilization is often left out: John Seager, undated letter mailed to constituents, Population Connection, 2120 L Street, NW, Suite 500, Washington, DC 20037.

"we must stop population growth": John Seager, undated letter mailed to constituents, Population Connection, 2120 L Street, NW, Suite 500, Washington, DC 20037.

seventy billion metric tons: See "Family Planning and Education" in Project Drawdown, "Table of Solutions," accessed May 10, 2023, https://www.drawdown.org/solutions/table-of-solutions.

between 33 and 40 percent of all food grown: Moira Borens et al., "Reducing Food Loss: What Grocery Retailers and Manufacturers Can Do," *The Daily Read*, McKinsey, September 7, 2022, https://www.mckinsey.com/industries/consumer-packaged-goods/our-insights/reducing-food-loss-what-grocery-retailers-and-manufacturers-can-do. "Food loss" refers to what happens on the farm or during processing and transport. "Food waste" happens during the different phases of retail sales and consumption.

lost or wasted edible food releases: Borens et al., "Reducing Food Loss."

two billion tons of food a year: Borens et al., "Reducing Food Loss."

cooperation across the food supply chain: Borens et al., "Reducing Food Loss."

food system occupies billions of acres: Hannah Ritchie, "If the World Adopted a Plant-Based Diet We Would Reduce Global Agricultural Land Use from 4 to 1 Billion Hectares," Our World in Data, March 4, 2021, https://ourworldindata.org/land-use-diets.

15 percent of total GHG releases: Paul Hawken, ed., *Drawdown: The Most Comprehensive Plan Ever Proposed to Reverse Global Warming* (New York: Penguin Books, 2017), 39.

impacts will be even less sustainable: Damian Carrington, "Huge Reduction in Meat-Eating 'Essential' to Avoid Climate Breakdown," *Guardian* (London), October 18, 2018, https://www.theguardian.com/environment/2018/oct/10/huge-reduction-in-meat-eating-essential-to-avoid-climate-breakdown.

more than fifty billion animals for food every year: "Annual U.S. Animal Death Stats," Animal Clock, accessed March 9, 2023, https://animalclock.org/.

as much manure as eighty million humans: Mark Bittman, "Meat Is Murder. But You Know That Already," *New York Times Book Review*, October 13, 2019, https://www.nytimes.com/2019/09/17/books/review/we-are-the-weather-jonathan-safran-foer.html.

raising livestock takes more water and energy: Frances Moore Lappé, *Diet for a Small Planet* (New York: Ballantine Books, 1971).

40 percent of all corn grown: Economic Research Service, U.S. Department of Agriculture, "Feed Grains Sector at a Glance," January 27, 2023, https://www.ers.usda.gov/topics/crops/corn-and-other-feed-grains/feed-grains-sector-at-a-glance/.

without reducing beef, pork, and lamb: Marco Springmann et al., "Options for keeping the food system within environmental limits," *Nature* 562 (October 25, 2018): 519–24.

offers major health benefits: Michael A. Clark et al., "Multiple Health and Environmental Impacts of Foods," *Proceedings of the National Academy of Sciences* 116, no. 46 (October 28, 2019): 23357–62.

a trillion dollars in annual health care costs: Hawken, ed., *Drawdown*, 39.

fooling the public for decades: Berger, *Climate Myths*, 8, 10–11, 13–15, 20–21, 25–26, 28–29.

the IRA will diminish U.S. emissions: J. D. Jenkins et al., "Preliminary Report: The Climate and Energy Impacts of the Inflation Reduction Act of 2022," REPEAT Project, Princeton University, Princeton, NJ, August 2022.

50 to 52 percent by 2030: "Fact Sheet: President Biden Sets 2030 Greenhouse Gas Pollution Reduction Target Aimed at Creating Good-Paying Union Jobs and Securing U.S. Leadership on Clean Energy Technologies," White House Briefing Room, April 22, 2021, https://www.whitehouse.gov/

briefing-room/statements-releases/2021/04/22/fact-sheet-president-biden-sets-2030-greenhouse-gas-pollution-reduction-target-aimed-at-creating-good-paying-union-jobs-and-securing-u-s-leadership-on-clean-energy-technologies/.

CODA: RECOMMENDED CLIMATE POLICIES AND STRATEGIES— THE TO-DO LIST

more than seventy are currently: "Declared National Emergencies Under the National Emergencies Act: A Running List of Presidential Emergency Declarations Under the National Emergencies Act of 1976," Brennan Center for Justice, last updated February 17, 2023, https://www.brennancenter.org/our-work/research-reports/declared-national-emergencies-under-national-emergencies-act.

cost the U.S. Treasury $7 trillion: Natasha Sarin, "The Case for a Robust Attack on the Tax Gap," U.S. Department of the Treasury, September 7, 2021, https://home.treasury.gov/news/featured-stories/the-case-for-a-robust-attack-on-the-tax-gap.

strengthening their power grids: They might do so by (a) improving interconnections between regions, (b) implementing advanced demand management systems, (c) constructing surplus renewable-energy capacity for use during periods of exceptional demand, and (d) adding energy storage to flatten out demand peak.

average interconnection time has almost doubled: Emily Pontecorvo, "Renewables Are Growing— But a Backlog of Projects Is Holding Up a Greener Grid," Grist, April 20, 2022, https://grist.org/energy/renewables-are-growing-but-a-backlog-of-projects-is-holding-up-a-greener-grid/; Shannon Osaka, "This Little-Known Bottleneck Is Blocking Clean Energy for Millions," *Washington Post*, December 20, 2022, https://www.washingtonpost.com/climate-environment/2022/12/20/clean-energy-bottleneck-transmission-lines/.

About the Author

John J. Berger, PhD, is an environmental science and policy specialist, prizewinning author, and journalist. A graduate of Stanford University and the University of California, he has produced ten other books and more than one hundred articles on climate change and the clean-energy transition. His work has been published by *Scientific American*, *HuffPost*, the *Los Angeles Times*, the *Boston Globe*, the *Christian Science Monitor*, *USA Today Magazine*, and others. He has been a consultant to the National Research Council of the National Academies of Sciences, Engineering, and Medicine; corporations; utilities; and the U.S. Congress, as well as a newspaperman, an editor, and a professor at the University of Maryland. He cofounded the Nuclear Information and Resource Service to assist citizen safe-energy groups and founded Restoring the Earth, a nonprofit organization devoted to bringing environmental restoration to national attention. He lives in El Cerrito, California, and is the father of Daniel and Michael Berger and the husband of Nancy Gordon. His hobbies are hiking, West Coast Swing and other partner dances, and instrumental guitar.

About the Contributor

Russ Feingold is the president of the American Constitution Society. He served as a Wisconsin state senator from 1983 to 1993 and a U.S. senator from Wisconsin from 1993 to 2011. From 2013 to 2015, he served as the U.S. Special Envoy for the Great Lakes Region of Africa and the Democratic Republic of the Congo.

During his eighteen years in the U.S. Senate, Feingold was ranked sixth in the Senate for bipartisan voting. He is a recipient of the John F. Kennedy Profile in Courage Award and cosponsored the Bipartisan Campaign Reform Act (McCain–Feingold Act), the only major piece of campaign finance reform legislation passed into law in decades. He was the only senator to vote against the initial enactment of the USA PATRIOT Act during the first vote on the legislation and was well known for his opposition to the Iraq War and as the Senate's leading opponent of the death penalty. He served on the Judiciary, Foreign Relations, Budget, and Intelligence Committees. Feingold was also chairman or ranking member of the Constitution Subcommittee.

Feingold has taught law and public policy extensively at various prominent American universities. He is the honorary ambassador for the Campaign for Nature, a global effort calling on policymakers to address the growing biodiversity crisis. He is the author of *While America Sleeps: A Wake-Up Call for the Post-9/11 Era*; contributes regularly to publications such as the *New York Times*, the *Washington Post*, and the *Guardian*, and appears frequently on MSNBC and

CNN. He holds a BA from the University of Wisconsin–Madison, a BA and an MA from the University of Oxford (as a Rhodes Scholar), and a JD from Harvard Law School—all degrees awarded with honors.

Index